Deep Learning for Biomedical Data Analysis

Mourad Elloumi
Editor

Deep Learning for Biomedical Data Analysis

Techniques, Approaches, and Applications

Springer

Editor
Mourad Elloumi
Computing and Information Technology
The University of Bisha
Bisha, Saudi Arabia

ISBN 978-3-030-71675-2 ISBN 978-3-030-71676-9 (eBook)
https://doi.org/10.1007/978-3-030-71676-9

This Springer imprint is published by the registered company Springer Nature Switzerland AG
The registered company address is: Gewerbestrasse 11, 6330 Cham, Switzerland

Contents

Part I Deep Learning for Biomedical Data Analysis

1-Dimensional Convolution Neural Network Classification Technique for Gene Expression Data 3
Samson Anosh Babu Parisapogu, Chandra Sekhara Rao Annavarapu, and Mourad Elloumi

Classification of Sequences with Deep Artificial Neural Networks: Representation and Architectural Issues 27
Domenico Amato, Mattia Antonino Di Gangi, Antonino Fiannaca, Laura La Paglia, Massimo La Rosa, Giosué Lo Bosco, Riccardo Rizzo, and Alfonso Urso

A Deep Learning Model for MicroRNA-Target Binding 61
Ahmet Paker and Hasan Oğul

Recurrent Neural Networks Architectures for Accidental Fall Detection on Wearable Embedded Devices 81
Mirto Musci and Marco Piastra

Part II Deep Learning for Biomedical Image Analysis

Medical Image Retrieval System Using Deep Learning Techniques 101
Jitesh Pradhan, Arup Kumar Pal, and Haider Banka

Medical Image Fusion Using Deep Learning 129
Ashif Sheikh, Jitesh Pradhan, Arpit Dhuriya, and Arup Kumar Pal

Deep Learning for Histopathological Image Analysis 153
Cédric Wemmert, Jonathan Weber, Friedrich Feuerhake, and Germain Forestier

Innovative Deep Learning Approach for Biomedical Data Instantiation and Visualization ... 171
Ryad Zemouri and Daniel Racoceanu

Convolutional Neural Networks in Advanced Biomedical Imaging Applications ... 197
Daniel A. Greenfield, Germán González, and Conor L. Evans

Part III Deep Learning for Medical Diagnostics

Deep Learning for Lung Disease Detection from Chest X-Rays Images ... 239
Ebenezer Jangam, Chandra Sekhara Rao Annavarapu, and Mourad Elloumi

Deep Learning in Multi-Omics Data Integration in Cancer Diagnostic ... 255
Abedalrhman Alkhateeb, Ashraf Abou Tabl, and Luis Rueda

Using Deep Learning with Canadian Primary Care Data for Disease Diagnosis ... 273
Hasan Zafari, Leanne Kosowan, Jason T. Lam, William Peeler, Mohammad Gasmallah, Farhana Zulkernine, and Alexander Singer

Brain Tumor Segmentation and Surveillance with Deep Artificial Neural Networks ... 311
Asim Waqas, Dimah Dera, Ghulam Rasool, Nidhal Carla Bouaynaya, and Hassan M. Fathallah-Shaykh

Index ... 351

Part I
Deep Learning for Biomedical Data Analysis

1-Dimensional Convolution Neural Network Classification Technique for Gene Expression Data

Samson Anosh Babu Parisapogu, Chandra Sekhara Rao Annavarapu, and Mourad Elloumi

Abstract In the field of bioinformatics, the development of computational methods has drawn significant interest in predicting clinical outcomes of biological data, which has a large number of features. DNA microarray technology is an approach to monitor the expression levels of sizable genes simultaneously. Microarray gene expression data is more useful for predicting and understanding various diseases such as cancer. Most of the microarray data are believed to be high dimensional, redundant, and noisy. In recent years, deep learning has become a research topic in the field of *Machine Learning* (ML) that achieves remarkable results in learning high-level latent features within identical samples. This chapter discusses various filter techniques which reduce the high dimensionality of microarray data and different deep learning classification techniques such as 2-*Dimensional Convolution Neural Network* (2D- CNN) and 1-*Dimensional* CNN (1D-CNN). The proposed method used the fisher criterion and 1D-CNN techniques for microarray cancer samples prediction.

Keywords Gene expression data · Deep learning · Convolution neural network · Machine learning · Classification

1 Introduction

Computational molecular biology is an interdisciplinary subject that includes different fields as biological science, statistics, mathematics, information technology, physics, chemistry and computer science. The analysis of biological data involves the study of a wide range of data generated in biology. This biological data is gener-

---

S. A. B. Parisapogu · C. S. R. Annavarapu (✉)
Department of Computer Science and Engineering, Indian Institute of Technology, Dhanbad, Jharkhand, India
e-mail: samson.enosh.17dr000327@cse.ism.ac.in; acsrao@iitism.ac.in

M. Elloumi
Faculty of Computing and Information Technology, The University of Bisha, Bisha, Saudi Arabia

M. Elloumi (ed.), *Deep Learning for Biomedical Data Analysis*,
https://doi.org/10.1007/978-3-030-71676-9_1

Table 1 Gene expression data format

	Sample 1	...	Sample j	...	Sample M
Gene 1	a_{11}	...	a_{1j}	...	a_{1M}
...	...	...	...	...	...
Gene i	a_{i1}	...	a_{ij}	...	a_{iM}
...	...	...	...	...	...
Gene N	a_{N1}	...	a_{Nj}	...	a_{NM}

ated from different sources, including laboratory experiments, medical records, etc. Different types of biological data include nucleotide sequences, gene expression data, macromolecular 3D structure, metabolic pathways, protein sequences, protein patterns or motifs and medical images [1]. Unlike a genome, which provides only static sequence information, microarray experiments produce gene expression patterns that provide cell functions dynamic information. Understanding the biological intercellular and intra-cellular processes underlying many diseases is essential for improving the sample classification for diagnostic and prognostic purposes and patient treatments.

Biomedical specialists are attempting to find relationships among genes and disease or formative stages, as well as relationships between genes. For example, an application of microarrays is the revelation of novel biomarkers for cancer, which can give increasingly exact determination and monitoring tools for early recognition of a specific subtype of disease or assessment of the viability of a particular treatment protocol. Different technologies are used to interpret these biological data. For example, microarray technology is useful for measuring the expression levels of a large number of genes under different environmental conditions, and *Next Generation Sequencing* (NGS) Technology for massively parallel DNA sequencing. This kind of experiments on a large amount of biological data leads to an absolute requirement of collection, storage and computational analysis [2].

In the last decade, biological data analytics has improved with the development of associated techniques such as *Machine Learning* (ML), *Evolutionary Algorithms* (EA) and *Deep Learning* (DL). These techniques are capable of handling more complex relationships in the biological data. For example, the prediction of cancer disease from the microarray data can be carried out using different ML algorithms (classification and clustering). While dealing with the microarray datasets, which has high dimensionality, are usually complex and noisy makes the classification task inconvenient [3, 4]. Table 1 presents the gene expression data format.

Due to this high dimensionality and redundancy, usual classification methods became challenging to apply on gene expression data efficiently. To reduce the problem of high dimensionality, improving learning accuracy and removing irrelevant data from gene expression data, many *filter* [48] and *wrapper* [49] approaches were applied. The *filter* method selects feature subsets independently of any learning algorithm and relies on various measures of the general characters of the training data. The wrapper method uses the predictive accuracy of a predetermined learning algorithm to determine the goodness of the selected subsets and is computationally expensive. Some of the filter methods are as follows: *Z-Score*, *Minimum Redundancy and Maximum Relevance* (mRMR) [5], T-Test [6], *Information Gain* [7], *Fisher criterion* [8] and *K-Means-Signal-to-Noise Ratio* (KM-SNR) ranking [9] etc. The effective gene selection aims to select the small subset of essential genes that are highly predictive and maximizing the ability of classifiers to classify the samples accurately. The task of finding reducts is reported to be NP-hard [10, 11].

Artificial intelligence (AI) is the reproduction of human intelligence by machine and computer systems. This evolution of reproduction done by learning—rules to use information, reasoning—rules to reach proper execution and self-correction—take appropriate actions based on learning, reasoning procedures. ML is made use of computational methods to get better machinery performances by detecting influential patterns and inconsistency information. ML makes decisions based on what they memorize or learn from data.

DL solves problems that were difficult with ML. DL uses *Artificial Neural Networks* (ANNs) [33]. An ANN acts very much like a human brain to increase computational work and provide accurate results. At present, AI is getting smarter by changing the way of its branches, namely ML and DL with enormous computational powers. *Convolutional Neural Network* (CNN) mimics brain function in processing information [19].

In this chapter, 1D-CNN, which is a multilayered DL algorithm, is proposed to classify microarray cancer data to recognize the kind of disease. CNN has the capacity in managing with insufficient data and boosting classification performance. Furthermore, CNN is also influential in detecting latent characteristics of cancer from comparable types. The organization of the chapter is as follows. Section 2 describes the related works of gene expression data classification. The preliminaries, which includes various filter and DL classification techniques, are explained in Sect. 3. Section 4 provides the details of the selected datasets. Section 5 describes the proposed approach, and Sect. 6 presents the experimental results. Finally, Sect. 7 concludes the chapter.

2 Related Works

The accessibility of open repositories of data that are appropriate for AI research is highly essential to the field of biomedical informatics. The *University of California, Irvine* (UCI) ML repository (https://archive.ics.uci.edu/ml/index.php), which contains a collection of databases, and data generators, which are used by the ML community for the empirical analysis of ML algorithms. It has enabled many numbers of analysts to demonstrate the performance of new statistical and ML algorithms. ML has excellent accuracy results when it has pre-processed data, but in real-time applications, it is not so easy to obtain pre-processed data.

Microarray information has been used in past researches to perform malignant growth, such as cancer classification by utilizing ML methodologies. *Decision Tree* (DT) was the most primitive ML method acquainted in comparing human proteins to informative genes in proteins containing ailment [29]. In analyzing microarray data, two conventional techniques which have been engaged are *classification* and *clustering*. Moreover, there are various procedures which have been executed beforehand in classifying microarray data, which includes *k-Nearest Neighbours* (KNN) [30], *Support Vector Machines* (SVM) [31], *Multilayer Perceptron* (MLP) [32], and variants of ANNs [33].

The breast cancer and leukemia datasets in performing selection of informative genes from microarray data reasoned that KNN classifier performed superiorly than random forest in terms of accurate classification [34]. From a hybrid classifier with a combination of *Particle Swarm Optimization* (PSO) and adaptive KNN in choosing useful genes recognizes a bunch of genes that meet the criteria of classification [35].

Deep Neural Network (DNN) has multiple neural layers where each layer describes hidden information from real-world raw data [12–14]. DL has been applied in many fields of computational biology. In [15], the author designed a deep learning-based framework for the prediction of drug-target interaction. In [16, 17], deep learning was applied for the prediction of eukaryotic protein subcellular localization. In [18], deep learning algorithm based on the 2D-CNN was applied for classification of microarray gene expression data.

3 Preliminaries

In this section, we present the details of three various *filter* approaches that can apply on microarray gene expression data, deep learning and CNN.

3.1 Forward Selection and Minimum Redundancy: Maximum Relevance Criterion Filter Approach

The *Minimum Redundancy - Maximum Relevance* (MRMR) criterion *filter* approach can be used to rank the genes based on their relevance and redundancy values. This criterion has been explained in [20–22] with forward selection search strategy. Given a set X_S of selected variables, the method updates X_S with the variable $X_i \in X_R$ that maximizes $v_i - z_i$, where v_i represents the relevance terms and z_i represents the redundancy terms. In detail, v_i is the relevance of X_i to the output of Y alone, and z_i is the average redundancy of X_i to each selected variables $X_j \in X_S$:

$$v_i = I(X_i; Y) \tag{1}$$

$$z_i = \frac{1}{|X_S|} \sum_{X_j \in X_S} I(X_i; X_j) \tag{2}$$

$$X_i^{MRMR} = \arg \max_{X_i \in X_R} \{v_i - z_i\} \tag{3}$$

In every step, this method chooses the variable having the best trade-off between relevance and redundancy. This selection approach is fast and efficient. At step d of forwarding search, the algorithm computes $n - d$ evaluations where each evaluation requires the estimation of $d + 1$ bivariate densities (one for each already selected variable and one for with the output). Therefore, MRMR avoids the estimation of multivariate densities by using multiple bivariate densities.

In [20], the authors justification of MRMR is as follows:

$$I(X; Y) = H(X) + H(Y) - H(X, Y) \tag{4}$$

with

$$R(X_1; X_2; \ldots; X_n) = \sum_{i=1}^{n} H(X_i) - H(X) \tag{5}$$

and

$$R(X_1; X_2; \ldots; X_n; Y) = \sum_{i=1}^{n} H(X_i) + H(Y) - H(X, Y) \tag{6}$$

Hence

$$I(X; Y) = R(X_1; X_2; \ldots; X_n; Y) - R(X_1; X_2; \ldots; X_n) \tag{7}$$

where:

- The minimum of the 2nd term $R(X_1; X_2; \ldots; X_n)$ is reached for independent variables since, in that case, $H(X) = \Sigma_i H(X_i)$ and $R(X_1; X_2; \ldots; X_n) = \Sigma_i H(X_i) - H(X) = 0$. Hence, if X_S is already selected, a variable X_i should have a minimal redundancy $I(X_i; X_S)$ with the subset. However, according to the authors the approximation of $I(X_i; X_S)$ is with $\frac{1}{|S|} \Sigma_{j \in S} I(X_i; X_j)$.
- The maximum of the 1st term $R(X_1; X_2; \ldots; X_n; Y)$ is attained for maximally dependent variables.

Qualitatively, in a sequential setting where a selected subset X_S is given, independence between the variables in X is achieved by minimizing $\frac{1}{|X_S|} \Sigma_{X_j \in X_S} I(X_i; X_j) \simeq I(X_i; X_S)$ and maximizing dependency between the variables of X and Y, that is, by maximizing $I(X_i; Y)$.

3.2 Fisher Criterion Filter Approach

In recent studies, the focus has been on the Fisher ranking measure [23], and this metric has proven its robustness against data scarcity [24] in eliminating the weak features, in this work, we used Fisher criterion to rank the genes. The Fisher criterion calculation for genes is as followed in Eq. 8.

$$FC(j) = \frac{(\mu_{j1} - \mu_{j2})^2}{\sigma_{j1}^2 - \sigma_{j2}^2} \tag{8}$$

Where, μ_{jc} is the sample mean of gene j in class c and σ_{jc}^2 is variance of gene j in c. The top N genes possessing the highest Fisher value are to be selected for the next step.

3.3 k-Means and Signal-to-Noise Ratio Filter Approach

Using this *k-Means and Signal-to-Noise Ratio* (KM-SNR) *filter* method [25], initially, all the genes are grouped into different clusters by k-means (KM) algorithm. Then genes in each cluster are positioned separately using the *Signal-to-Noise Ratio* (SNR) ranking method. These two methods are applied to overcome the redundancy issue of the gene selection process and to decrease the search space complexity. Best ranked genes from each cluster then sent to the next stage.

3.3.1 k-Means Algorithm

The *k-Means* clustering is the most utilized unsupervised learning approach, based on limiting a formal objective function and also minimizing the maximum distance from its closest center to every point. One of the excellent heuristics for understanding *k-Means* problem depends on finding a locally minimal solution [26]. The *k-Means* clustering aims to divide n data points of d dimensional space into k groups in which every data point belongs to the group with the closest mean (i.e., minimizing the within-cluster sum of squares), filling in as a model of the group. These data points clustered based on feature similarity. The data points inside a partition belong to the cluster. These partitions are also called *Voronoi partitions* [50], which is a process of segmenting a space into regions, around a set of points called seeds. The following Algorithm 1, represents the pseudocode of *k-Means* clustering algorithm.

Algorithm 1: *k*-Means clustering algorithm

Input: k = Number of clusters to be made and
S = Set of features or Dataset
Output: k clusters
// Initialization:
1: Set the cluster value k
2: Randomly select k features as initial centers from the dataset
// Loop process
3: **while** not convergence **do**
4: Assign each data point to its nearest centroid according to squared Euclidean *distance*
5: update *mean* of each cluster
6: **end while**

3.3.2 Signal-to-Noise-Ratio Ranking Approach

The *Signal-to-Noise-Ratio* (SNR) is a measurement of the expression patterns relationship between signal strength and noise of the features. It is also known as the proportion of signal to the noise power. Here, signal strength refers to the maximal distinction in the mean expression between two classes and noise refers to the minimal standard deviation of expression within each category [27, 28]. By using this method, we rank the features depending on SNR score value. The SNR score value is mathematically calculated using the following Eq. 9.

$$S_v = (\mu_1 - \mu_2)/(\sigma_1 + \sigma_2) \tag{9}$$

Here, S_v represents the SNR value, μ_1 and μ_2 are the means of $class_1$ and $class_2$ respectively. And, σ_1 and σ_2 are the standard deviations of $class_1$ and $class_2$ respectively.

3.4 Deep Learning

DL begins to administer our day by day life, exhibiting such arrangements that must be envisioned in science fiction movies just a decade earlier. Indeed, the presentation of the *AlexNet*, maybe one can think about, has started with the pivotal article published in the journal, Science, in 2006 by Hinton and Salakhutdinov [36], which described the importance of "the depth" of an ANN in ML. It fundamentally calls attention to the way that ANNs with few hidden layers can have an amazing learning capacity, that further improve with increasing depth—or equivalently the number of hidden layers. Thus comes the term "Deep" learning, a specific ML branch, which can handle intricate patterns and objects in enormous size datasets.

DL is a way to deal with ML that has drawn slowly on our knowledge of the human mind, statistics and applied math as it developed over the past several decades. In recent years, it has seen enormous growth in its popularity and usefulness, due in large part to all-powerful computers, more substantial datasets and techniques to train deeper ANNs. DL has solved increasingly complicated applications with increasing accuracy over time.

Because of the capacity of learning on multilayered representations, DL is prevalent in drawing results from complex issues. In this sense, DL is the most progressive way to be utilized in collecting and processing abstract information from several layers. Such attributes present DL as an appropriate way to be considered in dissecting and contemplating gene expression information. The ability to learning multilayered representations makes DL a flexible procedure in creating progressively accurate outcomes in a speedier time. Multi-layered representation is a component that structures the general architecture of DL [37].

ML and DL contrast in terms of performance relying upon the amount of data. For low dimensionality dataset, DL works inefficiently, as it requires data consisting of high dimensionality to comprehend learning to be carried out [38].

3.5 Convolutional Neural Network

During the most recent decade, *Convolutional Neural Networks* (CNNs) [51] has turned into the de facto standard for different Computer Vision and ML tasks. CNNs are feed-forward *Artificial Neural Networks* (ANNs) [52] with alternating convolutional and subsampling layers. Profound 2D-CNNs with many hidden layers and with many parameters can learn intricate patterns giving that they train on a gigantic size visual database with ground-truth labels. Figure 1 visualizes the pipeline of usual CNN architecture.[1]

1D-CNNs have been proposed in a few applications, for example, customized biomedical data classification and early finding, structural health observing,

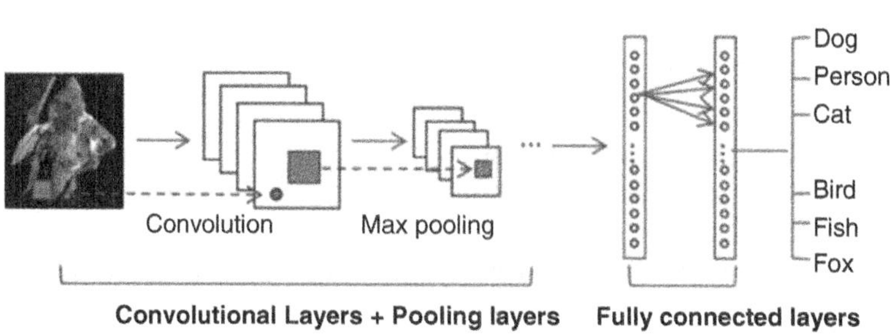

Fig. 1 The pipeline of usual CNN architecture [39]

[1]Image taken from https://www.sciencedirect.com/science/article/pii/S0925231215017634#f0010.

anomaly detection and identification in motor fault detection. Moreover, the real-time and minimum-cost hardware usage is plausible because of the reduced and straightforward configuration of 1D-CNNs that perform only 1D convolutions. The following subsections present a comprehensive review of the general architecture and principals of 2D-CNNs and 1D-CNNs [40]. The CNNs has three main components, such as the input layer, hidden layer, and latent layers. These latent (hidden) layers may categorize as a fully-connected layer, a pooling layer, or a convolutional layer. The definitions and details are as follows [39, 41]:

Convolution layer is the first layer in CNN architecture. The procedure of convolution deals with the iterative execution of explicit function towards the output of a different function. This layer comprises of various maps of neurons, portrayed as maps of features or filters. It is moderately indistinguishable in size to the dimensionality of the input. Neural reactivity deciphers through quantifying discrete convolution of receptors. The quantification manages to figure absolute neural weights of input and assigning activation function

Max pooling layer concerns with delivering a few grids from the splitting convolution layers output. In matrices, most of the grid values used to be sequenced. Operators are used in performing the calculation on every matrix to quantify average or maximize value.

A fully connected layer is a practically complete CNN, involving 90% of architectural parameters of CNN. The layer enables input to be transmitted in the ANN with preset vector lengths. A layer changes dimensional data before classification. The convolutional layer also undergoes a transformation, which enables the retaining of data integrity.

3.5.1 2D Convolutional Neural Networks

Although it has been right around thirty years after the first CNN proposed, present-day CNN structures still share the underlying properties with the absolute initial one, for example, convolutional and pooling layers. To begin with, the ubiquity and the broad scope of utilization areas of deep CNNs can ascribe to the following advantages:

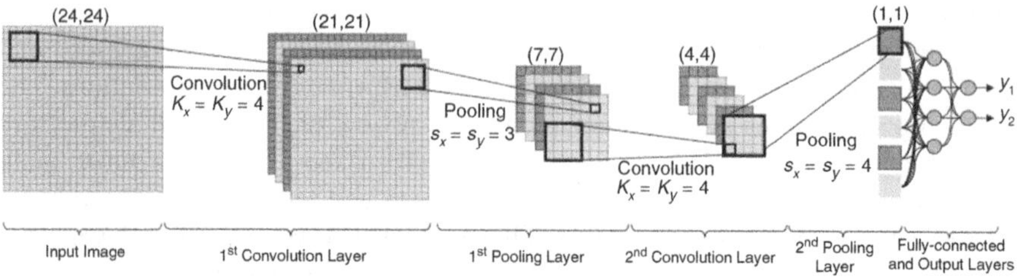

Fig. 2 The sample illustration of CNN with two convolution and one fully-connected layers [40]

1. CNNs intertwine the feature extraction and classification procedures into a single learning body. They can learn to optimize the features during the training stage legitimately from the raw input.
2. As the CNN neurons are connected sparsely along with tied weights, CNNs can process more inputs with an extraordinary computational proficiency compared with the regular fully connected *Multi-Layer Perceptrons* (MLP) networks.
3. CNNs are resistant to little changes in the input information, including translation, scaling, skewing, and distortion.
4. CNNs can adapt various sizes of inputs.

In a conventional MLPs, each hidden neuron contains scalar weights, input and output. In any case, because of the 2D nature of pictures, every neuron in CNN contains 2-D planes for weights, known as the kernel, and input and outputs which are known as a feature map. The classification of a 24×24 pixel grayscale image of two categories by conventional CNN is shown in Fig. 2 [40]. This sample CNN consists of two convolution layers and two pooling layers. The output of the second pooling layer handled by a fully-connected layer and followed by the output layer that produces the classification result.

The interconnections assigned with the weighting filters (w) and a kernel size of (K_x, K_y), which feeds the convolutional layers. As the convolution happens inside the boundary limits of the image, the feature map dimension is decreased to $(K_x - 1, K_y - 1)$ pixels from the width and height, respectively. The values (S_x, S_y) initialized in pooling layers as subsampling factors. In the sample Fig. 2, the kernel sizes of the two convolution layers assigned as $K_x = K_y = 4$, while the subsampling elements set as $S_x = S_y = 3$ for the first pooling layer and $S_x = S_y = 4$ for the subsequent one. Note that these values purposely are chosen so that the last pooling layer (i.e. the input of fully-connected layer) outputs are scalars (1×1). The output layer comprises of two fully-connected neurons relating to the number of classes to which the image is categorized. The following steps show a complete forward-propagation process of the given example CNN:

1. For the CNN, a grayscale 24×24-pixel image fed as the input layer.
2. Every neuron of the 1st convolution layer performs a linear convolution between the image and related filter to create the input feature map of the neuron.

3. Then each neurons input feature map is passed through the activation function to produce the output feature map of the neuron of the convolution neuron.
4. Inside the pooling layer, every neuron's feature map is made by pulverizing the output feature map of the past neuron of the convolution layer. In the given example image, 7 × 7 feature maps made in the 1st pooling layer.
5. Steps 3 and 4 repeated and the outputs of the 2nd pooling layer become the inputs of the fully-connected layers, which are indistinguishable from the layers of a conventional MLP.
6. The scalar outputs are forward-propagated through the accompanying fully-connected and output layers to create the last output that results in the classification of the input image.

CNNs are prevalently trained in a regulated way by a stochastic gradient descent technique or the *Back-Propagation* (BP) method. During the iterations of BP, the gradient magnitude (or sensitivity) of each ANN parameter such as the weights of convolution and fully-connected layers computed. The parameter sensitivities are then utilized to update the CNN parameters until a specific criterion attains iteratively. A detailed explanation of the BP in 2D-CNNs can found in [42].

3.5.2 1D Convolutional Neural Networks

In recent years, an alternative version of 2D-CNNs called 1D *Convolutional Neural Networks* (1D-CNNs) had been created [42]. As per the studies, specific applications of 1D-CNNs are beneficial and accordingly desirable over their 2D partners in dealing with 1D signals because of the following reasons:

- Except for matrix operations, *Forward-Propagation* (FP) and BP in 1D-CNNs require basic array operations. That implies the computational complexity nature of 1D-CNNs is substantially lower than 2D-CNNs.
- 1D-CNNs with moderately shallow architectures (for example, less number of hidden layers and neurons) are capable in learning challenging tasks including 1D signals. Then again, 2D-CNNs, for the most part, require deeper architectures to deal with such kind of tasks. ANNs with shallow architectures are a lot simpler to train and implement.
- Typically, preparing deep 2D-CNNs requires unique hardware arrangement (for example, cloud computing or GPUs). Then again, any CPU execution over a standard computer is plausible and generally quick for training compact 1D-CNNs with few hidden layers (i.e., 2 or less) and neurons (for example, less than 50).
- Because of the low computational requirements of 1D-CNNs, these are well-suited for real-time and low-cost applications, particularly on mobile or hand-held devices.

A sample 1D-CNN configuration with 3 CNN, and 2 MLP layers shown in Fig. 3. Two distinct layers proposed here: (1) the "CNN-layers" where both 1D

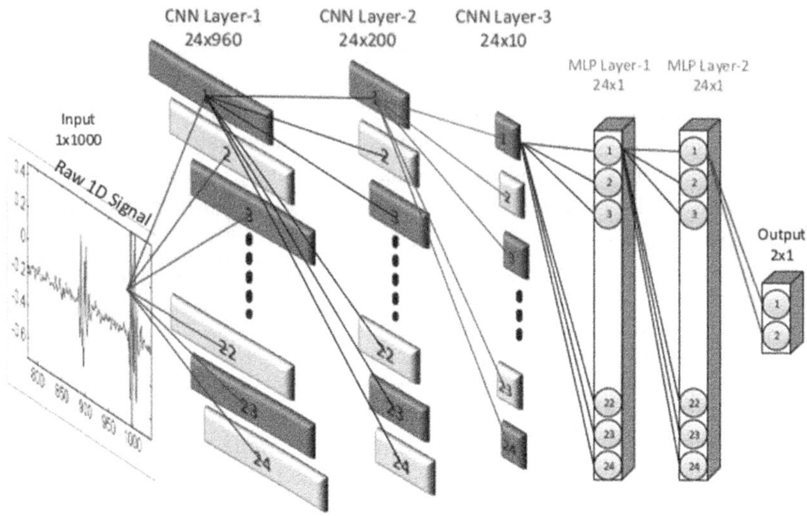

Fig. 3 A sample 1D-CNN configuration with 3 CNN and 2 MLP layers [40]

convolutions and sub-sampling (pooling) happens, and (2) Fully-connected layers that are indistinguishable from the layers of the *Multi-Layer Perceptron* (MLP), also called as "MLP-layers". The following hyper-parameters shape the design of a 1D-CNN:

1. The number of hidden CNN and MLP layers/neurons (Fig. 3, consists of 3 hidden CNN and 2 MLP layers).
2. The CNN layer filter (kernel) size (Fig. 3 has 41 as filter size for all hidden CNN layers).
3. Each CNN layers subsampling factor (In Fig. 3, the subsampling factor assigned to 4).
4. The activation and pooling functions decision.

As per the conventional 2D-CNNs, the input layer is a passive layer that gets the raw 1D signal and the output layer is an MLP layer with the number of neurons equivalent to the number of classes. Three sequential CNN layers of a 1D-CNN exhibited in Fig. 4. As appeared in this figure, the 1D filter kernels have size 3, and the sub-sampling factor is 2 where the kth neuron in the hidden CNN layer, l, first performs a grouping of convolutions, then the sum of which passed through the activation function, f, followed by the sub-sampling activity.

In reality, the primary distinction between 1D and 2D-CNNs, where 1D arrays replace 2D matrices for both kernels and feature maps. In the subsequent step, the CNN layers process the raw 1D data and "learn to extract" such features which use in the classification task of MLP-layers. As an outcome, both the operations, feature extraction and classification intertwined into one process that can be optimized to performance classification maximization. It is an essential advantage of 1D-CNNs that bring about a low computational complexity. Since the CNN topology permit

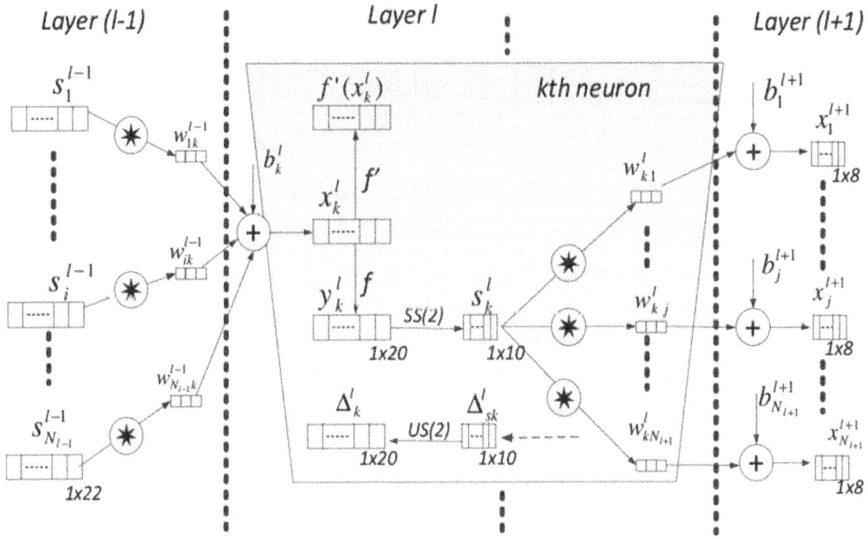

Fig. 4 1D-CNN representation with three consecutive hidden CNN layers [43]

the varieties in the input layer dimension so that the sub-sampling factor of the output CNN layer is tuned adaptively, known as adaptive implementation. The more details, along with FP and BP of CNN layers, are exhibited in [40].

4 Dataset Details

The microarray gene expression datasets used in the experiment downloaded from https://www.gems-system.org. The five benchmark microarray datasets such as ALL-AML leukemia, Prostate Tumor, DLBCL, MLL leukemia, and SRBCT utilized in this experiment.

The ALL-AML Leukemia [9] dataset consists of two classes ALL (*Acute Lymphocytic Leukemia*) and AML (*Acute Myeloid Leukemia*). In this, each sample has 7129 genes. This dataset contains 47 ALL samples, and 25 AML class samples. The Prostate Tumor [44] is a two-class dataset having 12,600 genes. This dataset consists of 77 prostate samples and 59 normal samples. The DLBCL [45] consists of two classes named DLBCL (*Diffuse large B-cell lymphoma*) and FL (*Follicular Lymphoma*) and contains 5469 genes. In this dataset, 58 DLBCL, and 19 FL samples exist.

The MLL leukemia [46] has three classes named ALL, AML, and MLL (*myeloid/ lymphoid leukemia or mixed-lineage leukemia*). Each sample contains 12,582 genes. ALL contains 24 samples, MLL consists of 20 samples, and AML has 28 samples. The SRBCT (*Small-Round-Blue-Cell Tumor*) [47] is a four-class

Table 2 Gene expression dataset details

Dataset	#Genes	#Samples	#Classes
Prostate tumor	12,600	136	2
ALL-AML	7129	72	2
DLBCL	5469	77	2
MLL	12,582	72	3
SRBCT	2308	83	4

dataset having 2308 genes. This dataset contains a total of 29 EWS (*Ewing sarcoma*) samples, 25 RMS (*Rhabdomyosarcoma*) samples, 11 BL (*Burkitt lymphoma*) samples, and 18 NB (*Neuroblastoma*) samples. The description of these three benchmark microarray datasets with details such as dataset name, number of genes, sample size, and the total number of classes of each dataset construction presented in Table 2.

5 Proposed Approach

In this section, we introduce the proposed DL 1D-CNN classification technique, applied to the gene expression dataset. Figure 5, shows the summarized flowchart of the proposed work. The proposed workflow has following four steps.

1. Preprocessing step using Fisher ranking
2. Model creation
3. Model training
4. Model testing

5.1 Preprocessing Using Fisher Ranking Approach

In this stage, The Fisher ranking method, which explained in Sect. 3.2, was applied as a filter approach to reduce the high dimensionality of the input microarray data. As explained in Sect. 3.2, the genes in each cluster are ranked using the Fisher ranking method, and the top-ranked genes from the result are gathered to make a minimized microarray dataset. We consider that the reduced microarray dataset having the highest ranking is considered as less redundant and sent to the next step.

5.2 Model Creation

For the model creation, a total of eight layers have used for 1D-CNN, which includes one input layer, two 1D Convolutional layers followed by a dropout, and a Max Pooling layer, a flatten dimension layer is to the next of it. After this layer, a dense

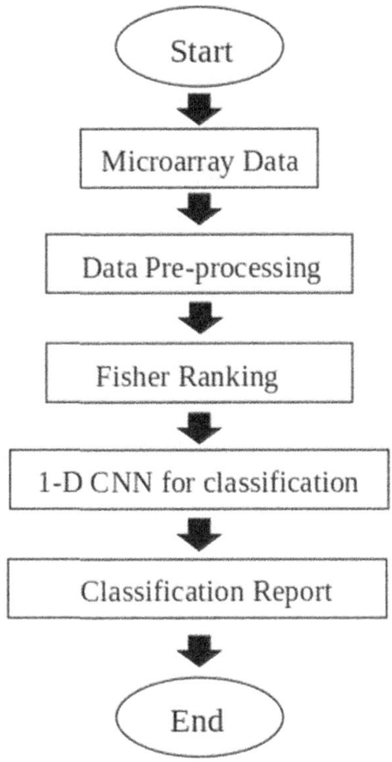

Fig. 5 The summarized proposed workflow of the deep learning 1D-CNN approach applied on microarray gene expression data

layer is followed, and then the final output layer. It is important to note that 8-layer 1D-CNN does not contain many parameters needed to be trained as compared to another 8-layer 2D-CNN with the same number of layers.

The size of the ANN input layer depends on the dataset used for the ANN, which ranges from 100, 150, and 200 to 250 features as per the output result of the preprocessing step. Although the complete data input of the microarray for which the number of samples is much less than the number of features, we applied the proposed method. The first 1D convolutional layer uses 100 kernels, each of size 20. Then, followed by a second 1D convolutional layer which uses 64 kernels each of size 10. The regularization method used here is Dropout. A dropout layer with a keep probability of 0.5 has applied after the convolutional layers. A Max-pooling layer is next with a kernel size of 5. To arrange the output tensor of the previous layer linearly, a flatten dimension layer is used, which converts an input tensor to a single dimension vector. It is followed by a dense layer with a size of 10 neurons.

Finally, we used an output layer with a soft-max activation function for multi-class gene expression datasets and a sigmoid activation function for binary class datasets. The rest of the layers uses *Rectified Linear Unit* (ReLu) as an activation function. The loss function which was used to calculate classification loss is the categorical cross-entropy for multi-class datasets, and *Mean Squared Error* (MSE)

```
Layer (type)                   Output Shape         Param #
===============================================================
C1 (Conv1D)                    (None, 231, 100)     2100
_______________________________________________________________
C2 (Conv1D)                    (None, 222, 64)      64064
_______________________________________________________________
dropout_layer1 (Dropout)       (None, 222, 64)      0
_______________________________________________________________
maxpool_layer (MaxPooling1D)   (None, 44, 64)       0
_______________________________________________________________
flatten_dimension (Flatten)    (None, 2816)         0
_______________________________________________________________
dense_53 (Dense)               (None, 10)           28170
_______________________________________________________________
dense_54 (Dense)               (None, 1)            11
===============================================================
Total params: 94,345
Trainable params: 94,345
Non-trainable params: 0
_______________________________________________________________
```

Fig. 6 An example 1D-CNN structure for the binary class dataset having 100 features as input data

for binary class datasets. An example 1D-CNN structure for the binary class dataset having 100 features as an input data is visualized in Fig. 6.

5.3 *Model Training*

The data is divided into two sets after random shuffling with 80% of the data belonging to the training set and 20% data of the test set. The Labels of each sample were converted to a one-hot encoded format for the multiclass dataset. The optimizer used for training the model is Adam (Adaptive Moment Estimation). The training process is done using the mini-batch technique with a batch size of 20 and 1000 epochs.

5.4 *Model Testing*

Testing is necessary to gauge the generalizability of the model. The model is tested using the test set, which contains examples unfamiliar to the model. The results of the classification task tabulated for the case of the training set, as well as the testing set for further analysis like f1-score calculation and confusion matrix. These different parameters help us to conclude on model performance on the dataset and its reliability on unseen data.

6 Experimental Results

We have implemented the proposed 1D-CNN DL classification technique with the model created in Sect. 5.2 for microarray gene expression data consisting of five different cancer datasets. The proposed model was implemented in *Python* using packages of *Keras* with *TensorFlow* backend and *Scikit* Learn. The experiments were carried out on system with following specifications:

- Processor: Intel Core i5-8300H CPU @ 2.30GHZ
- Storage: SSD- 128 GB and RAM- 8.00 GB
- OS and system type: Windows 10 64-bit operating system, x64-based processor
- Software and tools: Anaconda with jupyter notebook 5.7.8 and python 3.7.3

In the experiment, initially all the complete datasets of five microarray datasets which are explained in Sect. 4 have taken and ranked them using the fisher criterion ranking approach, which was explained in Sect. 3.2. After ranking, the top 100, 150, 200, and 250 genes were taken separately and applied the proposed 1D-CNN DL classification technique, explained in Sect. 3.5.2 with 80:20 train and test split, then 70:30 train and test splits, respectively.

The experiments repeated to calculate the average performance of the model while tested. Because the convergence also depends on how well the model weights were initialized randomly for each run. The training time of the model is dependent on the number of epochs they trained. For example, a simple 1D-CNN model (2 × 1D-Convolution layers, max-pool layer, dense layer, and output layer) trained for 100 epochs completed training in 208 seconds. Other model's training time varied and took approximately less than an hour to train. Memory usage is directly related to the number of weights (parameters) adjusted in the model. A simple model like the multi-layer ANN has a large number of parameters to be adjusted compared to models like CNN. For example, one of the CNN used in our experiments was a model with two 1D convolution layers followed by a max-pool layer (max-pool does not amount any weights) and then a dense layer followed by the output layer. This model had 358,058 numbers of parameters. The memory requirements for the models utilized were not large as compared to popular DNNs having a large number of layers and parameters which take hours to train and have large memory requirements.

The performance of the classifier model was evaluated after getting the *True Positive* (TP), *False Positive* (FP), *True Negative* (TN), and *False Negative* (FN) values.

The formulae to calculate the performance measures are as follows:

$$Positive\ Predictive\ Value(PPV)\ (or)\ Precision = \frac{TP}{TP+FP}$$

$$Recall\ (or)\ Sensitivity\ (or)\ True\ Positive\ Rate(TPR) = \frac{TP}{TP+FN}$$

$$False\ Positive\ Rate(FPR) = \frac{FP}{TN+FP}$$

$Accuracy = \frac{TP+TN}{TP+TN+FP+FN}$

$f1\text{-}score = 2 * \frac{Precision*Recall}{Precision+Recall}$

In Tables 3 and 4 the classification results of 1D-CNN on complete datasets with 70:30 and 80:20 train and test splits are visualized. In these Tables 3 and 4, the classification results containing Dataset name, number of genes, training accuracy, testing accuracy, precision, recall, f1-score measure, *False-Positive Rate* (FPR), *Area Under Curve* (AUC), and execution of training time are displayed.

Similarly, in Tables 5 and 6 the classification results of 1D-CNN on top 100 fisher ranked genes of all the datasets have visualized. In Table 5, the classification results containing dataset name, training accuracy, testing accuracy, precision, recall, f1-score measure, FPR, AUC, and execution of training time with 70:30 train and test split are displayed. Then in Table 6, the classification results containing dataset name, training accuracy, testing accuracy, precision, recall, f1-score measure, FPR, AUC with 80:20 train, and test splits are displayed.

Table 3 Classification of gene expression data using 1D-CNN on complete datasets with 70:30 train-test ratio

Dataset	#Genes	Training Acc.	Testing Acc.	Precision	Recall	f1-score	FPR	AUC	Trainin time
Prostate	12,600	60	48.78	0.24	0.49	0.32	0.51	0.49	6099.07
DLBCL	5469	100	95.83	0.96	0.96	0.96	0.04	0.96	1347.19
ALL-AML	7129	100	100	1	1	1	0	1	3636.95
MLL	12,582	100	100	1	1	1	0	1	5900.05
SRBCT	2308	100	95.99	0.96	0.96	0.96	0.01	0.97	1302.51

Table 4 Classification of gene expression data using 1D-CNN on complete datasets with 80:20 train-test ratio

Dataset	#Genes	Training Acc.	Testing Acc.	Precision	Recall	f1-score	FPR	AUC	Trainin time
Prostate	12,600	60.19	42.86	0.18	0.43	0.26	0.57	0.43	5870.41
DLBCL	5469	75.41	75	0.56	0.75	0.64	0.25	0.75	1683.89
ALL-AML	7129	100	95.83	0.96	0.96	0.96	0.04	0.96	245.11
MLL	12,582	100	86.67	0.87	0.87	0.87	0.06	0.9	6236.03
SRBCT	2308	100	100	1	1	1	0	1	1482.82

Table 5 Classification of gene expression data using 1D-CNN for top 100 fisher ranked genes with 70:30 train-test ratio

Dataset	Training Acc.	Testing Acc.	Precision	Recall	f1-score	FPR	AUC	Training time
Prostate	98.95	95.12	0.95	0.95	0.95	0.05	0.95	138.35
DLBCL	100	91.67	0.92	0.92	0.92	0.08	0.92	62.48
ALL-AML	100	100	1	1	1	0	1	61.05
MLL	100	90.91	0.92	0.91	0.91	0.05	0.93	62.47
SRBCT	100	87.99	0.90	0.88	0.88	0.04	0.92	65.46

Table 6 Classification of gene expression data using 1D-CNN for top 100 fisher ranked genes with 80:20 train-test ratio

Dataset	Training Acc.	Testing Acc.	Precision	Recall	f1-score	FPR	AUC
Prostate	100	89.28	0.91	0.89	0.89	0.11	0.89
DLBCL	100	87.50	0.89	0.88	0.86	0.12	0.87
ALL-AML	100	100	1	1	1	0	1
MLL	100	86.66	0.89	0.87	0.86	0.06	0.90
SRBCT	100	82.35	0.86	0.82	0.83	0.05	0.88

Table 7 Classification of gene expression data using 1D-CNN for top 150 fisher ranked genes with 70:30 train-test ratio

Dataset	Training Acc.	Testing Acc.	Precision	Recall	f1-score	FPR	AUC	Training time
Prostate	98.95	90.24	0.91	0.90	0.90	0.09	0.91	183.56
DLBCL	100	87.50	0.87	0.88	0.87	0.12	0.87	79.18
ALL-AML	100	100	1	1	1	0	1	76.24
MLL	100	90.91	0.92	0.91	0.91	0.05	0.93	79.63
SRBCT	100	80.00	0.88	0.80	0.81	0.07	0.87	85.87

Table 8 Classification of gene expression data using 1D-CNN for top 150 fisher ranked genes with 80:20 train-test ratio

Dataset	Training Acc.	Testing Acc.	Precision	Recall	f1-score	FPR	AUC
Prostate	99.07	96.42	0.97	0.96	0.96	0.03	0.96
DLBCL	98.36	93.75	0.94	0.94	0.93	0.06	0.93
ALL-AML	100	100	1	1	1	0	1
MLL	100	86.66	0.89	0.87	0.86	0.06	0.90
SRBCT	100	70.58	0.46	0.53	0.48	0.09	0.81

From Tables 7 and 8 we displayed the classification results of 1D-CNN for the top 150 fisher ranked genes of all the datasets. In Table 7, the classification results containing dataset name, training accuracy, testing accuracy, precision, recall, f1-score, FPR, AUC, and execution of training time with 70:30 train and test split are displayed. Then in Table 8, the classification results containing dataset name, training accuracy, testing accuracy, precision, recall, f1-score measure, FPR, AUC with 80:20 train, and test splits are displayed.

In Tables 9, 10, 11, and 12 we displayed the classification results of 1D-CNN for the top 200, and top 250 fishers ranked genes of all the datasets. In Tables 9 and 11, the classification results containing dataset name, training accuracy, testing accuracy, precision, recall, f1-score, FPR, AUC, and execution of training time with 70:30 train and test split are displayed. Then in Tables 10 and 12, the classification results containing dataset name, training accuracy, testing accuracy, precision, recall, f1-score measure, FPR, AUC with 80:20 train, and test splits are displayed.

Table 9 Classification of gene expression data using 1D-CNN for top 200 fisher ranked genes with 70:30 train-test ratio

Dataset	Training Acc.	Testing Acc.	Precision	Recall	f1-score	FPR	AUC	Training time
Prostate	98.95	92.68	0.94	0.93	0.93	0.07	0.93	235.23
DLBCL	100	95.83	0.96	0.96	0.96	0.04	0.96	145.82
ALL-AML	100	100	1	1	1	0	1	92.54
MLL	100	90.91	0.92	0.91	0.91	0.05	0.93	96.18
SRBCT	100	80.00	0.77	0.80	0.77	0.07	0.87	153.78

Table 10 Classification of gene expression data using 1D-CNN for top 200 fisher ranked genes with 80:20 train-test ratio

Dataset	Training Acc.	Testing Acc.	Precision	Recall	f1-score	FPR	AUC
Prostate	99.07	89.28	0.91	0.89	0.89	0.11	0.89
DLBCL	100	100	1	1	1	0	1
ALL-AML	100	100	1	1	1	0	1
MLL	100	93.33	0.95	0.93	0.93	0.03	0.95
SRBCT	100	76.47	0.75	0.76	0.74	0.07	0.84

Table 11 Classification of gene expression data using 1D-CNN for top 250 fisher ranked genes with 70:30 train-test ratio

Dataset	Training Acc.	Testing Acc.	Precision	Recall	f1-score	FPR	AUC	Training time
Prostate	100	97.56	0.98	0.98	0.98	0.02	0.98	281.18
DLBCL	100	95.83	0.96	0.96	0.96	0.04	0.96	157.45
ALL-AML	100	95.45	0.96	0.95	0.95	0.04	0.95	158.45
MLL	100	95.45	0.96	0.95	0.96	0.02	0.97	159.82
SRBCT	100	75.99	0.76	0.76	0.74	0.08	0.84	158.25

Table 12 Classification of gene expression data using 1D-CNN for top 250 fisher ranked genes with 80:20 train-test ratio

Dataset	Training Acc.	Testing Acc.	Precision	Recall	f1-score	FPR	AUC
Prostate	99.07	89.28	0.91	0.89	0.89	0.11	0.89
DLBCL	100	100	1	1	1	0	1
ALL-AML	100	100	1	1	1	0	1
MLL	100	93.33	0.95	0.93	0.93	0.03	0.95
SRBCT	100	76.47	0.75	0.76	0.74	0.07	0.84

By observing the Tables 4, 5, 6, 7, 8, 9, 10, 11 and 12, we can say that the classification performance of the proposed 1D-CNN DL model performed efficiently for both train and test split ratios of 70:30 and 80:20. Although the classification performance of the proposed 1D-CNN model performed efficiently on both the train and test split ratios, the performance of 1D-CNN for 70:30 train and split ratio was shown better results on all the datasets when compared with 80:20 train and test the split ratio. As we already know that the DL techniques are highly efficient for usage in image classification and high dimensional data. With the experiment of the application of 1D-CNN of DL technique on microarray gene expression data, proved that DL techniques could also show efficient results on small sample size datasets.

7 Conclusion

In this chapter, we proposed a method for classifying microarray gene expression data using a 1D-CNN, which is a DL classification technique. This chapter explained the details of microarray preprocessing filter techniques such as forward selection and MRMR approach, *Fisher Criterion* approach, and *k-Means* and SNR approach. Then the DL classification techniques such as 2D-CNN and 1D-CNN have explored. Initially, we applied the fisher ranking filter approach to rank the genes and taken top 100, 150, 200, and 250 genes. On the selected top gene subsets, we applied the proposed 1D-CNN DL technique to classify the microarray samples. The datasets split into 70:30, and 80:20 ratios, and the proposed classification technique showed efficient performance results on 70:30 train and test split compared with 80:20 train and test split of datasets. With this, we can conclude that the deep learning classification techniques could apply to small dimensional datasets. In the future, we can also apply various deep learning classification techniques on the biological datasets to classify the disease samples.

References

1. Bhatia, D. (2010). Medical informatics: A boon to the healthcare industry. Chronicles of Young Scientists, 1(3), 26.
2. Roh, S. W., Abell, G. C., Kim, K. H., Nam, Y. D., & Bae, J. W. (2010). Comparing microarrays and next-generation sequencing technologies for microbial ecology research. Trends in biotechnology, 28(6), 291–299.
3. Ghorai, S., Mukherjee, A., Sengupta, S., & Dutta, P. K. (2010, December). Multicategory cancer classification from gene expression data by multiclass NPPC ensemble. In 2010 International Conference on Systems in Medicine and Biology (pp. 41–48). IEEE.
4. Annavarapu, C. S. R., Dara, S., & Banka, H. (2016). Cancer microarray data feature selection using multi-objective binary particle swarm optimization algorithm. EXCLI journal, 15, 460.
5. Ding, C., & Peng, H. (2005). Minimum redundancy feature selection from microarray gene expression data. Journal of bioinformatics and computational biology, 3(02), 185–205.
6. Jafari, P., & Azuaje, F. (2006). An assessment of recently published gene expression data analyses: reporting experimental design and statistical factors. BMC Medical Informatics and Decision Making, 6(1), 27.
7. Wang, Z. (2005). Neuro-fuzzy modeling for microarray cancer gene expression data. First year transfer report, University of Oxford.

8. Yassi, M., & Moattar, M. H. (2014). Robust and stable feature selection by integrating ranking methods and wrapper technique in genetic data classification. Biochemical and biophysical research communications, 446(4), 850–856.
9. Golub, T. R., Slonim, D. K., Tamayo, P., Huard, C., Gaasenbeek, M., Mesirov, J. P., ... & Bloomfield, C. D. (1999). Molecular classification of cancer: class discovery and class prediction by gene expression monitoring. science, 286(5439), 531–537.
10. Alshamlan, H. M., Badr, G. H., & Alohali, Y. A. (2015). Genetic Bee Colony (GBC) algorithm: A new gene selection method for microarray cancer classification. Computational biology and chemistry, 56, 49–60.
11. Skowron, A., & Rauszer, C. (1992). The discernibility matrices and functions in information systems. In Intelligent decision support (pp. 331–362). Springer, Dordrecht.
12. Hatcher, W. G., & Yu, W. (2018). A survey of deep learning: platforms, applications and emerging research trends. IEEE Access, 6, 24411–24432.
13. Sze, V., Chen, Y. H., Yang, T. J., & Emer, J. S. (2017). Efficient processing of deep neural networks: A tutorial and survey. Proceedings of the IEEE, 105(12), 2295–2329.
14. Gheisari, M., Wang, G., & Bhuiyan, M. Z. A. (2017, July). A survey on deep learning in big data. In 2017 IEEE International Conference on Computational Science and Engineering (CSE) and IEEE International Conference on Embedded and Ubiquitous Computing (EUC) (Vol. 2, pp. 173–180). IEEE.
15. Wen, M., Zhang, Z., Niu, S., Sha, H., Yang, R., Yun, Y., & Lu, H. (2017). Deep-learning-based drug?target interaction prediction. Journal of proteome research, 16(4), 1401–1409.
16. Wei, L., Ding, Y., Su, R., Tang, J., & Zou, Q. (2018). Prediction of human protein subcellular localization using deep learning. Journal of Parallel and Distributed Computing, 117, 212–217.
17. Almagro Armenteros, J. J., Snderby, C. K., Snderby, S. K., Nielsen, H., & Winther, O. (2017). DeepLoc: prediction of protein subcellular localization using deep learning. Bioinformatics, 33(21), 3387–3395.
18. Zeebaree, D. Q., Haron, H., & Abdulazeez, A. M. (2018, October). Gene Selection and Classification of Microarray Data Using Convolutional Neural Network. In 2018 International Conference on Advanced Science and Engineering (ICOASE) (pp. 145–150). IEEE.
19. Zeng, T., & Ji, S. (2015, November). Deep convolutional neural networks for multi-instance multi-task learning. In 2015 IEEE International Conference on Data Mining (pp. 579–588). IEEE.
20. Peng, H., Long, F., & Ding, C. (2005). Feature selection based on mutual information: criteria of max-dependency, max-relevance, and min-redundancy. IEEE Transactions on Pattern Analysis & Machine Intelligence, (8), 1226–1238.
21. Tourassi, G. D., Frederick, E. D., Markey, M. K., & Floyd Jr, C. E. (2001). Application of the mutual information criterion for feature selection in computer?aided diagnosis. Medical physics, 28(12), 2394–2402.
22. Meyer, P. E., & Bontempi, G. (2013). Information?Theoretic Gene Selection In Expression Data. Biological Knowledge Discovery Handbook: Preprocessing, Mining, and Postprocessing of Biological Data, 399–420.
23. Sharbaf, F. V., Mosafer, S., & Moattar, M. H. (2016). A hybrid gene selection approach for microarray data classification using cellular learning automata and ant colony optimization. Genomics, 107(6), 231–238.
24. Yassi, M., & Moattar, M. H. (2014). Robust and stable feature selection by integrating ranking methods and wrapper technique in genetic data classification. Biochemical and biophysical research communications, 446(4), 850–856.
25. Hengpraprohm, S., & Chongstitvatana, P. (2007, October). Selecting Informative Genes from Microarray Data for Cancer Classification with Genetic Programming Classifier Using K-Means Clustering and SNR Ranking. In 2007 Frontiers in the Convergence of Bioscience and Information Technologies (pp. 211–218). IEEE.
26. Forgey, E. (1965). Cluster analysis of multivariate data: Efficiency vs. interpretability of classification. Biometrics, 21(3), 768–769.

27. Sahu, B., Dehuri, S., & Jagadev, A. K. (2017). Feature selection model based on clustering and ranking in pipeline for microarray data. Informatics in Medicine Unlocked, 9, 107–122.
28. Cuperlovic-Culf, M., Belacel, N., & Ouellette, R. J. (2005). Determination of tumour marker genes from gene expression data. Drug discovery today, 10(6), 429–437.
29. Liao, Q., Jiang, L., Wang, X., Zhang, C., & Ding, Y. (2017, December). Cancer classification with multi-task deep learning. In 2017 International Conference on Security, Pattern Analysis, and Cybernetics (SPAC) (pp. 76–81). IEEE.
30. Li, C., Zhang, S., Zhang, H., Pang, L., Lam, K., Hui, C., & Zhang, S. (2012). Using the K-nearest neighbor algorithm for the classification of lymph node metastasis in gastric cancer. Computational and mathematical methods in medicine, 2012.
31. Furey, T. S., Cristianini, N., Duffy, N., Bednarski, D. W., Schummer, M., & Haussler, D. (2000). Support vector machine classification and validation of cancer tissue samples using microarray expression data. Bioinformatics, 16(10), 906–914.
32. Wang, Z., Wang, Y., Xuan, J., Dong, Y., Bakay, M., Feng, Y., ... & Hoffman, E. P. (2006). Optimized multilayer perceptrons for molecular classification and diagnosis using genomic data. Bioinformatics, 22(6), 755–761.
33. Asyali, M. H., Colak, D., Demirkaya, O., & Inan, M. S. (2006). Gene expression profile classification: a review. Current Bioinformatics, 1(1), 55–73.
34. Kumar, C. A., Sooraj, M. P., & Ramakrishnan, S. (2017). A comparative performance evaluation of supervised feature selection algorithms on microarray datasets. Procedia computer science, 115, 209–217.
35. Kar, S., Sharma, K. D., & Maitra, M. (2015). Gene selection from microarray gene expression data for classification of cancer subgroups employing PSO and adaptive K-nearest neighborhood technique. Expert Systems with Applications, 42(1), 612–627.
36. Hinton, G. E., & Salakhutdinov, R. R. (2006). Reducing the dimensionality of data with neural networks. science, 313(5786), 504–507.
37. Bianchini, M., & Scarselli, F. (2014). On the complexity of neural network classifiers: A comparison between shallow and deep architectures. IEEE transactions on neural networks and learning systems, 25(8), 1553–1565.
38. Wang, H., Meghawat, A., Morency, L. P., & Xing, E. P. (2017, July). Select-additive learning: Improving generalization in multimodal sentiment analysis. In 2017 IEEE International Conference on Multimedia and Expo (ICME) (pp. 949–954). IEEE.
39. Guo, Y., Liu, Y., Oerlemans, A., Lao, S., Wu, S., & Lew, M. S. (2016). Deep learning for visual understanding: A review. Neurocomputing, 187, 27–48.
40. Kiranyaz, S., Avci, O., Abdeljaber, O., Ince, T., Gabbouj, M., & Inman, D. J. (2019). 1D Convolutional Neural Networks and Applications: A Survey. arXiv preprint arXiv:1905.03554.
41. Zeebaree, D. Q., Haron, H., & Abdulazeez, A. M. (2018, October). Gene Selection and Classification of Microarray Data Using Convolutional Neural Network. In 2018 International Conference on Advanced Science and Engineering (ICOASE) (pp. 145–150). IEEE.
42. Kiranyaz, S., Ince, T., & Gabbouj, M. (2015). Real-time patient-specific ECG classification by 1-D convolutional neural networks. IEEE Transactions on Biomedical Engineering, 63(3), 664–675.
43. Kiranyaz, S., Gastli, A., Ben-Brahim, L., Alemadi, N., & Gabbouj, M. (2018). Real-time fault detection and identification for MMC using 1D convolutional neural networks. IEEE Transactions on Industrial Electronics.
44. Singh, D., Febbo, P. G., Ross, K., Jackson, D. G., Manola, J., Ladd, C., ... & Lander, E. S. (2002). Gene expression correlates of clinical prostate cancer behavior. Cancer cell, 1(2), 203–209.
45. Alizadeh, A. A., Eisen, M. B., Davis, R. E., Ma, C., Lossos, I. S., Rosenwald, A., ... & Powell, J. I. (2000). Distinct types of diffuse large B-cell lymphoma identified by gene expression profiling. Nature, 403(6769), 503.
46. Armstrong, S. A., Staunton, J. E., Silverman, L. B., Pieters, R., den Boer, M. L., Minden, M. D., ... & Korsmeyer, S. J. (2002). MLL translocations specify a distinct gene expression profile that distinguishes a unique leukemia. Nature genetics, 30(1), 41–47.

47. Pati, S. K., Das, A. K., & Ghosh, A. (2013, December). Gene selection using multi-objective genetic algorithm integrating cellular automata and rough set theory. In International Conference on Swarm, Evolutionary, and Memetic Computing (pp. 144–155). Springer, Cham.
48. Lazar, C., Taminau, J., Meganck, S., Steenhoff, D., Coletta, A., Molter, C., ... & Nowe, A. (2012). A survey on filter techniques for feature selection in gene expression microarray analysis. IEEE/ACM Transactions on Computational Biology and Bioinformatics (TCBB), 9(4), 1106–1119.
49. Hira, Z. M., & Gillies, D. F. (2015). A review of feature selection and feature extraction methods applied on microarray data. Advances in bioinformatics, 2015.
50. Kotu, V., & Deshpande, B. (2014). Predictive analytics and data mining: concepts and practice with rapidminer. Morgan Kaufmann.
51. Saha, S. (2018). A comprehensive guide to convolutional neural networks? the ELI5 way.
52. Hopfield, J. J. (1988). Artificial neural networks. IEEE Circuits and Devices Magazine, 4(5), 3–10.

Classification of Sequences with Deep Artificial Neural Networks: Representation and Architectural Issues

Domenico Amato, Mattia Antonino Di Gangi, Antonino Fiannaca, Laura La Paglia, Massimo La Rosa, Giosué Lo Bosco, Riccardo Rizzo, and Alfonso Urso

Abstract DNA sequences are the basic data type that is processed to perform a generic study of biological data analysis. One key component of the biological analysis is represented by sequence classification, a methodology that is widely used to analyze sequential data of different nature. However, its application to DNA sequences requires a proper representation of such sequences, which is still an open research problem. *Machine Learning* (ML) methodologies have given a fundamental contribution to the solution of the problem. Among them, recently, also *Deep Neural Network* (DNN) models have shown strongly encouraging results. In this chapter, we deal with specific classification problems related to two biological scenarios: (A) metagenomics and (B) chromatin organization. The investigations have been carried out by considering DNA sequences as input data for the classification methodologies. In particular, we study and test the efficacy of (1) different DNA sequence representations and (2) several *Deep Learning* (DL) architectures that process sequences for the solution of the related supervised classification problems. Although developed for specific classification tasks, we think that such architectures

D. Amato
Dipartimento di Matematica e Informatica, Università degli studi di Palermo, Palermo, Italy
e-mail: domenico.amato01@unipa.it

M. A. Di Gangi
Fondazione Bruno Kessler, Università degli Studi di Trento, Trento, Italy
e-mail: mattia.digangi@alumni.unitn.it

A. Fiannaca · L. La Paglia · M. La Rosa · R. Rizzo · A. Urso
ICAR-CNR, National Research Council of Italy, Palermo, Italy
e-mail: antonino.fiannaca@icar.cnr.it; laura.lapaglia@icar.cnr.it; massimo.larosa@icar.cnr.it; riccardo.rizzo@icar.cnr.it; alfonso.urso@icar.cnr.it

G. Lo Bosco (✉)
Dipartimento di Matematica e Informatica, Università degli studi di Palermo, Palermo, Italy

Dipartimento di Scienze per l'Innovazione Tecnologica, Istituto Euro-Mediterraneo di Scienza e Tecnologia, Palermo, Italy
e-mail: giosue.lobosco@unipa.it

M. Elloumi (ed.), *Deep Learning for Biomedical Data Analysis*,
https://doi.org/10.1007/978-3-030-71676-9_2

could be served as a suggestion for developing other DNN models that process the same kind of input.

Keywords Deep neural network · Sequence classification · Bacteria classification · Nucleosome identification · Metagenomics

1 Introduction

Biological sequence classification has many fields of application as structural bioinformatics, genomics, transcriptomics, epigenomics and metagenomics [7]. Classification of genomic or proteomic sequences can be very useful to extract phylogenetic information [64]. Biological data is produced nowadays in large quantities by using the new advanced *Next Generation Sequencing* (NGS) methodologies, which has raised the need for new state-of-the-art bioinformatics aids [16] to properly analyze them. Indeed, one of the biggest challenges in Bioinformatics is to efficiently annotate the massive volume of biological sequences, both nucleotides and amino acids [61]. Actually, specific patterns of nucleotides or amino acids, as sequence motifs, can be compared to reconstruct the phylogenetic tree of different genomes. The importance of studying sequence patterns relies on their linkage with their biological function [12]. Here we present two distinct biological scenarios that well explain the relevance of classifying sequences: (1) metagenomics, and (2) chromatin organization. The metagenome is the genetic material belonging to a multitude of bacterial populations in a specific environmental sample. The advantage of metagenomics investigation is that it can be performed by directly analyzing biological samples, avoiding all the laboratory steps required to culture, isolate and keep bacteria in cellular cultures [52, 70]. The analysis of microbial communities conceptually relies on species richness and differential abundance [28, 60]. It is important to consider both features because bacteria could have an equal number of species (species richness) but different abundances [9]. Metagenomics has been investigated especially in the biomedical field. Specific patterns of microbial flora are often linked to the prediction, onset or outcome of different pathologies [8, 49, 65]. Because of this, it is important to well characterize different sequences belonging to different bacterial species to investigate their potential role in these pathologies.

When applied to human microbiota, metagenomics is a study of the totality of nucleic acids belonging to all microbes present in the human organism.

Human microbiota may have relevant functional properties, such as conferring host protection against invading pathogen, regulation of diverse host physiological functions including metabolism, development and homeostasis of immune and the nervous system. Thus, microbiota imbalance could lead to host dysfunction in a multitude of processes [26]. An interesting study of Chaput et al. [3] evidences how the microbiome can play a relevant role in cancer treatment. Its profile can predict clinical response and colitis in melanoma patients treated with target therapy. Frankel et al. [15] report that metabolomic profiling is strongly associated with specific microbiota profiles of the human gut, and they associate with immune checkpoint therapy efficacy in melanoma patients. For these reasons, many research groups tried to characterize human microbiome through meta-omics projects including both 16S *ribosomal* RNA (rRNA) and *Whole-Genome Shotgun* (WGS) approaches. Bioinformatics plays a key role in this kind of analysis, because of both the large amount of data produced and the difficulty to deeply classify different bacterial species. To this end, what we propose here, is a supervised classifier, based on a *Deep Neural Network* (DNN), which can classify 16S sequences as belonging to different bacterial species.

The second biological scenario discussed in this chapter is chromatin organization [74]. The eukaryote genome is packed as chromatin [51], the fundamental unit of packaging is called *nucleosome*, and it consists of DNA wrapped around a protein core (*octamer*). Nucleosomes are separated from each other by sequences of DNA called *linker DNA*. Starting from this low-level organization, chromatin is coiled into Higher-Order Structures to finally form the chromosomes. Nucleosome positioning studies are relevant for two main reasons: first, it indicates the physical packaging of DNA; this phenomenon drives the determination of the final architecture of chromatin in the cell [68, 69] both trough the DNA sequence itself, both through the interaction of different epigenetic factors, including remodelling proteins [56]. Second, nucleosome positioning drives gene regulation through a well structured and nonrandom sequence of events linked to post-translational modification of histones and DNA bounded to regulatory proteins [2, 23, 57, 58]. Furthermore, nucleosomes regulate the accessibility of different regulative element to DNA [41, 63], and they are critical for other biological processes such as replication [36] and recombination [48]. For these reasons, nucleosome analysis is an important area of study in biology that is producing interesting findings. For sure there is clear evidence that the DNA sequence is responsible for nucleosome positioning [25]. Following this consideration, several computational models that use DNA sequence as input have been proposed so far for nucleosome prediction [47]. Also for this case, we propose DNN classifier for the problem of nucleosome prediction starting from the DNA sequence information.

Deep Learning (DL) is nowadays a successful paradigm for big data classification [30]. Their implementation is almost accessible to all, due to the low-level cost of GPU computing cards. For sure, DL techniques represent now state of the art for the supervised and unsupervised classification task.

Most of the contributions of DL methods to bioinformatics have been in genomic medicine and medical imaging research fields [42]. DL contribution to sequence classification is at present an active research field [34].

In the following sections, we will show how the problems related to these two scenarios were approached using DNNs. In Sect. 2 we introduce the most common sequence representations; Sect. 3 presents an introduction to deep neural architectures for sequence classification, while their experimental results to DNA sequences are reported and discussed in Sect. 4. Finally, Sect. 5 concludes this chapter.

2 Sequences Representation

Machine Learning (ML) algorithms require numeric input variables as input and can not process label data directly. As such, DNA sequences need to be converted to numerical representations. Formally, a DNA sequence s of length $l(s)$ is a string of symbols from a finite alphabet, this alphabet is often $\Sigma = \{A, T, C, G\}$ but it can also contain other symbols that represent the "ambiguity" of a combination of bases (for example in the IUPAC notation W stands for A or T, S stands for C or G, and so on). The numerical representation of a sequence can be seen as a mapping ϕ of the sequence s into a numerical multidimensional feature vector x_s of fixed size.

2.1 *Representation of Fixed-Length Sequences*

Let m be the number of labels in the alphabet Σ, one can set an *integer encoding*, i.e. a bijection between Σ and the set of labels $\{1, .., m\}$. The adoption of integer encoding as a numerical sequence representation generates numerical data which inherits the natural order property of the numbers. This could represent a useless property regardless of the problem to solve, that could be potentially learnt by a supervised learning algorithm. To avoid such kind of problem, the solution is the use of the so-called *one hot encoding*. It is a binary representation of the labels by a binary vector whose bit length is equal to the number of different labels. In the case of a DNA sequence s of length $l(s)$, it is transformed into a multidimensional vector x_s of size $4 \times l(s)$. Each column j of the matrix has all zero entries except for a single one at position i. The position corresponds to the integer chosen to represent the label. The results of the representation by one-hot encoding lead to a sparse representation of the input, and this could be an issue regarding the informative content of the representation. Conversely, an advantage is that the required space complexity is fixed, and does not depend on any parameter except $l(s)$. Another advantage of this representation is that it naturally maintains the sequence order of symbols along the DNA string. This is an important feature when it is important to take into account ordinal features of the sequence, such as its periodicity. An example of a DNA sequence coded by one hot is represented in Fig. 1 (top).

2.2 *Spectral Representations*

A simple mapping of sequence s to a real values array in the space $\Re^m$ can be obtained using a set of finite set of pre-selected words $P = \{p_i, \ldots, p_m\}$ and enumerate the occurrence of the words in the sequence s. The set P is often constituted by k-mers, a set containing any string of length k whose symbols are

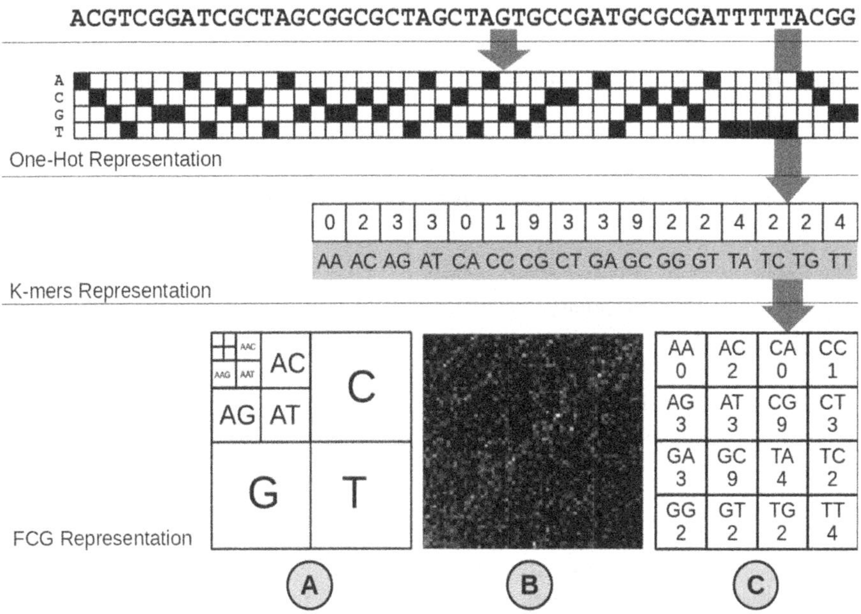

Fig. 1 Construction of the sequence representations. The sequence s in the upper section of the figure is represented as a simple one-hot coding visualized as a black and withe array. In the middle part is represented the k-mers representation with $k = 2$, and this spectral representation is rearranged in the corresponding *Frequency Chaos Game Representation* (FCGR). In this last row A is the graphical depiction of the organizing schema of the matrix, B is a fingerprint image example for a sequence representation with $k = 6$, C is the FCG Representation of the sequence s in the top row

taken in the alphabet Σ. In this case the sequence s is mapped to a vector $x_s \in \Re^{4^k}$ and the component x_s^i counts the occurrences of the i-th k-tuple into the string s. The counting process uses a window of length k that runs through the sequence s, from position 1 to $l(s) - k + 1$. An example of this coding is reported in Fig. 1 (middle) with $k = 2$. The main advantage of this representation is that the vector length depends only on the number of k-mers and not on the sequence length $l(s)$. By varying the value of k it is possible to change the space dimensionality. This means that the representing vectors can be dense if k is small and very sparse for large values of k. A large value of k has also an impact on the complexity of the ML classifier and to the processing time. The spectral representation has shown to be effectively for several sequence classification applications [37], even for barcoding sequences [13, 14, 53]. It is also important to remark that this representation involves an exponential space complexity with respect to the value of k. To this purpose, several solutions for the selection of relevant k-mers have been proposed [10, 40, 45, 46]. Moreover, with the adoption of the spectral representation, the positional information of each nucleotide symbols along the DNA sequence is lost. It is also necessary to recall that the spectral representation permits to compute sequence similarity in a *alignment free* way. The representation of a sequence by a numerical vector gives the possibility to compute sequence similarities by the adoption of standard distances between vector, such as the euclidean one.

2.3 Frequency Chaos-Game Representations

The *Frequency Chaos Game Representation* (FCGR) [67] is a rearrangement of the values of the k-mers frequency counting in a matrix and then in a greyscale image. The arrangement of the elements in the matrix follows the schema in the lower left part of Fig. 1 (label A). For each sequence s a matrix A^s with dimension $2^k \times 2^k$ is obtained, and normalized as:

$$\overline{A^s} = \frac{4^k * A^s}{\sum_{i,j} a^s_{i,j}} \tag{1}$$

where a^s_{ij} is the element of the matrix A^s: the counting frequency of the i, j element and k is the length of the k-mers. An example of the matrix A^s is reported in Fig. 1 lower right. The matrix $\overline{A^s}$ is then translated in a grey-scale image that acts as a "fingerprint" for the sequence s; an example is in the lower centre part of Fig. 1 (label B). In this example image, obtained by using $k = 6$, some of the pixels are black because the corresponding k-mers are not present in the sequence.

Considering that the FCGR is the same as the Spectral Representation, the main characteristics described in Sect. 2.2 are maintained. The only additional characteristic is that the values are packed in a matrix or image so that a single value has 8 neighbourhoods instead of 2 as in a vector. This makes sense when using a 2D *convolutional layer* in classifiers.

3 Deep Learning Architectures for Sequences Classification

Deep neural networks (DNNs) are complex neural structures with many layers between input and output layers. They can be implemented combining many kinds of layers with different characteristics and functions. A well known DNN is the *Convolutional Neural Network* (CNN) whose *convolutional layers* can be trained to extract features from input vectors [31]. Other kinds are *Recurrent Neural Networks* (RNN) [24], that can process sequences of features. A recurrent layer can be useful when the sequence representation contains feature sequence information, as one-hot coding does. Among these recurrent layers, one of the most effective is the *Long Short-Term Memory* (LSTM) layer [22].

Deep Belief Networks (DBN) is a class of DNNs which are made of multiple layers of graphical models having both directed and undirected links [21]. It is composed of multiple layers of hidden units, with connections between the layers but not between units within each layer. Those layers are *Restricted Boltzmann Machines* [19].

3.1 *Restricted Boltzmann Machines Layers in Artificial Neural Networks*

The *Restricted Boltzmann Machines* (RBM) [19] are *Artificial Neural Networks* (ANNs) with two layers, commonly referred to as visible and hidden layer, respectively. A pair of nodes from each of the two groups of layers may have symmetric connections, whereas there are no connections between nodes within a group. Since inputs of all visible nodes are connected to all hidden nodes, an RBM can be considered a symmetric bipartite graph. Symmetric means that each visible node is connected to each hidden node. Bipartite stands for the two layers. A property related to this structure is that hidden $h = \{h_1, h_2, \ldots, h_i\}$ and visible $v = \{v_1, v_2, \ldots, v_j\}$ units are conditionally independent, according to Eqs. (2) and (3):

$$p(h|v) = \sum_i p(h_i|v) \tag{2}$$

$$p(v|h) = \sum_j p(v_j|h) \tag{3}$$

Usually, for binary units (i.e., RBM where h_i and $v_j \in \{0, 1\}$), it is possible to consider a probabilistic version of the activation function as the following:

$$P(h_i = 1|v) = sigm(c_i + W_i \cdot v) \tag{4}$$

$$P(v_i = 1|h) = sigm(b_j + W_j \cdot h) \tag{5}$$

where b and c represent the offset of visible and hidden layers, and W is the matrix of weights connecting h and v units.

RBMs are well-suited for dimensionality reduction, classification, regression, collaborative filtering, feature learning and topic modelling [43]. The main purpose of RBM is to represent input data in a lower-dimensional space. If the reduced representation is, in turn, considered as the input of the RBM, it is possible to obtain an estimate of the probability distribution of the original signal. The error between the estimated and actual input data is minimised by considering their *Kullback–Leibler Divergence* [29].

3.2 Convolutional Layers in Artificial Neural Networks

Convolutional layers are used in DNNs because they can extract features from complex input data [32].

In convolutional layers, each neuron is not connected to the whole set of input signals (features in an input vector) but it has its receptive field that scans the whole input vector. For example, a neuron that receives an input from a layer of 10,000 neurons do not have 10,000 connections and weights, but a small subset (for example 25) and these connections are shifted to cover the whole input while building the neuron output. This operation is called *convolution* and the set of weight associated with the neuron is called *convolutional kernel*.

The shift movement of the kernel on the input vector can be of 1 step at a time or more and is called *stride*. Stride is used to reducing the output dimensions and stride values greater than 3 steps are rarely adopted. Sometimes it is necessary to extend the input matrix to a convolutional layer to have the desired input size. This operation is called *padding*. In this work, we considered only 1D or 2D input matrices.

In order to briefly explain the mechanism of a convolutional layer assume that the neuron i has a non-linearity ϕ and $2 * n + 1$ connections with the input vector x (so that its weights are $W_i^l \in \Re^{2*n+1}$); its output can be calculated in two steps; the first step is the convolution with the input:

$$q_k^i = \sum_{u=-n}^{n} w_u^i * x_{k-u} \tag{6}$$

In Eq. (6) q_k^i is the component k of the i-th output vector and w_u^i is the component u of the i-th kernel vector. To obtain the neuron output signal a bias term b^i is added and a non-linear function is should be applied to q_k^i value:

$$h_k^i = \phi(q_k^i + b^i) \tag{7}$$

The vector h_k^i, where $K = 1, 2, \ldots D^1$is is the number of neurons in layer 1, is the output of the convolutional layer. The non-linearity ϕ can be a sigmoid or tanh; often is considered a *Rectified Linear Units* (ReLU) function, that is easier to calculate [44]. The calculations for the layers other than the first are similar.

3.3 Recurrent Layers for Sequence Classification

A DNA sequence is a string of symbols from a finite alphabet where the order of the symbols matters. Taking this into account, it could be useful to adopt computational methodologies able to catch the ordinal dependencies between symbols. When the representation of the sequence is the one-hot representation, described in Sect. 2.1, the sequence of nucleotides is represented and this information can be caught by a recurrent layer.

An RNN layer is a processing unit that processes sequences of data. It is characterized by an *hidden state* h that is updated after each time step computation. The new hidden state at time t is a function g of the old hidden state at time $t-1$ and the input at t:

$$h_t = g(W_h x_t + U_h h_{t-1} + b_h)$$

where W_h and U_h are respectively the weight matrices for the input and the hidden state, b_h is a bias vector, and g is a nonlinear activation function. The hidden state h_t is a summary of the vectors seen until time step t, thus the last hidden state h_T should contain a summary of the entire sequence. The main drawback of RNN is that it adopts a learning algorithm called *back propagation through time* which lead to loose long term dependencies between the vectors composing the sequence.

An LSTM layer [22] is a variant of a recurrent layer explicitly designed to alleviate their main issues, by selecting the inputs that are relevant for updating the hidden state. This is achieved by the use of *gates*. Gates allow the LSTM to regulate the removal or addition of information to the cell state, establishing an effective mechanism to let information through. The LSTM gates operations are defined by the following equations:

$$f_t = \sigma(W^f x_t + U^f h_{t-1} + b^f)$$

$$i_t = \sigma(W^i x_t + U^i h_{t-1} + b^i)$$

$$o_t = \sigma(W^o x_t + U^o h_{t-1} + b^o)$$

$$c_t = \tanh(W^c x_t + U^c h_{t-1} + b^c)$$

$$s_t = f_t \odot s_{t-1} + i_t \odot c_t$$

$$h_t = \tanh(s_t) \odot o_t$$

where i_t, f_t, o_t are respectively, the input, forget and output gates, $\odot$ represents the element-wise multiplication and σ is the sigmoid function. We can consider the gates as vectors that, assuming values in the range [0, 1], decide, component by component, which part of the input of the previous hidden state and of the candidate output should flow through the ANN.

3.4 Other Useful Layers in Neural Networks

DNN architectures can also be conveniently composed of some additional layers that, in addition to the ones previously described, can improve their performance.

For instance, the so-called *embedding layer*, that is a lookup table of weights, is sometimes used as a learned representation of the input symbols. It is part of the set of trainable weights of a generic DNN. Let V be the set of symbols occurring in the training set, then $E \in R^{d\times|V|}$ is the embedding layer containing the input representation of size d for the $|V|$ symbols. The lookup matrix is indexed by the input symbols, each initially represented as a one-hot vector, and its values are fed as input to the ANN.

Also, it can be taken into account some other types of layers that do not contain neurons, so that they do not need training, but are useful for regularization or add a non-linearity. Some of these layers are *max-pooling*, *dropout* and *softmax*.

The *max-pooling* [71] is a non-linear down-sampling layer. In these processing layers, the input vector is partitioned into a set of non-overlapping regions (for example of 2×2 elements in the 2D architectures) and, for each sub-region, the maximum value is considered as output. This processing layer reduces the complexity for the higher layers and operates a sort of translational invariance. Convolution and max-pooling are usually considered together.

Dropout layers [62] are used to avoid overfitting of ANNs during training. With dropout, a random fraction of the units of a layer is ignored, and their weights do not change. Using dropout, a different subset of neural units is considered for training and the training is more "noisy". Adding this noise, the ANN is forced to use fewer neurons to produce the correct output. The parameter used in dropout layers represents the probability of a unit in the layer to be "used" (or not "dropped out"). During the test phase, all the units of the ANN are working.

Softmax layers are used as an output layer in DNNs. A *softmax* output layer allows to interpret the ANN output as probabilities; if the number of the output neurons is N and x_i $i = 1, 2, \ldots N$ the output value for each neuron in the layer preceding the softmax one, the output of the softmax will be:

$$s_i = \frac{e^{x_i}}{\sum_{k=1}^{N} e^{x_k}} \tag{8}$$

3.5 Layers Assembly in a Deep Neural Network

Many DNNs are composed of two parts: the first part is aimed to pre-processing, for example, to extract features from the input vector, the second part is made of fully connected layers and aimed to the classification task. This structure is depicted in Fig. 2.

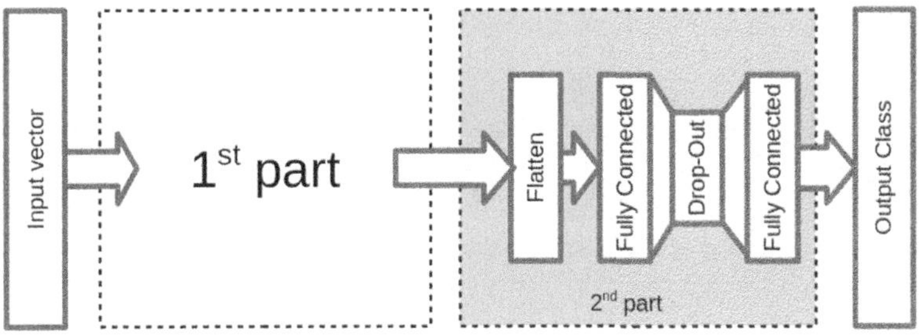

Fig. 2 Architecture of the discussed ANN. The first part can be a set of convolutional layers or any other layer capable to extract features from the input data

The features used for classification in the second part of the ANN are extracted from raw data by the computational layers in the first part, input raw data can be for example images [33], where it is difficult to decide a priori the right features for a classification task. Assuming that a convolutional layer can be useful for feature extraction, the first part of the ANN in Fig. 2 can be substituted with a set of convolutional layers if the feature sequence is not meaningful.

This kind of consideration is the one that guided the work in [54] where the k-mers features from sequences are calculated, and the CNN has the task of detection k-mers co-occurrence and frequency. Assuming that the k-mer frequency representation is suitable for CNN, it is necessary to design it. The ANN design has two aspects: the ANN architecture that is related to the number and kind of layers in the first part of the ANN, the kind of non-linearity involved and so on, and the number of ANN parameters (for example the number of neural units) that are tuned during the training phase of the ANN. These two aspects are interconnected and varying the number of layers can produce an effect similar to the variation of the number of parameters of the ANN.

An analysis of the CNN for sequence classification is reported in Zeng et al. [73]. In this chapter, CNNs are studied by varying the number of kernels and the number of layers. The authors find that increasing the number of convolutional layers gives only a small improvement in the ANN performances, and a more complex architecture needs more training time and more training data to obtain a little gain in the classification results. Starting from these observations, the authors conclude that CNN performances do not scale easily with the complexity of the ANN. Following this consideration, a CNN architecture with just two convolutional layers in Fig. 2 was used to classify sequences represented with a k-mer frequency vector. The obtained ANN is represented in Fig. 3.

The same approach was used in [55] for the classification of sequences represented with FCGR.

Other architecture can be useful for the case when sequences are represented with one-hot encoding. In this case, an LSTM layer can find the positional relationships of the features extracted by the first convolutional layer. The number of neurons in LSTM layer is related to its capability to recognize sequence fragments.

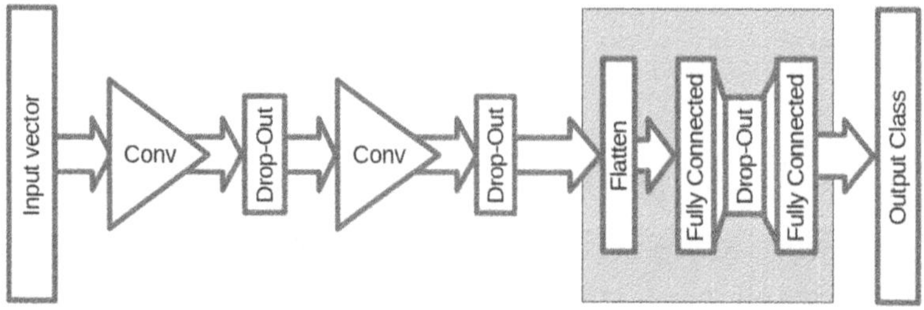

Fig. 3 Architecture of the discussed CNN

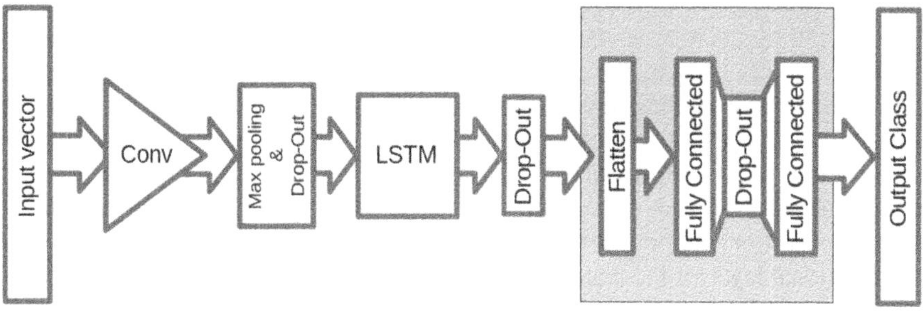

Fig. 4 DNN architecture with an LSTM layer

Figure 4 shows the obtained ANN by adding the LSTM layer in place of the second convolutional layer; the resulting ANN is constituted by a convolutional layer, a max-pooling layer, a dropout layer, an LSTM layer, and two fully connected layers. The convolutional layer main role is the feature extraction from the input data x of $4 \times l(s)$ binary values, where $l(s)$ is the length of the sequence.

The convolutional layer extracts a set of simple features from the sequences, and this representation is processed by a max-pooling layer that reduces the size of the representing vector. The dropout layer before the LSTM prevents overfitting during the training phase.

The LSTM layer scans the output of the sequential features of the previous layer and outputs its hidden state at each time step. The purpose is to find long-range relations between the time steps along all the sequence. The outputs from all the LSTM time steps are then concatenated in a single vector.

The second part of the ANN is the same for the two architectures and is made by two fully connected layers, the first with a ReLU non-linearity, the second with a sigmoid activation, and a dropout layer. The number of units in these layers change with the kind of classification task.

Another architecture one can exploit is the DBNs. As said before DBNs are generative probabilistic models consisting of several layers of stochastic and latent variables [20, 21]. Latent variables are also known as hidden units or feature extractors and usually assume binary values. As shown in Fig. 5, it can be defined as a stacked RBM (see Sect. 3.1), and the learning procedure is made up of two steps. In the first step, also-called *pre-training*, RBM layers training is carried out

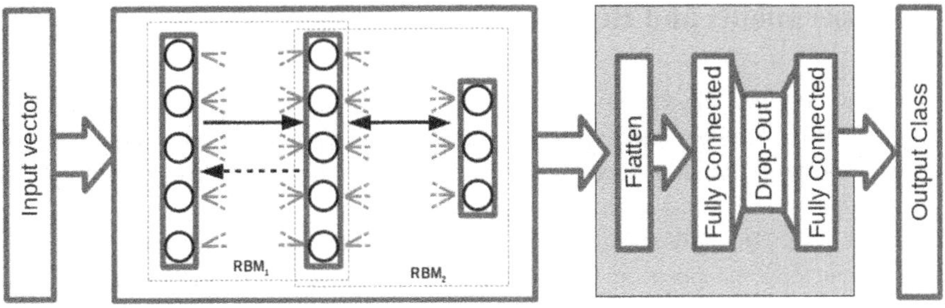

Fig. 5 DBN architecture

in an unsupervised way to represent the original input in a new dimensional space featuring a lower dimension. The second step, also known as *fine-tuning*, can be carried out by adding a layer of variables representing the desired outputs to apply supervised learning via the back-propagation algorithm.

DBNs are graphical models that learn to extract a deep hierarchical representation of data. DBNs model the joint distribution between the observed vector x and the l hidden layers h_k according to:

$$P(x, h_1, \ldots, h_l) = \prod_{k=0}^{(l-2)} P(h_k|h_{(k+1)})P(h_{(l-1)}, h_l)$$

where $x = h_0$, $P(h_{(k-1)})|h_k)$ is a conditional distribution for the visible unit conditioned by hidden units of RBM at the level k and $P(h_{(l-1)}|h_l)$ is the joint visible-hidden distribution in the RBM of the last level.

Unsupervised training can be applied to the DBN according to five steps:

1. Build the first layer with an RBM modelling the input $x = h_0$ as a visible level.
2. Use this level to achieve a representation of the input which is the input data of the second level. This representation can be considered as the mean activation $P(h_{(1)} = 1|h_{(0)})$ or samples of $P(h_{(1)}|h_{(0)})$.
3. Build the second level with an RBM by using transformed data (samples or mean activation) as samples for the training of the visible RBM layer.
4. Iterate steps 2. and 3. as many times as the desired number of layers, every time by bottom-up propagation of samples or mean values.
5. Tune and optimize all parameters (i.e., fine-tuning step) by a supervised training algorithm as back-propagation. Fine-tuning is applied by the controlled gradient descent of the negative log-likelihood cost function

The architecture described so far, needs a proper *learning algorithm* to find the ANN parameters necessary to perform the supervised or unsupervised classification for a real test case. Many algorithms exist in the literature, and their adoption depends also on the used neural architecture. For a deep survey about such kind of algorithms, the interested reader can take the book by Goodfellow et al. [17].

4 Experiments and Results

The most common classification DNNs are suitable for fixed-length input vectors (e.g. images). This explains why the k-mers representation is very common for variable or fixed-length sequences, while the one-hot representation is useful if all the sequences in the data set have the same length.

In the following subsections, two classification problems are discussed: prediction of nucleosome positioning and bacteria classification using 16S sequences. These are the two classification problems that belong respectively to the two biological scenarios tackled in this chapter, i.e. chromatin organization and metagenomics. Sequences processed in nucleosome positioning problem are fixed length so that they can be represented by using both one-hot coding or k-mer representation.

4.1 Prediction of Nucleosomes

For the specific problem of nucleosome prediction, we have experimented our architecture on three datasets, each one collecting DNA sequences underlying nucleosomes and not nucleosomes (called *linkers*) classes from a specific organism. Three organisms are considered, *Homo sapiens*(hsa), *Caenorhabditis elegans* (cel) and *Drosophila melanogaster*(Dmel). The details about the extraction and filtering phase of such data are described in the original paper by Guo et al. [18]. Each sequence of a dataset is 147 *base pairs* (bp) long. The distribution of the elements among the two classes of Nucleosomes and Linkers is balanced, for each dataset. The number of total sequences is 4573 for hsa, 5750 for Dmel and 5175 for cel. The classification problem we study in the following regards the classification of sequences among the two classes, starting from the DNA sequence information only, represented by different codings.

Nucleosome-Linker Classification Using k-mers Representation We have first investigated in the adoption of the k-mers representation and CNN using the datasets described above [39]. The used architecture for these experiments, in the one in Fig. 3. The focus of the study was about the efficacy of the representation for different values of k. We found that the k-mers length has an effect on the classification accuracy and we noticed that the results are quite the same for k ranging from 4 to 5. This is probably due to the sparsity of the obtained representation, considering that the feature number goes from 256 (4^4) to 1024 (4^5). Representation with $k > 5$ affect significantly the training time, causing an increase from $\approx$30 min to more than $\approx$2 h. It is also interesting that the impact of the convolution kernel size on the classification accuracy.

In these new experiments, we found that the accuracy can be increased using larger kernels. Figure 6 reports the results obtained by varying the size of the kernels in the convolutional layers, using a different number of training epochs. The new

Table 1 Details of the CNN architecture used for Nucleosome-Linkers classification with $k = 4$ k-mers representation

Parameter		Conv. Layer 1	Conv. Layer 2	Fully connected part
Kernel dimensions	(D)	3, 5, 7, 9	3, 5, 7, 9	–
	(D - 2*D)	3, 5, 7, 9	6, 10, 14, 18	–
Number of kernels (N)		32	64	–
Max pooling		2	2	–
Drop out		0.5	0.5	0.5
Number of units 1st layer		–	–	1024
Number of units 2nd layer		–	–	1

experiments have been carried out using a 10-fold cross-validation procedure. The plots show also an obvious dependence of the accuracy on the number of training epochs. The architecture details of the used CNN are reported in Table 1.

The convolutional layers assemble the features in the input vectors to build more complex features, and this process goes on layer by layer as many of the works in image classification show. This is, in some way, confirmed by the increasing performances with the kernel dimension in Fig. 6.

To capture more complex features, we doubled the dimension of the second layer in the second set of experiments. As reported in Fig. 7 this did not increase the classification performances as expected, while the training time is roughly the same (Figs. 8 and 9).

Nucleosome-Linker Classification Using One-Hot Representation There is strong evidence that specific kind of *periodicities* are observable in nucleosome related sequences [35]. This consideration leads to the suggestion of adopting a recurrent layer for the ANN architecture. To investigate the effectiveness of a recurrent layer, one has to consider a representation able to maintain the sequence of symbols in the DNA string. For this purpose, the one-hot representation has been adopted. Concerning the architecture, we have considered an LSTM layer after a convolution one. The idea is to let the convolutional layer to extract the most relevant local features, and then use the LSTM to find the relations of these features along the sequence [1, 5, 6]. Figure 4 shows the adopted architecture. The first layer from left to right is a convolutional layer whose main role is the feature extraction from the input data x of 4×147 binary values. It is characterized by a bank of $n = 50$ 1D convolutions [32] between the kernel vectors w^l $l = 1, 2, \ldots n$ and the input sequence x. The subsequent max-pooling layer, with width and stride values, equals to 2, helps to capture the most salient features extracted by the convolution and also reduces the output size of the input vectors. Then, the dropout layer with probability $p = 0.5$ is used to prevent overfitting. The LSTM layer is composed of 50 hidden memory units, and it scans sequentially the data. The role of this layer is to catch long-range relations between the symbols along all the sequence. The convolutional and LSTM layers use L_2 regularization with $\lambda = 0.001$. The outputs of all the LSTM hidden units are concatenated in a single vector, which is fed

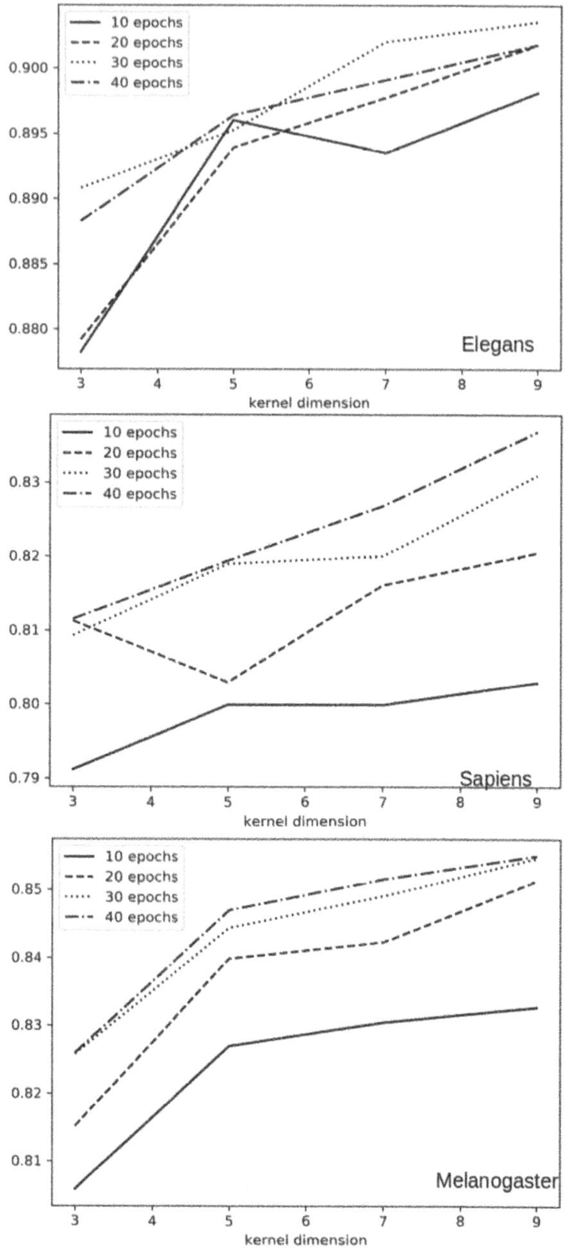

Fig. 6 *Accuracy* results for the three data sets: *Elegans*, *Melanogaster* and *Sapiens*. In the x-axis is reported the dimension D of the kernel in the first layer, that is the same of the second layer. The architecture is reported in Fig. 8

to 2 subsequent fully-connected layers that reduce its length first to 150 and then to 1. The first fully connected layer adopts a ReLU activation function, while the second a sigmoid causing the output of the ANN belongs in the interval [0, 1]. A summary of the architecture details are reported in Table 2, while Table 3 reports

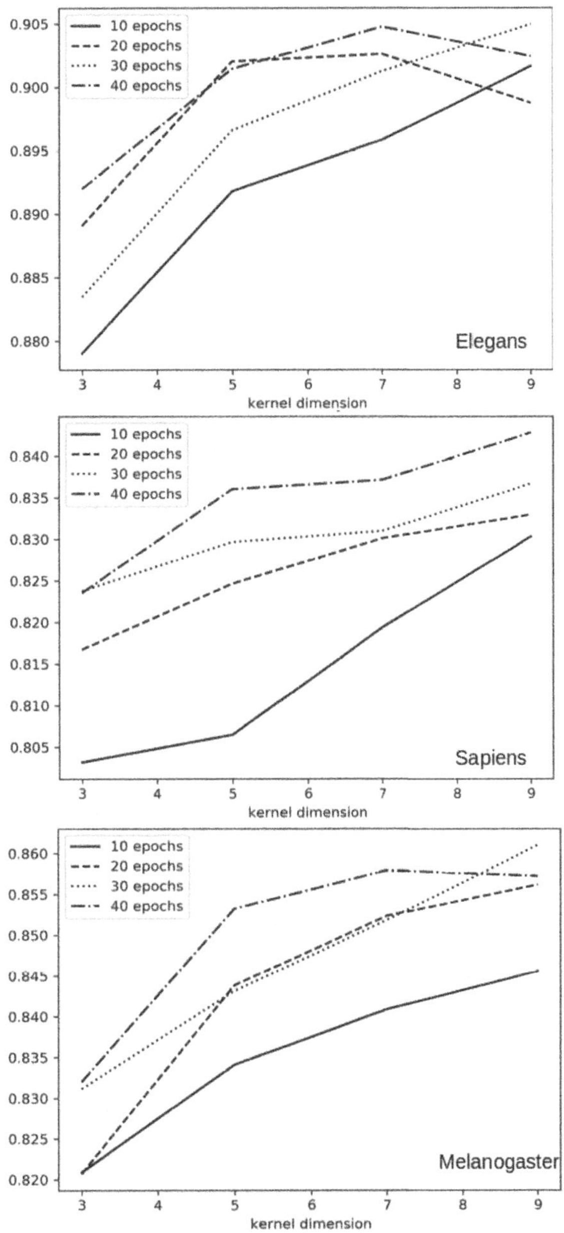

Fig. 7 *Accuracy* results for the three data sets: *Elegans*, *Melanogaster* and *Sapiens*. In the x-axis is reported the dimension of the kernel in the first layer D, the kernel dimension of the second layer is 2*D. The architecture is reported in Fig. 9

the results obtained by this architecture named CLSTM. We have reported mean values of accuracy, sensitivity and specificity of the three datasets computed by two different versions of CLSTM, named CLSTM-3 and CLSTM-5. Their difference is in the size of the convolutional kernel, i.e. 3 for the first and 5 for the second. The

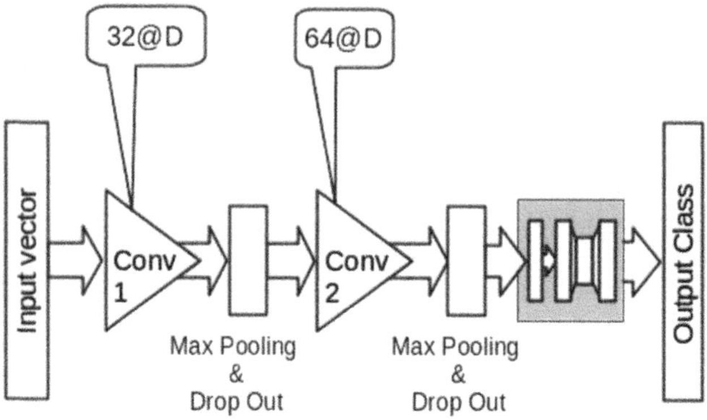

Fig. 8 ANN architecture corresponding to the results of Fig. 6

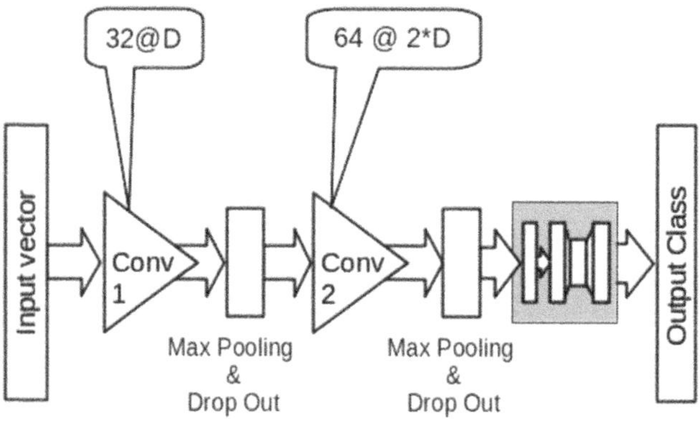

Fig. 9 ANN architecture corresponding to the results of Fig. 7

experiments have been conducted adopting a 10-fold cross-validation schema. A 10% of the dataset is selected among the training set as a validation step for early stopping. The predicted labels are obtained by thresholding the output value of the DLNN, that ranges in the interval [0, 1], with the value 0.5. Output values below 0.5 are classified as linkers otherwise as nucleosomes. Other details about the proposed neural models and further comparison with other methodologies can be found in other recent works [5, 6].

4.2 Bacteria Classification Using 16S Gene Sequences

In this section, we show the application of DL models for the taxonomic classification of bacteria considering only the 16S gene sequences. The problem is a multi-class classification and the number of output classes depends on the phyla

Table 2 Details of the CLSTM architecture used for Nucleosome-Linkers classification with one-hot representation

Parameter	Conv. Layer 1	LSTM	Fully connected part
Kernel dimensions (1-D)	3, 5	–	–
Number of kernels (N)	50	–	–
Max pooling	2	–	–
Drop out	0.5	–	0.5
Number of units 1st layer	–	50	150
Number of units 2nd layer	–	–	1

Table 3 10-fold cross-validation performances nucleosome datasets. cel, Dmel, hsa refers to the species; CLSTM refers to the DLNN proposed in this chapter and -3 or -5 refers to the kernel dimension in the first convolutional layer of the net

	Accuracy		Sensitivity		Specificity	
Method (species)	μ	σ	μ	σ	μ	σ
CLSTM-3(cel)	89.60	0.8	93.36	1.27	85.93	2.13
CLSTM-3(Dmel)	85.54	1.13	87.60	2.55	83.42	2.65
CLSTM-3(hsa)	84.65	2.16	89.67	2.83	79.64	4.29
CLSTM-5(cel)	89.62	2.45	93.04	3.68	86.34	5.54
CLSTM-5(Dmel)	85.60	0.75	87.81	2.79	83.33	2.74
CLSTM-5(hsa)	85.37	1.91	88.34	1,82	82.29	4.86

and taxonomic classification level. In the first and second scenario, the classification is carried on considering full-length sequences (about 1400 bp); in this case, the data set used is made of 3000 16S ribosomal RNA sequences downloaded from the *Ribosomal Database Project* II (RDP) [4]. The sequences were high quality with a length of 1200–1400 nucleotides from both uncultured and isolated sources, checked by the RDP quality system. From each of the most popular phyla of bacterias (*Actinobacteria*, *Firmicutes* and *Proteobacteria*) were randomly selected 1000 sequences. Table 7 shows the structure of the taxonomic categories. In the third scenario, the classification is based on short reads simulating the output of an NGS machine, to stay as much as possible close to the real-life metagenomic problem. The short-read sequences were generated according to [50, 72] and more details are in the original work [11]. In this case, *Proteobacteria phylum* was only considered and to obtain a balanced dataset, we selected 100 genera with 10 species for each genus. Two datasets, according to simulated sequencing technology, were obtained: *shotgun* (SG) and *amplicon* (AMP).

Bacteria Classification Using Full-Length 16S Sequences and One Hot Representation The bacteria classification is hierarchical and is composed of many levels (taxa levels). In the experiments described in the following, we used 5 levels of classification from *Phylum* to *Genus*. The classifier structure is the one in Fig. 10: There is one ANN for each taxa level to have the same number of training sequences for each ANN.

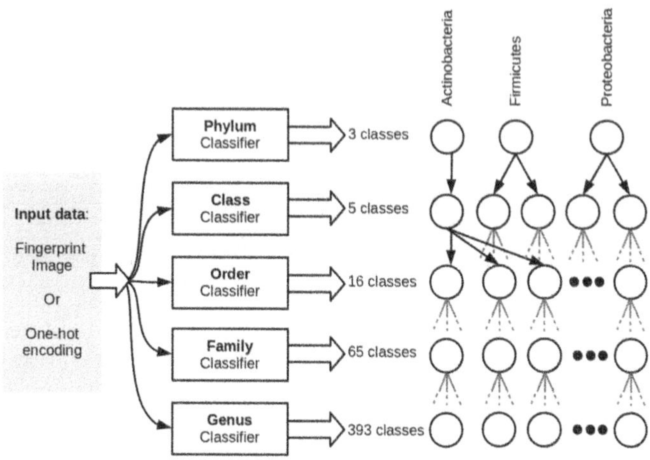

Fig. 10 Structure of the classifier for the 16S sequences (left) and the representation of the hierarchical classification of the considered three phyla (right). The input data is the representation of a 16S full-length sequence and each classifier is a trained ANN: in the case of full-length 16S represented with one-hot encoding were compared a CNN and an RNN, in the case of FCGR a CNN was used

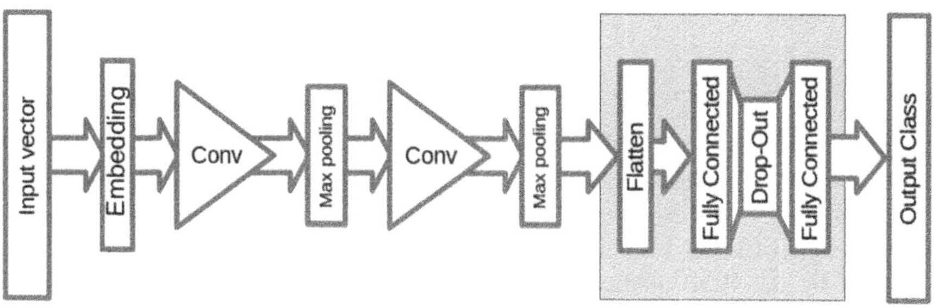

Fig. 11 Architecture of the ANN based on convolutional layers

In bacteria classification, our goal is to look for the more suitable neural architecture. To this purpose, we provide here a comparison of CNNs and RNNs DL architectures. The architecture for the CNN is a variant of the *LeNet* network [31], the RNN is an LSTM.

The ANN structures are reported in Figs. 11 and 12.

The first layer of the CNN acts as an embedding layer, takes as an input 16-dimensional one-hot encoding of sequence characters and produces as output a 10 dimensional continuous vector. Note that the one-hot representation size is different from 4 due to the nature of the input dataset, which is defined in the IUPAC alphabet. Finally, the output of the embedding layer is a $10 \times l$ matrix of real values, where $l = 1400$ is the length of the 16S sequences. On top of the input, we have two 2-D convolutional layers, each followed by a max-pooling layer. The two convolutional

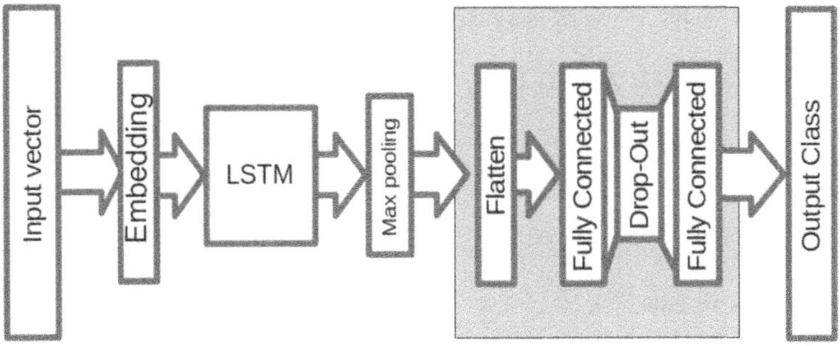

Fig. 12 Architecture of the ANN based on LSTM recurrent layers

Table 4 Average accuracy of the proposed models over 10-fold

	Phylum		Class		Order		Family		Genus	
	μ	σ	μ	σ	μ	σ	μ	σ	μ	σ
CNN	0.995	0.003	0.993	0.006	0.937	0.012	0.893	0.019	0.676	0.065
LSTM	0.982	0.028	0.977	0.022	0.902	0.028	0.857	0.034	0.728	0.030

layers use 10 and 20 filters respectively, each of size 5. The width and the stride of the pooling layers are both equal to 5.

The two convolutional layers are then stacked with two fully connected layers. The first one is composed of 500 units and uses a $tanh$ activation function. The second one is the classification layer, and uses the $softmax$ activation (Table 4).

This ANN is similar to that proposed in [54], except for the addition of an embedding layer and the absence of preprocessing of the sequences, achieved using the one-hot encoding. The details of the CNN architecture are summarized in Table 5. The RNN is a 6-layered ANN. The first layer is the embedding, and it is followed by a max-pooling layer of width and stride equal to 2. The max-pooling reduces the computation for the following layer, and at the same time gives some capability of translational invariance to the ANN. The subsequent LSTM layer processes the data from left to right and produces an output vector of size 20 at each time step. The output is then sub-sampled by another max-pooling layer and in the top are stacked two fully-connected layers like those of the CNN version. The basic details of the proposed LSTM architecture are summarized in Table 6. The same results are also reported in Fig. 13. For each taxonomic rank, a 10-Fold cross-validation has been performed. We have chosen 15 epochs for each fold, without using early stopping validation. This choice has been motivated by our experimental observations. In Table 4 we report the results of mean (μ) and standard deviation (σ) of the accuracy of the two ANNs computed on 10 test folds, for both ANNs (CNN and LSTM). Other details about this experiment can be found in the paper [38].

Table 5 Details of the CNN architecture used for 16s sequence classification with one-hot representation

Parameter	Embedding	Conv. Layer 1	Conv. Layer 2	Fully connected part
Size	10	–	–	–
Kernel dimensions (2-D)	–	5 × 5	5 × 5	–
Number of kernels (N)	–	10	20	–
Max pooling	–	5	5	–
Drop out	–	–	–	0.5
Number of units 1st layer	–	–	–	500
Number of units 2nd layer	–	–	–	3–393

Table 6 Details of the LSTM architecture used for 16s sequence classification with one-hot representation representation

Parameter	Embedding	LSTM	Fully connected part
Size	10	–	–
Kernel dimensions (D)	–	–	–
Number of kernels (N)	–	–	–
Max pooling	2	2	–
Drop out	–	–	0.5
Number of units 1st layer	–	20	500
Number of units 2nd layer	–	–	3–393

Bacteria Classification Using Full-Length 16S Sequences and Frequency Chaos Game Representation The classification of sequences using the FCGR is very similar to image classification, and a first application was presented in [55]. The classification of the same dataset with a k-mer representation and similar results was reported in [54]. The taxonomic categories from *Phylum* to *Genus* are reported in Table 7.

The classifier architecture is reported in Fig. 10. Even in this case, there is one classifier for each taxa level (the classifier details are in Table 8).

The classification results reported in the original paper [55] confirmed that the classification results are better than the *Support Vector Machine* (SVM) classifier [59] and that there is an increment of the performance if the representation goes from $k = 5$ to $k = 6$, but there is only a slightly better performance with a more complex representation ($k = 7$). As noticed before for each increase of k the fingerprint image of the sequence the number of pixels is multiplied by 4, and consequently the training time.

Following the same line of experiments, we wanted to test an increased kernel dimension and made the same experiment with $D = 5$. A larger kernel should be able to capture more large structure and patterns in the input fingerprint image. The obtained results are in Fig. 13, and in the same figure is reported the $D = 3$ results of the original paper for comparison. It is possible to notice that there is not a noticeable increase in performances.

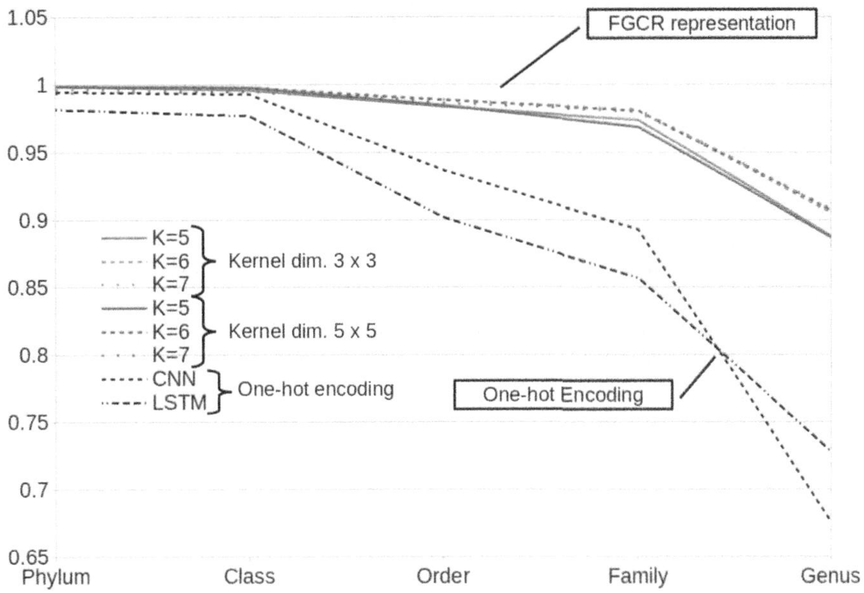

Fig. 13 *Accuracy* results for the three bacteria phyla classification using the 16S full length in the two cases: one-hot encoding and FCGR representation. The results obtained with the FCGR using k-mers with $k = 5$ are the same for ANNs with kernel dimensions equal to 3×3 or 5×5. The results are slightly better with a representation obtained with k-mers of dimension 6 or 7, regardless of the kernel dimension used in the ANN. The one-hot encoding gives worse results

Table 7 The 16S bacteria data set structure. The first three rows report the number of taxonomic categories for each taxa level. The last row the number of classes for each classifier in the architecture

	Phylum	Class	Order	Family	Genus
Actinobacteria	1	1	3	12	79
Frimicutes	1	2	3	19	110
Proteobacteria	1	2	13	34	204
Total number of classes	3	5	16	65	393

Bacteria Classification Using Short Reads 16S Sequences and Spectral Representation The dataset analysis leads to take into account several features. Representation of short reads given by the simulator using k-mers length ranging from 3 to 7 give hints to design the classifier. Each representing vector is modelled as a list of frequency values ordered using the natural order of k-mers.

Fig. 3 shows the ANN architecture.

As for CNN, we started from an initial configuration and then performing a grid search to find a trade-off between results and processing time. The initial configuration features a first convolution layer of 10 kernels and a kernel size of 5; a second layer with 20 kernels of the same dimension; the non-linearity is the ReLU;

Table 8 Network structure for the FCGR experiments. The number of output is variable depending on the taxa level of the classifier

Parameter	Conv. Layer 1	Conv. Layer 1	Fully connected part
Kernel dimensions	$3 \times 3, 5 \times 5$	$3 \times 3, 5 \times 5$	–
Number of kernels	10	20	–
Max pooling	2	2	–
Drop out	–	–	–
Number of units 1st layer	–	–	500
Number of units 2nd layer	–	–	3–393

Table 9 Network structure for spectral representation of short reads. The number of output is variable depending on the taxa level of the classifier

Parameter	Conv. Layer 1	Conv. Layer 2	Fully connected part
Kernel dimensions	5	5	–
Number of kernels	5	10	–
Max pooling	2	2	–
Drop out	–	–	0.5
Number of units	–	–	500
Number of units 2nd layer	–	–	3–100

a fixed pooling size of 2 and the last hidden layer with 500 units. The learning algorithm for CNN is the Adam optimization algorithm [27]. Classification results slightly depended on kernel size and kernels number (less than 1%). Therefore, the ANN configuration of Table 9 is considered.

As for the DBN parameters, in the two RBM layers, the same number of units is selected. The number of input features, which is strongly related to k-mer size, define the number of hidden units. Indicating with k the k-mer size, the number of input features is equal to 4^k. Consequently, the number of hidden units is equal to $4^{(k-1)}$ for $k = 3, 4, 5$ and 4^4 for $k = 6, 7$ to speed up the processing time. Figure 5 shows the DBN model. In this case, the optimization algorithm for DBN learning is the *Contrastive Divergence* (CD) method [19].

Both CNN and DBN are tested by a 10-fold cross-validation procedure and, consequently, results are averaged. Tests are carried out under several sizes of k-mer length ranging from 3 to 7 in order to find the minimum k-mer length providing the most of the information required for classification. Classification performances are evaluated in terms of accuracy, precision, recall and *F1* score. Figures 14 and 15 show the accuracy score, at different taxonomic levels, of classification using CNN and AMP and SG dataset, respectively, as a function of k-mer size. The trend using the DBN is very similar and it is not shown. Regardless of the ANN type and taxonomic level, the highest accuracy is achieved with a 7 k-mer size, corresponding to its maximum value. Accuracy scores range from 99% at *Class* taxon and 80% at *Genus* level with SG dataset. In Figs. 16 and 17 we make a comparison between classification scores among CNN and DBN, considering AMP and SG dataset

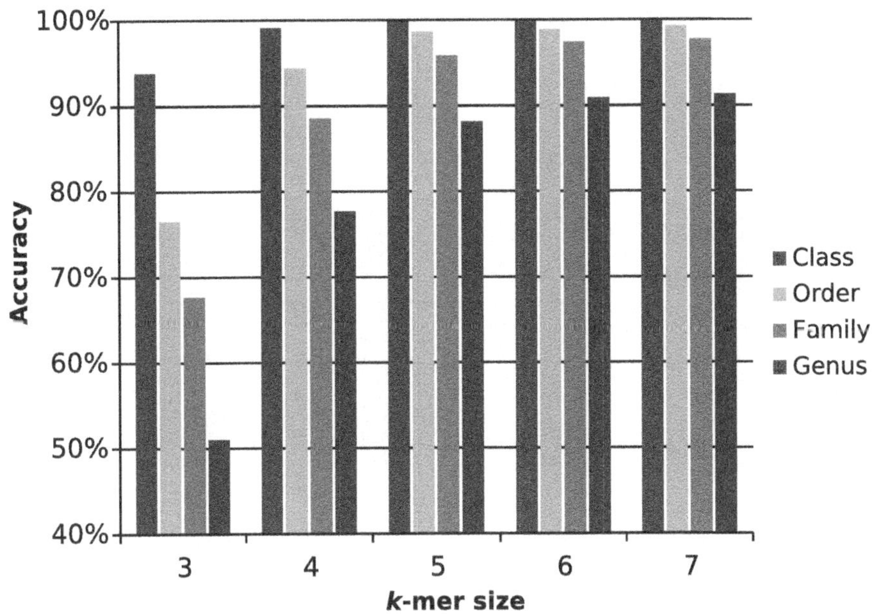

Fig. 14 *Accuracy* results, using CNN and AMP dataset, from *Class* to *Genus* level for increasing value of *k*-mer size

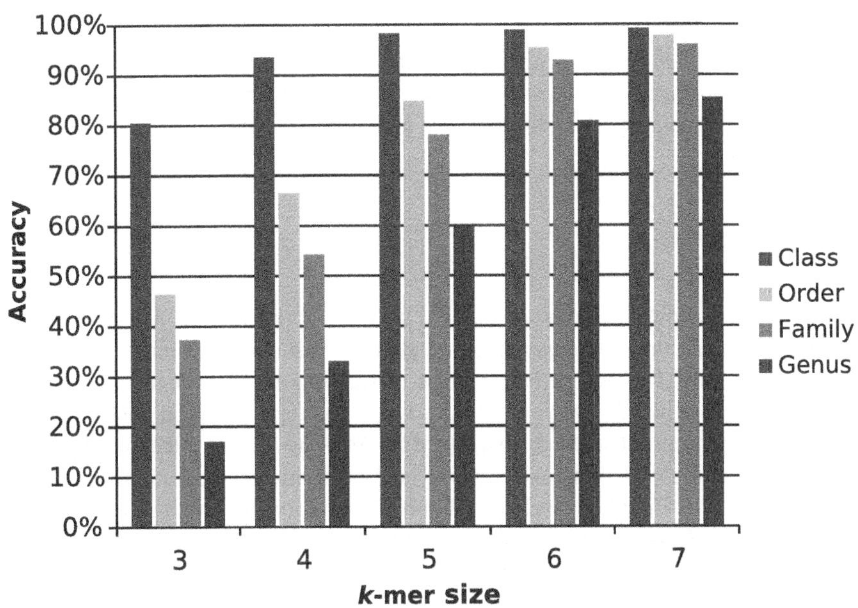

Fig. 15 *Accuracy* results, using CNN and SG dataset, from *Class* to *Genus* level for increasing value of *k*-mer size

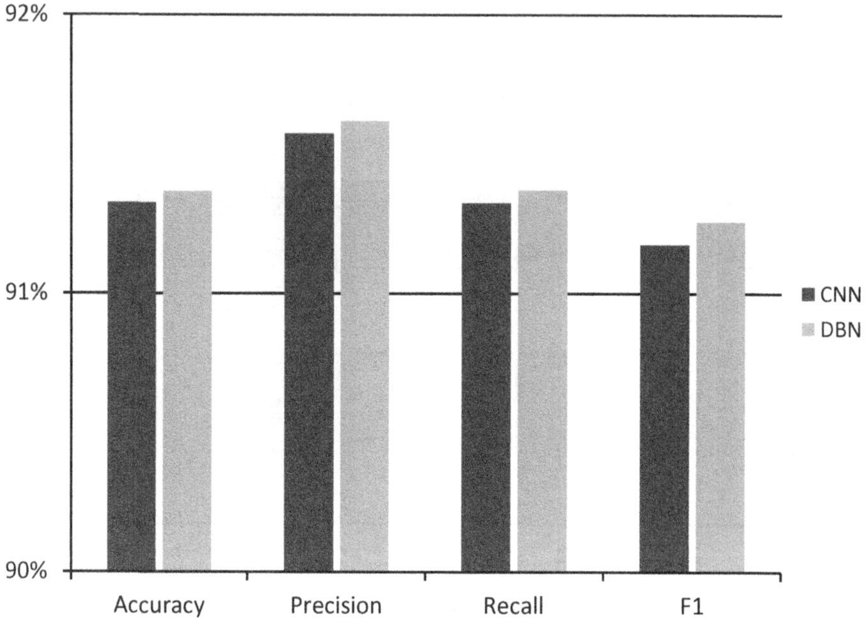

Fig. 16 *Accuracy, Precision, Recall* and F1 scores, using CNN and DBN, on AMP dataset, at *Genus* level and k-mer with $k = 7$

respectively, at *Genus* level with $k = 7$. The results are quite similar between the two ANN models, and the best scores are reached with the AMP dataset (about 90%). With SG dataset, the best scores are about 85.5%.

4.3 Discussions

The sequence representations considered in this work are one hot and k-mers encoding. The second one depends on a parameter k and suffers form exponential space complexity. Moreover, the k-mers counting is a task linear in time. This is an issue that should be taken into account when this representation is needed. The space complexity problem can be narrowed by adopting feature selection strategies, and to the specific classification problems here tackled several solutions have been proposed [10, 40, 45]. Conversely, one-hot representation does not need any feature extraction or selection process. For the classification of nucleosomes and linkers sequences specific architectures was used (see Figs. 3 and 4). In both cases the first convolution layer was used to extract simple features from the input data, that should be collected by the second convolution layer. The first set of experiments was aimed to understand if a larger kernel in the second layer would be more effective and capable to assemble more complex features from the output of the

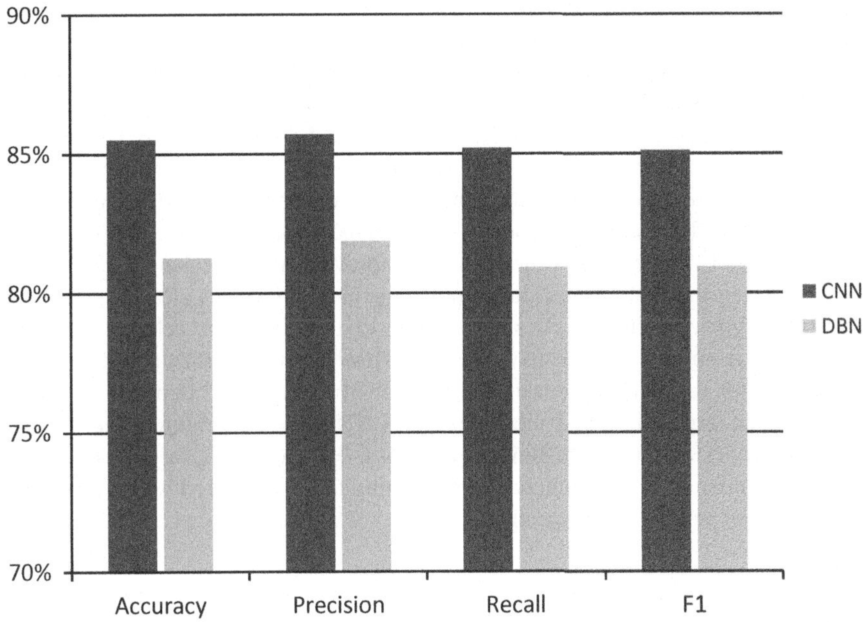

Fig. 17 *Accuracy, Precision, Recall* and F1 scores, using CNN and DBN, on SG dataset, at *Genus* level and k-mer with $k = 7$

first layer. Plots in Figs. 6 and 7 show that the use of different k learning epochs does not involve a significant accuracy improvement. The best accuracy values for the CNN architecture, shown in Figs. 6 and 7, are 0.9, 0.85, 0.84 respectively for *Elegans*, *Melanogaster*, *Sapiens*. The other architecture which adopts the one-hot representation, and the LSTM layer in place of the second convolutional layer, seems to be more effective for the case of *Sapiens* (see Table 3). This architecture can exploit the ordinal information in the sequence, a feature that seems useful for the case of nucleosome classification. Indeed the sequence contains useful information that in the k-mers representation is naturally discarded. For sure, for the CNN to be comparable to the LSTM, larger k is needed. This involves a very huge preprocessing phase in terms of space that does not motivate the adoption of this architecture. A deeper study of the LSTM architecture, involving huge sequence datasets and the comparison with state of the art methodologies for nucleosome identification, is reported in [5].

Conversely, for the case of 16s classification, the adoption of one hot representation did not lead to successful results neither using convolutional, nor recurrent layers. The used architectures are shown in Figs. 11 and 12. For sure, several studies on bacteria classification has shown the effectiveness of using the k-mer representation for 16s sequences [66]. Another successful representation for 16s bacteria classification was the chaos game one. We recall that the CNNs were originally developed for image classification, and the FCGR representation generates image data. In the reported experiments the dimension of the obtained fingerprint images varies with the dimension of k-mers this means that larger images are more sparse. Probably this is the reason why there is no improvement in the classification results when the k-mers dimension ranges from 6 to 7. Moreover, the results are the same for the two kernel dimensions used, possibly because the features are small and a larger kernel is not useful. For the classification of short-reads of 16S sequences, key performance parameter is the k-mer size. The size of input representation strongly depends on the k-mer size k, being equal to 4^k. As shown in Fig. 14 for the CNN, the accuracy improves with increasing k-mer. At the *Genus* taxonomic level, the accuracy improvement with k-mer is clearer since there are 100 categories to classify. As shown in Fig. 15, the CNN approach performances are boosted from $k = 5$ to $k = 6$, whereas the DBN approach shows a stable growth trend. Yet, for the DBN approach, the number of hidden units depends on the k-mer size. For large values of k-mer size (6, 7) CNN and DBN feature a very similar trend. Since performances are almost stable for $k = 6$ and $k = 7$, larger k-mer sizes are not considered. A huge amount of processing time is required with 65,536 and 262,144 input vector size, respectively.

4.4 Execution Times

All the experiments were carried out on a cluster with 24 nodes and the following configuration:

- **OS:** CentOS 6.3
- **CPU:** 1 X Intel®Xeon®CPU E5-2670 @ 2.6 GHz
- **HDD:** 1 TB SATA
- **RAM:** 128 GB DDR3 @ 1.6 GHz
- **GPU:** 48 x GPU NVIDIA KEPLER K20

The execution time (s), for the training phase are summarized in Tables 10 and 11. We took into account the size of the k-mers, the ANN architecture and, in case of nucleosome case study, the species.

Table 10 Training times, in seconds, for nucleosome prediction case study, considering k-mers size, ANN architecture and species

	Cel		Dmel		Hsa	
ANN						
k-mer	CNN (Fig. 8)	CNN (Fig. 9)	CNN (Fig. 8)	CNN (Fig. 9)	CNN (Fig. 8)	CNN (Fig. 9)
3	300	304	325	335	261	268
5	380	347	360	379	290	312
7	420	441	456	495	367	360
9	440	490	510	536	413	436

Table 11 Training times, in seconds, for bacteria classification case study, considering *k*-mers size, ANN architecture and sequence length

ANN	Full sequences		Short reads	
k-mer	CNN (Fig. 11)	LSTM (Fig. 12)	DBN (Fig. 5)	CNN (Fig. 3)
3	732	878	7288	686
4	825	990	8170	1256
5	1171	1422	11,875	3091
6	2044	2452	20,346	8021
7	37,054	44,553	37,161	24,204

5 Conclusions

The goal of this chapter is to present sequence representations and DL architectures useful for DNA sequence classification problems related to *metagenomics* and *chromatin* organization.

We show the difference in performance between one-hot encoding and *k*-mers representation in several tasks characterized by different sequence lengths. The two different sequence encodings call for different DL-based features extractors. First of all the use of spectral representation, which produces a fixed-length representation, justifies the use of convolutional layers. In this case, a large number of layers is not required, as also reported in [73]. The chaos game representation does not improve significantly the performances and can have a long training time.

One-hot encoding preserves the sequential nature of the data, but it can be infeasible to train ANN on very long sequences. If the classification problem involves small sequences that make feasible the one-hot encoding, an LSTM layer can be useful, as it takes advantage of the sequential structure and can find more complex patterns than a bag of *k*-mers. Classification of long sequences in a complex taxonomic order as 16S bacteria classification can be difficult especially in the metagenomic approach, but also for this experiment, CNN with one-dimensional kernels can have good performances.

Our experimental results and analysis on different DNA sequence classification tasks can be used as a starting point and strong baselines to develop new DL techniques that can enhance the state of the art of this field.

Acknowledgments Additional support to Giosué Lo Bosco and Domenico Amato has been granted by Project INdAM - GNCS "Computational Intelligence methods for Digital Health".

References

1. Amato, D., Di Gangi, M.A., Lo Bosco, G., Rizzo, R.: Recurrent deep neural networks fornucleosome classification. In: Raposo, M., Ribeiro, P., Sério, S., Staiano, A., Ciaramella, A. (eds.) Computational Intelligence Methods for Bioinformatics and Biostatistics. pp. 118–127. Springer International Publishing, Cham (2020)

2. Cairns, B.R.: Chromatin remodeling complexes: strength in diversity, precision through specialization. Current opinion in genetics & development **15**(2), 185–190 (2005)
3. Chaput, N., Lepage, P., Coutzac, C., Soularue, E., Le Roux, K., Monot, C., Boselli, L., Routier, E., Cassard, L., Collins, M., *et al.*: Baseline gut microbiota predicts clinical response and colitis in metastatic melanoma patients treated with ipilimumab. Annals of Oncology **28**(6), 1368–1379 (2017)
4. Cole, J.R., Wang, Q., Cardenas, E., Fish, J., Chai, B., Farris, R.J., Kulam-Syed-Mohideen, A., McGarrell, D.M., Marsh, T., Garrity, G.M., *et al.*: The ribosomal database project: improved alignments and new tools for rrna analysis. Nucleic acids research **37**(suppl_1), D141–D145 (2008)
5. Di Gangi, M., Lo Bosco, G., Rizzo, R.: Deep learning architectures for prediction of nucleosome positioning from sequences data. BMC Bioinformatics **19**(14), 418 (Nov 2018)
6. Di Gangi, M.A., Gaglio, S., La Bua, C., Lo Bosco, G., Rizzo, R.: A deep learning network for exploiting positional information in nucleosome related sequences. In: Rojas, I., Ortuño, F. (eds.) Bioinformatics and Biomedical Engineering: 5th International Work-Conference, IWBBIO 2017, Granada, Spain, April 26–28, 2017, Proceedings, Part II, pp. 524–533. Springer International Publishing (2017)
7. Durbin, R., Eddy, S.R., Krogh, A., Mitchison, G.: Biological Sequence Analysis: Probabilistic Models of Proteins and Nucleic Acids. Cambridge University Press (1998)
8. Escobar-Zepeda, A., Vera-Ponce de León, A., Sanchez-Flores, A.: The road to metagenomics: from microbiology to dna sequencing technologies and bioinformatics. Frontiers in genetics **6**, 348 (2015)
9. Escobar-Zepeda, A., Vera-Ponce de León, A., Sanchez-Flores, A.: The Road to Metagenomics: From Microbiology to DNA Sequencing Technologies and Bioinformatics. Frontiers in Genetics **6**(348) (2015)
10. Ferraro Petrillo, U., Sorella, M., Cattaneo, G., Giancarlo, R., Rombo, S.E.: Analyzing big datasets of genomic sequences: fast and scalable collection of k-mer statistics. BMC Bioinformatics **20**(4), 138 (Apr 2019)
11. Fiannaca, A., La Paglia, L., La Rosa, M., Renda, G., Rizzo, R., Gaglio, S., Urso, A., *et al.*: Deep learning models for bacteria taxonomic classification of metagenomic data. BMC bioinformatics **19**(7), 198 (2018)
12. Fiannaca, A., La Rosa, M., La Paglia, L., Rizzo, R., Urso, A.: nrc: non-coding rna classifier based on structural features. BioData mining **10**(1), 27 (2017)
13. Fiannaca, A., La Rosa, M., Rizzo, R., Urso, A.: Analysis of dna barcode sequences using neural gas and spectral representation. In: Iliadis, L., Papadopoulos, H., Jayne, C. (eds.) Engineering Applications of Neural Networks, Communications in Computer and Information Science, vol. 384, pp. 212–221 (2013)
14. Fiannaca, A., La Rosa, M., Rizzo, R., Urso, A.: A k-mer-based barcode dna classification methodology based on spectral representation and a neural gas network. Artificial Intelligence in Medicine **64**(3), 173–184 (2015). https://doi.org/10.1016/j.artmed.2015.06.002
15. Frankel, A.E., Coughlin, L.A., Kim, J., Froehlich, T.W., Xie, Y., Frenkel, E.P., Koh, A.Y.: Metagenomic shotgun sequencing and unbiased metabolomic profiling identify specific human gut microbiota and metabolites associated with immune checkpoint therapy efficacy in melanoma patients. Neoplasia **19**(10), 848–855 (2017)
16. Giancarlo, R., Lo Bosco, G., Pinello, L., Utro, F.: The three steps of clustering in the post-genomic era: A synopsis. In: Rizzo, R., Lisboa, P.J.G. (eds.) Computational Intelligence Methods for Bioinformatics and Biostatistics. pp. 13–30. Springer Berlin Heidelberg, Berlin, Heidelberg (2011)
17. Goodfellow, I.J., Bengio, Y., Courville, A.: Deep Learning. MIT Press, Cambridge, MA, USA (2016), http://www.deeplearningbook.org
18. Guo, S.H., Deng, E.Z., Xu, L.Q., Ding, H., Lin, H., Chen, W., Chou, K.C.: inuc-pseknc: a sequence-based predictor for predicting nucleosome positioning in genomes with pseudo k-tuple nucleotide composition. Bioinformatics **30**(11), 1522–1529 (2014)

19. Hinton, G.E.: Training Products of Experts by Minimizing Contrastive Divergence. Neural Computation **14**(8), 1771–1800 (2002)
20. Hinton, G.E.: Reducing the Dimensionality of Data with Neural Networks. Science **313**(5786), 504–507 (2006)
21. Hinton, G.E., Osindero, S., Teh, Y.W.: A Fast Learning Algorithm for Deep Belief Nets. Neural Computation **18**(7), 1527–1554 (2006)
22. Hochreiter, S., Schmidhuber, J.: Long short-term memory. Neural Computation **9**(8), 1735–1780 (1997)
23. Jones, P.A., Baylin, S.B.: The epigenomics of cancer. Cell **128**(4), 683–692 (2007)
24. Jordan, M.I.: Attractor dynamics and parallelism in a connectionist sequential machine. In: Artificial neural networks: concept learning, pp. 112–127 (1990)
25. Kaplan, N., K Moore, I., Mittendorf, Y., J Gossett, A., Tillo, D., Field, Y., M LeProust, E., R Hughes, T., Lieb, J., Widom, J., Segal, E.: The dna-encoded nucleosome organization of a eukaryotic genome. Nature **458**, 362–6 (03 2009)
26. Kho, Z.Y., Lal, S.K.: The human gut microbiome–a potential controller of wellness and disease. Frontiers in microbiology **9** (2018)
27. Kingma, D., Ba, J.: Adam: A method for stochastic optimization. In: Proceedings of the 3rd International Conference on Learning Representations (ICLR) (2014)
28. Krebs, C.J.: Species diversity measures. Ecological methodology (1999)
29. Kullback, S., Leibler, R.A.: On Information and Sufficiency. The Annals of Mathematical Statistics **22**(1), 79–86 (1951)
30. LeCun, Y., Bengio, Y., Hinton, G.: Deep learning. Nature **521**, 436–444 (2015)
31. Lecun, Y., èon Bottou, L., Bengio, Y., Haffner, P.: Gradient-based learning applied to document recognition. In: Proceedings of the IEEE. pp. 2278–2324 (1998)
32. LeCun, Y., Bottou, L., Bengio, Y., Haffner, P.: Gradient-based learning applied to document recognition. Proceedings of the IEEE **86**(11), 2278–2324 (1998)
33. LeCun, Y., Bottou, L., Bengio, Y., Haffner, P.: Gradient-based learning applied to document recognition. Proceedings of the IEEE **86**(11), 2278–2324 (1998)
34. Li, Y., Huang, C., Ding, L., Li, Z., Pan, Y., Gao, X.: Deep learning in bioinformatics: Introduction, application, and perspective in the big data era. Methods (2019)
35. Liu, H., Lin, S., Cai, Z., Sun, X.: Role of 10–11bp periodicities of eukaryotic dna sequence in nucleosome positioning. Bio Systems **105**, 295–9 (06 2011)
36. Liu, M.J., Seddon, A.E., Tsai, Z.T.Y., Major, I.T., Floer, M., Howe, G.A., Shiu, S.H.: Determinants of nucleosome positioning and their influence on plant gene expression. Genome research **25**(8), 1182–1195 (2015)
37. Lo Bosco, G.: Alignment free dissimilarities for nucleosome classification. In: Computational Intelligence Methods for Bioinformatics and Biostatistics, Lecture Notes in Computer Science, vol. 9874, pp. 114–128 (2016)
38. Lo Bosco, G., Di Gangi, M.A.: Deep learning architectures for dna sequence classification. In: Petrosino, A., Loia, V., Pedrycz, W. (eds.) Fuzzy Logic and Soft Computing Applications. pp. 162–171. Springer International Publishing, Cham (2017)
39. Lo Bosco, G., Rizzo, R., Fiannaca, A., La Rosa, M., Urso, A.: A deep learning model for epigenomic studies. In: 12th International Conference on Signal-Image Technology Internet-Based Systems (SITIS). pp. 688–692. IEEE (2016)
40. Lo Bosco, G., Rizzo, R., Fiannaca, A., La Rosa, M., Urso, A.: Variable ranking feature selection for the identification of nucleosome related sequences. In: Benczúr, A., Thalheim, B., Horváth, T., Chiusano, S., Cerquitelli, T., Sidló, C., Revesz, P.Z. (eds.) New Trends in Databases and Information Systems. pp. 314–324. Springer International Publishing (2018)
41. Lu, Q., Wallrath, L.L., Elgin, S.C.: Nucleosome positioning and gene regulation. Journal of cellular biochemistry **55**(1), 83–92 (1994)
42. Min, S., Lee, B., Yoon, S.: Deep learning in bioinformatics. Briefings in Bioinformatics pp. 1–19 (2016)

43. Montúfar, G.: Restricted boltzmann machines: Introduction and review. In: Ay, N., Gibilisco, P., Matúš, F. (eds.) Information Geometry and Its Applications. pp. 75–115. Springer International Publishing, Cham (2018)
44. Nair, V., Hinton, G.E.: Rectified linear units improve restricted boltzmann machines. In: Proceedings of the 27th international conference on machine learning (ICML-10). pp. 807–814 (2010)
45. Pinello, L., Lo Bosco, G.: A new feature selection methodology for k-mers representation of dna sequences. In: Computational Intelligence Methods for Bioinformatics and Biostatistics, Lecture Notes in Computer Science, vol. 8623, pp. 99–108 (2015)
46. Pinello, L., Lo Bosco, G., Hanlon, B., Yuan, G.C.: A motif-independent metric for dna sequence specificity. BMC Bioinformatics **12** (2011)
47. Pinello, L., Lo Bosco, G., Yuan, G.C.: Applications of alignment-free methods in epigenomics. Briefings in Bioinformatics **15**(3), 419–430 (2014)
48. Pulivarthy, S.R., Lion, M., Kuzu, G., Matthews, A.G., Borowsky, M.L., Morris, J., Kingston, R.E., Dennis, J.H., Tolstorukov, M.Y., Oettinger, M.A.: Regulated large-scale nucleosome density patterns and precise nucleosome positioning correlate with v (d) j recombination. Proceedings of the National Academy of Sciences **113**(42), E6427–E6436 (2016)
49. Qin, J., Li, Y., Cai, Z., Li, S., Zhu, J., Zhang, F., Liang, S., Zhang, W., Guan, Y., Shen, D., *et al.*: A metagenome-wide association study of gut microbiota in type 2 diabetes. Nature **490**(7418), 55 (2012)
50. Ramazzotti, M., Berná, L., Donati, C., Cavalieri, D.: riboframe: an improved method for microbial taxonomy profiling from non-targeted metagenomics. Frontiers in genetics **6**, 329 (2015)
51. Ridgway, P., Almouzni, G.: Chromatin assembly and organization. Journal of cell science **114**(15), 2711–2712 (2001)
52. Rinke, C., Schwientek, P., Sczyrba, A., Ivanova, N.N., Anderson, I.J., Cheng, J.F., Darling, A., Malfatti, S., Swan, B.K., Gies, E.A., Dodsworth, J.A., Hedlund, B.P., Tsiamis, G., Sievert, S.M., Liu, W.T., Eisen, J.A., Hallam, S.J., Kyrpides, N.C., Stepanauskas, R., Rubin, E.M., Hugenholtz, P., Woyke, T.: Insights into the phylogeny and coding potential of microbial dark matter. Nature **499**(7459), 431–437 (2013)
53. Rizzo, R., Fiannaca, A., La Rosa, M., Urso, A.: The general regression neural network to classify barcode and mini-barcode dna. In: Computational Intelligence Methods for Bioinformatics and Biostatistics, Lecture Notes in Computer Science, vol. 8623, pp. 142–155 (2015)
54. Rizzo, R., Fiannaca, A., La Rosa, M., Urso, A.: A deep learning approach to dna sequence classification. In: International Meeting on Computational Intelligence Methods for Bioinformatics and Biostatistics. pp. 129–140. Springer (2015)
55. Rizzo, R., Fiannaca, A., La Rosa, M., Urso, A.: Classification experiments of dna sequences by using a deep neural network and chaos game representation. In: Proceedings of the 17th International Conference on Computer Systems and Technologies 2016. pp. 222–228. ACM (2016)
56. Sala, A., Toto, M., Pinello, L., Gabriele, A., Di Benedetto, V., Ingrassia, A.M., Lo Bosco, G., Di Gesù, V., Giancarlo, R., Corona, D.F.V.: Genome-wide characterization of chromatin binding and nucleosome spacing activity of the nucleosome remodelling atpase iswi. The EMBO Journal **30**(9), 1766–1777 (2011)
57. Schnitzler, G.R.: Control of nucleosome positions by dna sequence and remodeling machines. Cell biochemistry and biophysics **51**(2–3), 67–80 (2008)
58. Shahbazian, M.D., Grunstein, M.: Functions of site-specific histone acetylation and deacetylation. Annu. Rev. Biochem. **76**, 75–100 (2007)
59. Shawe-Taylor, J., Cristianini, N.: Support vector machines. An Introduction to Support Vector Machines and Other Kernel-based Learning Methods pp. 93–112 (2000)
60. Simpson, E.H.: Measurement of Diversity. Nature **163**(4148), 688–688 (1949)
61. Song, Y.J., Cho, D.H.: Classification of various genomic sequences based on distribution of repeated k-word. In: 2017 39th Annual International Conference of the IEEE Engineering in Medicine and Biology Society (EMBC). pp. 3894–3897. IEEE (2017)

62. Srivastava, N., Hinton, G., Krizhevsky, A., Sutskever, I., Salakhutdinov, R.: Dropout: a simple way to prevent neural networks from overfitting. The Journal of Machine Learning Research **15**(1), 1929–1958 (2014)
63. Svaren, J., Horz, W.: Transcription factors vs. nucleosomes: Regulation of the pho5 promoter in yeast. Trends in Biochemical Sciences **22**, 93–97 (1997)
64. Tekaia, F., Lazcano, A., Dujon, B.: The genomic tree as revealed from whole proteome comparisons. Genome research **9**(6), 550–557 (1999)
65. Turnbaugh, P.J., Ley, R.E., Mahowald, M.A., Magrini, V., Mardis, E.R., Gordon, J.I.: An obesity-associated gut microbiome with increased capacity for energy harvest. nature **444**(7122), 1027 (2006)
66. Vinje, H., Liland, K.H., Almøy, T., Snipen, L.: Comparing k-mer based methods for improved classification of 16s sequences. BMC Bioinformatics **16**(1), 205 (Jul 2015)
67. Wang, Y., Hill, K., Singh, S., Kari, L.: The spectrum of genomic signatures: from dinucleotides to chaos game representation. Gene **346**, 173–185 (2005)
68. Weiner, A., Hughes, A., Yassour, M., Rando, O.J., Friedman, N.: High-resolution nucleosome mapping reveals transcription-dependent promoter packaging. Genome research **20**(1), 90–100 (2010)
69. Whitehouse, I., Tsukiyama, T.: Antagonistic forces that position nucleosomes in vivo. Nature structural & molecular biology **13**(7), 633 (2006)
70. Wooley, J.C., Ye, Y.: Metagenomics: Facts and Artifacts, and Computational Challenges. Journal of Computer Science and Technology **25**(1), 71–81 (2010)
71. Wu, H., Gu, X.: Towards dropout training for convolutional neural networks. Neural Networks **71**, 1–10 (2015)
72. Yuan, C., Lei, J., Cole, J., Sun, Y.: Reconstructing 16s rrna genes in metagenomic data. Bioinformatics **31**(12), i35–i43 (2015)
73. Zeng, H., Edwards, M.D., Liu, G., Gifford, D.K.: Convolutional neural network architectures for predicting dna–protein binding. Bioinformatics **32**(12), i121–i127 (2016)
74. Zhang, J., Peng, W., Wang, L.: Lenup: learning nucleosome positioning from dna sequences with improved convolutional neural networks. Bioinformatics **34**(10), 1705–1712 (2018)

A Deep Learning Model for MicroRNA-Target Binding

Ahmet Paker and Hasan Oğul

Abstract MicroRNAs (miRNAs) are non-coding RNAs of ~21–23 bases length, which play critical role in gene expression. They bind the target mRNAs in the post-transcriptional level and cause translational inhibition or mRNA cleavage. Quick and effective detection of the binding sites of miRNAs is a major problem in bioinformatics. This chapter introduces a new technique to model microRNA-target binding using *Recurrent Neural Networks* (RNN) over a miRNA-target duplex sequence representation.

Keywords Deep Learning · Recurrent Neural Networks · Long-Short Term Memory · Sequence Alignment · miRNA · target prediction · miRNA target site

1 Introduction

MicroRNAs (miRNAs) are small and non-coding RNA molecules of ~21–23 bases length, which play an important role in gene expression. After transcription, they bind to target mRNAs and cause mRNA cleavage or translation inhibition in many living organisms. They bind their partial complementary target site and cause cleavage or posttranscriptional repression. They prohibit the genesis of peptides and output proteins [1, 2]. Recent research shown that gene regulation of psychiatric and neurodevelopmental disorders can be observable because of some miRNAs [3]. Since their function is usually elucidated and interpreted by the activities of their target mRNA molecules, rapid and efficient determination of the binding sites of

A. Paker
Department of Computer Engineering, Başkent University, Ankara, Turkey

H. Oğul (✉)
Faculty of Computer Sciences, Østfold University College, Halden, Norway
e-mail: hasan.ogul@hiof.no

M. Elloumi (ed.), *Deep Learning for Biomedical Data Analysis*,
https://doi.org/10.1007/978-3-030-71676-9_3

miRNAs is a major problem in molecular biology. Since experimental validation is usually long and cumbersome, computational techniques are needed to model target binding. The main problem is the elucidation of interaction between miRNAs with their target sites. It is known that this interaction is mediated through the sequences miRNA and mRNA binding site, although the mechanism behind this binding has not been completely understood yet. Therefore, computational prediction of miRNA targets is a challenging task to support global effort in understanding gene regulation [12, 13].

In this study, we introduce a new *Deep Learning* (DL) framework to predict miRNA-target binding given a mature miRNA sequence and a potential binding site of candidate target mRNA. The framework basically employs a *Long-Short Term Memory* (LSTM) network [18] fed by a duplex sequence obtained by complementary alignment of two sequences. New framework was tested in two different benchmark dataset and compared with existing methods. Finally, a web server is introduced. The web application gets a miRNA sequence from the user as a text input and shows all potential binding sites on the related mRNA sequences. The web server is available at https://mirna.atwebpages.com.

The rest of the chapter is organized as follows: In Sect. 2, we present a review of recent computational approaches using shallow and DL models. In Sect. 3, we introduce the components and hyperparameters of the used *Deep Neural Network* (DNN) [19] architecture. In Sect. 4, we present empirical results in different experimental setup. Finally, we discuss the future challenges and perspective.

2 Related Work

There exists a variety of algorithms and tools for miRNA target prediction. The approaches vary according to different assumptions about target selection mechanism such as perfect pairing of miRNA-mRNA seed match, seed match conservation across different species, number of predicted sites for the same miRNA on a certain 3'UTR, miRNA-target pair free energy outside the seed, target site accessibility and miRNA-target pair secondary structure.

RNAHybrid is one of the earliest methods to model miRNA–mRNA binding in terms of thermodynamic properties of resulting duplex [14]. It calculates the minimum free energy to form the duplex from predicted structure information. Several target prediction tools have utilized *RNAHybrid* methodology in their algorithms in addition to other factors [15]. *miRTif* is an advanced sequence-based model for miRNA-target binding [16]. The algorithm counts the frequencies of some predefined motifs in several lengths, defined over the complementary/non-complementary pairs in the duplex. These frequencies are used to feed a powerful *Machine Learning* (ML) method, called *Support Vector Machines* (SVM), which is able to learn an optimal *hyperplane* to separate positive and negative examples based on these features.

A probabilistic model was introduced to describe the binding preferences between a microRNA sequence and its target site [4]. They proposed a method which is based on a sequential probabilistic model to express miRNA sequences and its related target site. This model converts miRNA-binding site pair (duplex-sequence) to a new sequence. After that, to analyze a new sequence, they used *Variable Length Markov Chain* (VLMC) [9].

TargetScan v7.0 [8] uses the assumption that canonical sites are more functional than non-canonical sites to express miRNA binding sites. They extract 14 new features and train the data with multiple *linear regression* models [27]. *TarPmiR* [7] developed a random-forest-based approach to predict miRNA target sites. Their method is based on scanning miRNA on related mRNA sequence to get the perfect seed- matching sites. They used six conventional features and seven of their own features, together. Their method calculates the value of these features. As a result, *TarPmiR* selects the site which has the highest probability as target-site. *DeepMirTar* [5] is a recent method based on *Stacked de-noising Auto-encoder* (SdA) [19] DL method to predict human miRNA-targets on the site level. They used three different feature representations to express miRNA targets, i.e. high-level expert designed features, low-level expert designed features and raw-data-level designed features. Seed match, sequence composition, free energy, site accessibility, conservation, and hot-encoding are some of the examples of these features.

3 Materials and Methods

The classification of sequences is a modeling problem that you have a specific input sequence and predicts the target sequence. The difficulty of this problem is that the sequences can vary in length, consist of a large thesaurus of input characters and that the model should examine the long-term context or dependencies between characters in the input sequence. *Recurrent Neural Network* (RNN) [20] addresses this issue by adding the feedback mechanism which functions as a memory. Thus, the previous inputs in the model are kept in a kind of memory. LSTM expands this idea by creating a short term and long-term memory component. As a result, the LSTM model can give successful results in biological sequences which are made of repeating set of patterns. Here, we introduce a DL architecture to model miRNA-target binding by employing LSTM as shown in Fig. 1. Following sections details the layers in the architecture.

3.1 Alignment and Representation Layer

The input comprises two sequences which are questioned if they can perform a binding or not. Since LSTM can work on a single one-dimensional sequence input, we need a new representation of two inputs to feed an LSTM. To this end, we use

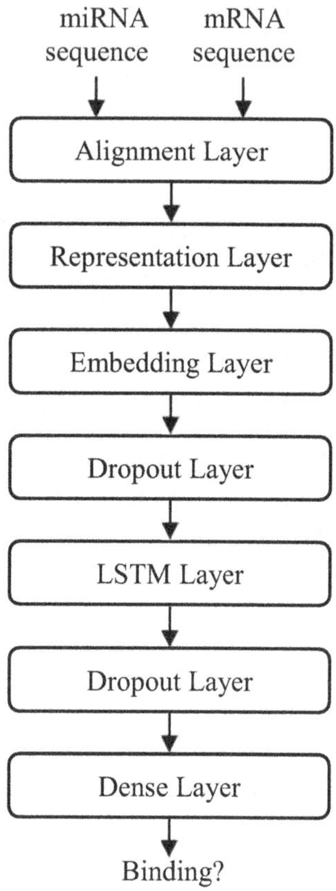

Fig. 1 DL architecture for miRNA-target binding prediction

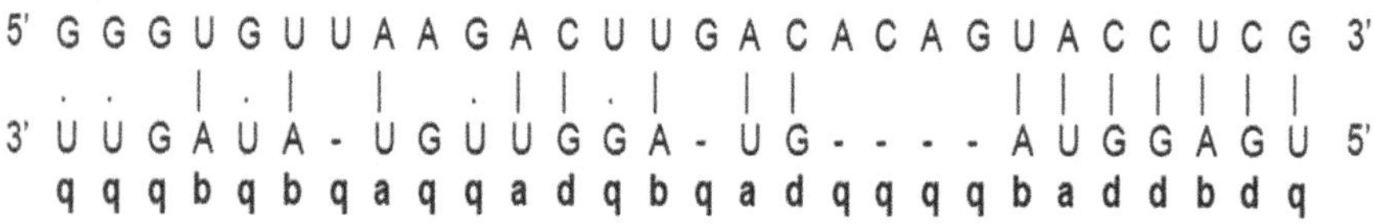

Fig. 2 Example alignment of mRNA binding site (top) and miRNA sequence (middle) converted to a new sequence of duplex formation (bottom)

a duplex sequence model offered by [4]. miRNA–mRNA duplex is constructed by complementary alignment of mature miRNA sequence with mRNA binding site. The alignment is performed using a dynamic programming algorithm with a penalty of 1 for both mismatches and gaps [6].

Resulting alignment is transformed into a new sequence defined over an alphabet of symbols representing distinct nucleotide pair types including mismatch and space in any other site (Fig. 2).

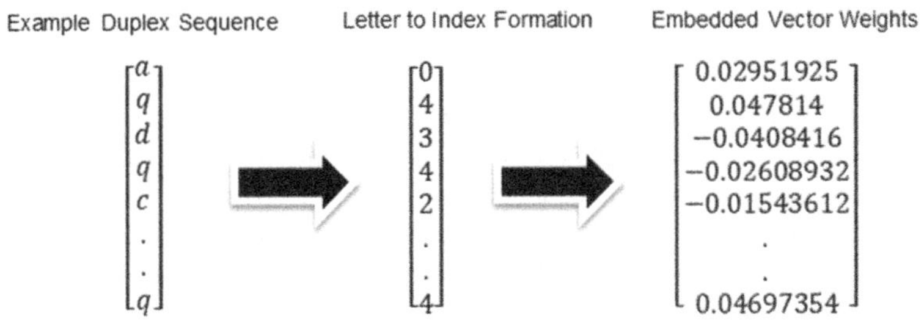

Fig. 3 Example of proposed embedded vector representation

3.2 *Embedding Layer*

The main idea behind the embedded word is that each word used in a language can be represented by a set of numerical values (vectors). Embedded words are N-dimensional vectors that try to capture word meaning and scope in their values.

First, each letter in duplex sequence is converted into an index. The character "a" converted index 0, Character "b" converted index 1, Character "c" converted index 2, Character "d" converted index 3, Character "q" converted index 4. The Embedded Word method uses the Euclidean distance to find the relationship between similar sequences. Once the dependencies between characters are found, an embedded vector is obtained (Fig. 3).

Before the data are learned and tested by the built *Deep Neural Network* (DNN) [19], the dimensions of each duplex sequence, letter to index and embedded vector weights are fixed in the form of a vector of length 32.

3.3 *Dropout Layer*

Dropping-out aims to drop or remove variables that are input to layers in the *Artificial Neural Network* (ANN). It has the effect of making the nodes in the ANN more robust for inputs and simulating multiple ANNs with different ANN structures. Technically, in each training step, the individual nodes are removed from the ANN with (1-*p*) probability, and a reduced ANN remains. The need for *Dropout* Layer is to reduce the possibility of overfitting data during training. A fully connected layer uses most of the parameters inefficient, and therefore, neurons develop interdependence between each other, which reduces the individual strength of each neuron during training and leads to overfitting of the training data.

Besides, *Dropout* is an approach in DNNs that relies on interdependent learning among smart neurons. *Dropout* approach is one of the most common regularization approaches used in DL.

3.4 LSTM Layer

Recurrent Neural Networks (RNNs) [20] are a family of ANNs to model sequential data. An RNN is a DNN where the links between the units form a directed loop. Since inputs are processed in sequence, the repetitive calculation is performed in hidden units with a cyclic connection. That's why memory is stored indirectly in hidden units called *state vectors*, and the output for the current input is calculated by considering all previous inputs using these state vectors.

Long Short-Term Memory (LSTM) [18] is an RNN, which can remember values at random intervals. Stored values are not changed when learned progress is saved. RNNs allow back and forth connections between neurons.

The RNN has a simpler structure than the LSTM and lacks the gating process. All RNNs have feedback loops at the repetitive layer. Thus, over time they are provided to keep the information "in memory". However, it is difficult to train standard RNNs to solve long-term interdependencies that require learning. The reason for this is that the gradient of the loss function gradually decreases over time. LSTM units contain a "memory cell" that can hold information in memory for a long time. A series of gates are used to control when the information is entered in memory, when it is exited and when it has been forgotten. This architecture allows them to learn long term dependencies. Also, RNN has a single layer (tanh) and LSTM has four interactive layers.

Figure 4 gives an overview of LSTM architecture. First, on the left, there is a new sequence value $\boldsymbol{X_t}$, which is combined with the previous output from cell h_{t-1}. The first step of this combined input is to crush it through a tanh layer. The second step is to pass this input through an input gate. An input gate is a layer of *sigmoid* active

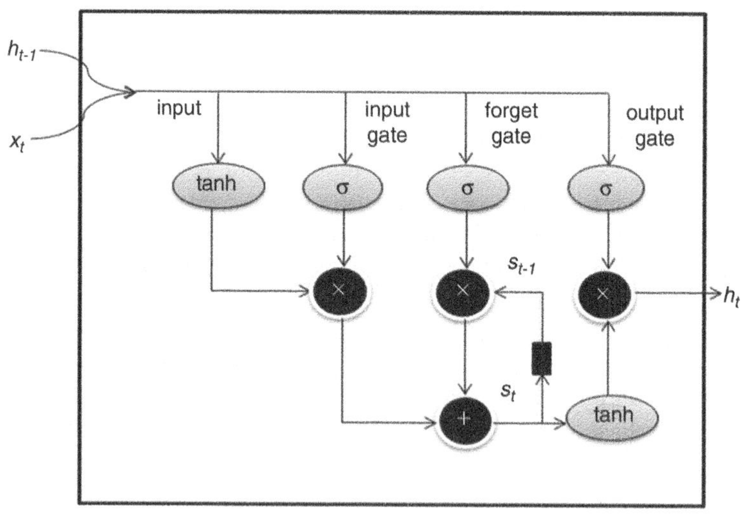

Fig. 4 LSTM Architecture

nodes whose output is multiplied by the squashed input. This *sigmoid* gate can move to destroy all unnecessary elements of the input vector. A *sigmoid* function returns values between 0 and 1 so that weights connecting the input to these nodes can be converted to output values close to zero (outputs close to "transition" from other values to "close" certain input values).

The next step in the LSTM network is the forget-gate. LSTM cells have an internal state that which is $\boldsymbol{S_t}$. This variable with a time-out delay is added to the input data to form an active iteration layer. Adding instead of multiplication helps reduce the risk of vanishing gradient. However, this iteration loop is controlled by a forget gate - it works in the same way as the input gate, but instead helps the ANN learn which status variables need to be "remembered" or "forgotten".

Lastly, an output gate specifies which values actually pass through the $\mathrm{h_t}$, cell as an output.

The mathematics of the LSTM cell is defined as below

The input is embedded between -1 and 1 with using tanh activation function.

$$g = \tanh\left(b^9 + x_t U^9 + h_{t-1} V^9\right)$$

The input gate and previous cell output are expressed with U^9 and V^9. b^9 is the input bias. This embedded input is multiplied by the output of the input gate which is defined as below:

$$i = \sigma\left(b^i + x_t U^i + h_{t-1} V^i\right)$$

Then the input section output will be as below:

$$g \circ i$$

The forget gate output is defined as:

$$f = \sigma\left(b^f + x_t U^f + h_{t-1} V^f\right)$$

Then the previous state and f will be multiplied. After that, output from forget gate will be expressed as:

$$s_t = s_{t-1} \circ f + g \circ i$$

The output gate is defined as below:

$$o = \sigma\left(b^o + x_t U^o + h_{t-1} V^o\right)$$

As a result, final output will be:

$$h_t = \tanh\left((S_t) \circ o\right)$$

3.5 Dense Layer

A *dense layer* is one of the neuron layers in the ANN. Each neuron receives input from all neurons in the previous layer so that it is densely bound. The layer has a weight matrix W, a bias vector b, and activations of the previous layer a. A *dense layer* is a fully connected ANN layer. The *dense layer* is used to modify the dimensions of the related vector. Mathematically, it applies a *scaling, rotation, translation, transformation* to the corresponding vector.

3.6 Optimization and Other Parameters

Usually, the performance of an ANN depends on several factors. Another factor that is neglected and contributed in the performance algorithm is the optimization method used to fit the model. In this section, one activation function, one loss function and three optimization methods are discussed.

Sigmoid Activation Function
Sigmoid functions are one of the most commonly used activation functions. In contrast to the linear function, in the *sigmoid* function, the output of the activation function will always be within range (0, 1). Then it can be used in binary classification problems.

Binary Cross-Entropy Loss Function
Binary Cross-Entropy Loss Function is used to measure the performance of a binary classification model whose output has a probability value between 0 and 1. As the predicted probability differs from the actual label, the loss of *cross-entropy* increases. Therefore, for example, when the actual observation label is 1, it will predict the probability of 0.006 and result in a high loss value. An excellent model would be 0 *log loss*.

Stochastic Gradient Descent Optimizer
Gradient Descent is a widely used optimization technique in ML and DL. It can be used with most learning algorithms. A *gradient* is the slope of a function; mathematically, they can be defined as partial derivatives of a range of parameters based on their input. *Gradient Descent* can be defined as an iterative method that tries to minimize the cost function and is used to find the values of the parameters of a function.

By "stochastic" is meant a system or a process associated with a random probability. Thus, in the *Stochastic Gradient Descent* (SGD), several random samples are selected instead of all the data set for each iteration. In SGD, it uses only one sample to perform each iteration. The sample is mixed randomly and selected to perform the iteration.

Adam Optimizer

Adam Optimizer [17] can be used instead of the classical SGD method to update recursive ANN weights based on training data. *Adam* is derived from adaptive moment estimation. *Adam* algorithm was presented by Diederik Kingmaand and Jimmy Ba in 2015. *Adam* has been created by combining the best features of the *AdaGrad* and *RMSProp* algorithms to provide an optimization algorithm that can handle the sparse transition in noisy problems. *Adam* is easy to use where most of the default configuration parameters are good for the problem. Relatively low memory requirements are an advantage of the *Adam* algorithm.

Adadelta Optimizer

Adadelta Optimizer [21] is derived from *Adagrad* and tries to reduce the aggression of *Adagrad*, reducing the learning rate monotonously. This is done by keeping the window of the past accumulated gradient limited to some fixed sizes w. At run time T, the average depends on the previous average and the current gradient.

3.7 Implementation Details

The experiments were done on the *Google Colab* environment. *Google Colab* tool is a *Jupyter* notebook-based system integrated with *Google Drive*. Features of the used machine: GPU: 1xTesla K80, 17.7 minutes compute time, having 2496 CUDA cores, 12GB GDDR5 VRAM, 33 GB memory space.

A 32 vector length is used to represent each word in the embedded layer. In addition, the parameter "500" is used to specify the size of the vector range that defines the words to be embedded. The size of the output layers for each word is defined in this layer (Table 1).

The second layer is a *Dropout* Layer. The main purpose of using this layer is eliminating useless and garbage data and also prevent overfitting. *Dropout* percentage is preferred as 20%. The third layer is the LSTM layer with 1000 memory units (smart neurons). The fourth layer is a *Dropout* Layer. Finally, since this is a classification problem, there has been used a dense output layer with a single neuron and sigmoid activation function to predict of 0 or 1 for two classes in the problem.

Table 1 Layers of proposed LSTM network

Layer Type	Output Shape
Embedding	(500,32)
Dropout_1	(500,32)
LSTM	1000
Dropout_2	1000
Dense	1

Since it is a binary classification problem, *log loss* is used as a loss function (binary cross entropy). In addition, the *Adam* algorithm was chosen as the optimizer. *Adam* is an optimization algorithm that recursively updates ANN weights in training data. In addition, the loss of validation is measured every 5 epochs to avoid overfitting. If the validation loss is increased from the previous one, the early stop function is activated. As a result, learning was interrupted.

The inputs (aligned duplex sequences) was extracted with *Python* 3. After that process, the learning step was developed in *Google Colaboratory* environment using the *Keras* library of python software language. In the web application, HTML5 and CSS3 were used in the frontend. To keep the data generated by the DL model, MySQL database was preferred. Also, PHP was used to manipulate data between the MySQL and the frontend. The web server is available at https://mirna.atwebpages.com.

4 Results

In this section, by performing various empirical analyzes on different datasets, the performance of the designed ANNs are compared based on various evaluation metrics.

4.1 Datasets

Two different datasets were used in this study. *DSet1* is taken from [9]. This dataset contains 283 positive and 115 negative miRNA-mRNA duplex sequences. There are 398 data in total. *DSet2* was obtained from the *DeepMirTar* repository [5]. In the first data set, 3915 positive data were collected, 473 of them were obtained from *mirMark* data [10] and 3442 of them were obtained from *CLASH* data [11]. 3905 negative data were generated using mock miRNAs. There are 7820 data in total. This set of data is preferred since there are many experimentally confirmed positive data.

4.2 Empirical Results

To evaluate the prediction performance we used accuracy, sensitivity, specificity, *Area Under Curve* (AUC). Given the number of *True Positives* (TP), *False Positives* (FP), *True Negatives* (TN) and *False Negatives* (FN), the evaluation metrics are calculated as follows:

$$Accuracy = \frac{\text{TP} + \text{TN}}{\text{TP} + \text{TN} + \text{FP} + \text{FN}}$$

$$Sensitivity \text{ (Recall)} = \frac{\text{TP}}{\text{TP} + \text{FN}}$$

$$Specificity = \frac{\text{TN}}{\text{TN} + \text{FP}}$$

$$\text{P}recision = \frac{\text{TP}}{\text{TP} + \text{FP}}$$

$$F1 \;\; Score = \frac{\text{Precision} * \text{Recall}}{\text{Precision} + \text{Recall}} * 2$$

Prediction Results of DSet1

34 different methods were used to test *DSet1*. From DS1_M1 to DS1_M30 method, performance of the DL system was measured according to different system hyper-parameters. Besides, DSet1 has been tried to be classified with some conventional shallow machine learning methods. SVM, DT, kNN and Random Forest were used in DS1_31, DS1_32, DS1_33 and DS1_M34, respectively.

Firstly, from DS1_M1 to DS1_M30 the dataset is randomly mixed. After that, LSTM model was evaluated by determining the test split size as 0.1. Hence, 40 randomly selected test data were gathered. 5 metrics were considered to evaluate success criterion: *Accuracy* (ACC), sensitivity, specificity, AUC and F1 score.

In Table 2, the performance criteria of the built LSTM model according to *Method Name, Batch Size, Neuron Size, Input Length, Loss Function*, and *Opt* (Optimizer system) parameters are given. As the results showed, DS1_M1 gave the best result with 4 batch size, 100 neuron size, 32 input length, *binary cross-entropy loss* functions, and *Adam* optimizer. Input length and loss function system hyperparameters have not been tested for different values. Because in the preprocessed dataset, the longest sized input sample is 32 in length. Furthermore, since the problem discussed is a binary classification problem, *binary cross-entropy loss* is preferred. According to Table 2, because the DS1_M1 method gives the best results, the increase in the number of neurons used in small size datasets decreases the success of the system. In other words, the increase in the complexity of the system affects the performance negatively. In general, the best results were obtained with *Adam* optimizer. Also, 4 batch size of gave better results than 8 (Table 2).

In Table 3, LSTM-based model is compared with commonly used traditional ML models. As shown, DL method cannot achieve a better performance when a small dataset is used.

Table 2 Methods applied based on different system hyperparameter configurations on *DSet1*

Method Name	Bat ch Size	Neuron Size	Input Length	Loss-Function	Opt	Acc	Sens	Spec	AUC
DS1_M1	**4**	**100**	**32**	**binary-ce**	***Adam***	**82,50**	**100**	**30,00**	**77,00**
DS1_M2	4	250	32	binary-ce	*Adam*	77,31	93,76	76,78	74,28
DS1_M3	4	500	32	binary-ce	*Adam*	77,61	83,14	77,12	74,79
DS1_M4	4	1000	32	binary-ce	*Adam*	55,44	86,61	61,26	71,77
DS1_M5	4	1500	32	binary-ce	*Adam*	51,72	85,88	60,60	59,53
DS1_M6	4	100	32	binary-ce	*Adad elta*	71,62	74,56	64,70	78,82
DS1_M7	4	250	32	binary-ce	*Adad elta*	74,99	71,14	71,23	76,70
DS1_M8	4	500	32	binary-ce	*Adad elta*	62,41	74,12	70,41	77,47
DS1_M9	4	1000	32	binary-ce	*Adad elta*	64,18	70,46	66,67	72,33
DS1_M10	4	1500	32	binary-ce	*Adad elta*	54,66	81,98	31,15	63,21
DS1_M11	4	100	32	binary-ce	SGD	64,77	55,12	51,16	64,88
DS1_M12	4	250	32	binary-ce	SGD	59,46	59,31	57,28	71,71
DS1_M13	4	500	32	binary-ce	SGD	57,86	68,82	53,33	66,53
DS1_M14	4	1000	32	binary-ce	SGD	54,44	69,08	33,76	61,38
DS1_M15	4	1500	32	binary-ce	SGD	53,99	77,71	35,62	66,22
DS1_M16	8	100	32	binary-ce	*Adam*	81,44	96,94	27,79	78,87
DS1_M17	8	250	32	binary-ce	*Adam*	77,46	81,61	80,92	78,15
DS1_M18	8	500	32	binary-ce	*Adam*	75,82	81,61	73,74	76,68
DS1_M19	8	1000	32	binary-ce	*Adam*	72,31	75,75	85,58	78,95
DS1_M20	8	1500	32	binary-ce	*Adam*	52,51	71,24	20,01	61,89
DS1_M21	8	100	32	binary-ce	*Adad elta*	72,25	84,85	71,81	81,66
DS1_M22	8	250	32	binary-ce	*Adad elta*	71,27	85,17	67,58	79,63
DS1_M23	8	500	32	binary-ce	*Adad elta*	63,76	85,17	61,26	75,55
DS1_M24	8	1000	32	binary-ce	*Adad elta*	64,85	81,81	66,80	71,52

Method Name	Bat ch Size	Neuron Size	Input Length	Loss-Function	Opt	Acc	Sens	Spec	AUC
DS1_M25	8	1500	32	binary-ce	*Adad elta*	67,67	86,09	45,16	74,32
DS1_M26	8	100	32	binary-ce	SGD	64,76	76,75	71,81	73,37
DS1_M27	8	250	32	binary-ce	SGD	64,76	76,75	71,81	73,37
DS1_M28	8	500	32	binary-ce	SGD	62,46	73,48	69,94	70,04
DS1_M29	8	1000	32	binary-ce	SGD	63,48	78,62	73,56	76,62
DS1_M30	8	1500	32	binary-ce	SGD	54,93	74,33	44,36	64,48

Table 3 Classification results of best mirLSTM (DS1_M1) model and other basic ML methods on *DSet1*

Method	Acc (%)	Sens (%)	Spec (%)	AUC (%)
mirLSTM [22]	82.50	100	30.00	77.00
SVM [23]	76.75	84.02	37.09	60.60
DT [24]	70.00	84.80	15.00	44.60
kNN [25]	85.50	89.30	64.51	77.80
RF [26]	85.00	100	0.005	81.10

Table 4 Comparison of best mirLSTM (DS1_M1) with previous methods in *DSet1*

Method	Acc (%)	Sens (%)	Spec (%)	AUC (%)
RNAHybrid [14]	63.70	64.10	60.50	71.00
miRTif [16]	81.90	83.60	73.70	89.00
probmiR [4]	84.60	86.70	73.70	94.00
mirLSTM [22]	82.50	100	30.00	77.00

Table 4 compares the mirLSTM method with available existing methods in terms of target binding prediction performance in *DSet2*. Again, mirLSTM has lower AUC compared with VLMC model used in *probmiR*.

Prediction Results of DSet2

In the first method named as DS2_M1, without doing any feature representation methods, raw data in [5] is given directly to our DL model to predict miRNA target sites. From DS2_M2 to DS2_M31, the *DSet2* was used. Also, in each method, the performance of the developed DL model was measured in terms of different system hyperparameters. Additionally, there is used data pre-processing and we intended to set up a classification model which is performed a miRNA target site prediction.

In Table 5, the performance criteria of the builded LSTM model according to *Method Name, Batch Size, Neuron Size, Input Length, Loss Function and Opt (Optimizer system)* parameters are given. As the results showed, DS2_M5 gave the best result with 64 batch size, 1000 neuron size, 32 input length, *binary cross-entropy loss* function and *Adam* optimizer. Input length and loss function system hyperparameters have not been tested for different values. Because in the preprocessed dataset, the longest sized input sample is 32 in length. Furthermore, since the problem discussed is a binary classification problem, binary cross-entropy loss is preferred. According to Table 1, because the DS2_M5 method gives the best results, the increase in the number of neurons used in large size datasets increases the success of the system. However, when the number of neurons was 1500, the performance of the model decreased because the model became more complex. In general, the best results were obtained with the *Adam* optimizer. In addition, the *Adadelta* optimizer is at least as successful as *Adam* optimizer. Also, 64 batch size of gave better results than 128.

In Table 6, LSTM-based model is compared with commonly used traditional machine learning models in *DSet2*. Here, we observe that LSTM can outperform other methods in terms of both accuracy and AUC.

Table 5 Methods applied based on different system hyperparameter configurations on *DSet2*

Method Name	Batch Size	Neuron Size	Input Length	Loss-Function	Opt	Acc	Sens	Spec	AUC
DS2_M2	64	100	32	binary- ce	*Adam*	84,02	82,38	85,60	91,00
DS2_M3	64	250	32	binary- ce	*Adam*	85,81	86,26	85,35	92,18
DS2_M4	64	500	32	binary- ce	*Adam*	60,61	33,41	87,12	80,73
DS2_M5	**64**	**1000**	**32**	**binary- ce**	***Adam***	**87,34**	**91,45**	**83,33**	**92,59**
DS2_M6	64	1500	32	binary- ce	*Adam*	55,12	93,78	17,42	58,65
DS2_M7	64	100	32	binary- ce	*Adadel ta*	78,26	71,50	84,84	84,78
DS2_M8	64	250	32	binary- ce	*Adadel ta*	85,29	83,41	87,12	89,96
DS2_M9	64	500	32	binary- ce	*Adadel ta*	84,14	79,01	89,14	89,74
DS2_M10	64	1000	32	binary- ce	*Adadel ta*	86,31	88,64	82,54	89,13
DS2_M11	64	1500	32	binary- ce	*Adadel ta*	60,19	93,56	0,22	54,61
DS2_M12	64	100	32	binary- ce	SGD	60,87	60,88	61,61	77,12
DS2_M13	64	250	32	binary- ce	SGD	60,10	68,13	52,52	78,67
DS2_M14	64	500	32	binary- ce	SGD	66,98	72,22	52,65	81,10
DS2_M15	64	1000	32	binary- ce	SGD	50,38	88,60	18,43	74,75
DS2_M16	64	1500	32	binary- ce	SGD	59,69	88,55	31,50	64,86
DS2_M17	128	100	32	binary- ce	*Adam*	82,23	83,93	80,55	90,15
DS2_M18	128	250	32	binary- ce	*Adam*	82,64	83,33	82,29	90,55
DS2_M19	128	500	32	binary- ce	*Adam*	70,08	44,00	88,47	78,75
DS2_M20	128	1000	32	binary- ce	*Adam*	61,13	38,60	83,08	55,69
DS2_M21	128	1500	32	binary- ce	*Adam*	57,05	91,44	10,80	63,66
DS2_M22	128	100	32	binary- ce	*Adadel ta*	78,52	81,08	78,78	83,99
DS2_M23	128	250	32	binary- ce	*Adadel ta*	86,11	87,81	80,94	90,68
DS2_M24	128	500	32	binary- ce	*Adadel ta*	79,67	82,80	89,33	90,95

(continued)

Table 5 (continued)

Method Name	Batch Size	Neuron Size	Input Length	Loss-Function	Opt	Acc	Sens	Spec	AUC
DS2_M25	128	1000	32	binary- ce	*Adadel ta*	83,58	88,77	81,08	90,95
DS2_M26	128	1500	32	binary- ce	*Adadel ta*	56,76	91,90	26,61	62,44
DS2_M27	128	100	32	binary- ce	SGD	66,34	57,57	66,66	71,58
DS2_M28	128	250	32	binary- ce	SGD	64,22	73,56	59,17	86,90
DS2_M29	128	500	32	binary- ce	SGD	62,94	77,46	58,87	83,80
DS2_M30	128	1000	32	binary- ce	SGD	51,56	84,57	22,11	72,21
DS2_M31	128	1500	32	binary- ce	SGD	64,69	88,76	35,53	60,41

Table 6 Classification results of best LSTM model (DS2_M5) and other basic ML methods on the *DSet2*

Method	Acc (%)	Sens (%)	Spec (%)	AUC (%)
mirLSTM [22]	87.34	91.45	83.33	92.59
SVM [23]	81.62	78.50	84.70	81.60
DT [24]	81.39	85.58	82.59	87.68
kNN [25]	86.71	74.80	98.50	88.40
RF [26]	83.25	81.73	94.20	85.40

Table 7 Comparison of mirLSTM (DS2_M5) with previous methods in *DSet2*

Method	Acc (%)	Sens (%)	Spec (%)	AUC (%)
TargetScan v7.0 [8]	58.00	60.20	59.20	67.00
TarPmiR [7]	74.40	73.60	76.50	80.00
DeepMirTar [5]	93.40	92.30	94.70	98.00
mirLSTM [22]	87.34	91.45	83.33	92.59

Table 7 compares the mirLSTM method with available existing methods in terms of target binding prediction performance in *DSet2*. mirLSTM can perform better than two methods, while *DeepMirTar* can still achieve highest accuracy.

In *DeepMirTar*, they represent the miRNA-mRNA pairs including 750 features. Some of these features are seed match, free energy, sequence composition, site accessibility, etc. On the other hand, the proposed methods represent the miRNA-mRNA pairs is based on a probabilistic approach. In the learning phase, In *DeepMirTar* they used SdA based on a DNN. They split dataset 60% training data, 20% validation data, 20% test data. Besides, in this work, the dataset divided into 90% training data and 10% test data. They optimized the hyperparameters via grid-search method. On the other hand, there has been optimized the hyperparameters with random search method. They chose the learning rate of 0.01 and batch size 10. There has been used the learning rate of 0.1 and batch size 64. Also, they used 1500 memory units (smart neurons) on the other hand there has been used 1000 smart neurons in the LSTM layer.

5 Conclusion

Discovery of miRNAs has significantly changed our understanding of gene regulation and genetic mechanisms of several diseases. Main issue in elucidating miRNA activities is to locate them in functional ANNs to explain how they mediate relevant pathways. Identifying individual targets is the key problem in this effort. Although the problem has been extensively studied in the last decade, current method still suffers from lack of consensus and having bias due to small datasets. Emergence of DL technologies with the increase amount the data has created opportunities to improve several pattern recognition tasks. However, we have not witnessed sufficient attempts so far in employing DL techniques in miRNA target prediction. In this chapter, we introduced a new DL architecture based on LSTMs to model binding between miRNA and putative target sequence of mRNA. The results have shown that the method can out-perform conventional shallow ML techniques when larger datasets are available. Future work may include the design of new sequence representation schemes to feed DL methods, automatic optimization of hyperparameters and adding attention layers into LSTM pipelines.

References

1. Bartel, D. (2009). MicroRNAs: Target Recognition and Regulatory Functions. Cell. 136(2), PP.215-233.
2. Bartel, D. (2004). MicroRNAs: Genomics, Biogenesis, Mechanism and Function. Cell. 116, PP.281-297
3. Xu, B., Hsu, P., Karayiorgou, M. and Gogos, J. (2012). MicroRNA dysregulation in neuropsychiatric disorders and cognitive dysfunction. Neurobiology of Disease, 46(2), pp.291- 301.
4. Oğul, H., Umu, S., Tuncel, Y. and Akkaya, M. (2011). A probabilistic approach to microRNA-target binding. *Biochemical and Biophysical Research Communications*, 413(1), pp.111-115.
5. Wen, M., Cong, P., Zhang, Z., Lu, H. and Li, T. (2018). DeepMirTar: a deep-learning approach for predicting human miRNA targets. *Bioinformatics*, 34(22), pp.3781-3787.
6. Needleman, S.B. and Wunsch, C.D. (1970). A general method applicable to the search for similarities in the amino acid sequences of two proteins. J. Mol. Biol., 48, 443–453.
7. Ding, J., Li, X., Hu, H.: TarPmiR: a new approach for microRNA target site prediction. Bioinformatics. 32, 2768-2775 (2016).
8. Agarwal, V., Bell, G., Nam, J., Bartel, D.: Predicting effective microRNA target sites in mammalian mRNAs. eLife. 4, (2015).
9. D. Ron, Y.Singer, N. Tishby, The power of amnesia: learning probabilistic automata with variable memory length, Mach. Learn 25 (1996) 117-149
10. Menor, M., et al. (2014) mirMark: a site-level and UTR-level classifier for miRNA target prediction. Genome biology,15,500
11. Helwak, A., et al. (2013) Mapping the human miRNA interactome by CLASH reveals frequent noncanonical binding. Cell, 153, 654-665
12. D Dede, H Oğul, TriClust: A Tool for Cross-Species Analysis of Gene Regulation, Molecular Informatics 33 (5), 382-387
13. H.Oğul, M.S.Akkaya, (2011), Data integration in functional analysis of microRNAs, Current Bioinformatics 6, 462-472.
14. M. Rehmsmeier, P. Steffen, M. Hochmann, et al., Fast and effective prediction of microRNA/-target duplexes, RNA 10 (2004) 1507–1517.
15. M. Hammell, Computational methods to identify miRNA targets, Semin. Cell Dev. Biol. 21 (2010) 738–744.
16. Y. Yang, Y.P. Wang, K.B. Li, miRTif: a support vector machine-based microRNA target interaction filter, BMC Bioinf. 9 (2008) S4.
17. Kingma, D.P., & Ba, J. (2014). Adam: A Method for Stochastic Optimization. *CoRR, abs/1412.6980.*
18. Hochreiter S, Schmidhuber J. Long short-term memory. Neural Comput. 1997;9(8):1735- 80
19. Hinton, G.E.; Salakhutdinov, R.R. Reducing the dimensionality of data with neural network. Science 2006, 7, 504–507.
20. Rumelhart, D. E., Hinton, G. E., and Williams, R. J. (1986). Learning representations by backpropagating errors. Nature, 323, 533–536.
21. M. D. Zeiler, Adadelta: an adaptive learning rate method. arXiv preprint (2012). arXiv:1212.5701

22. Paker, Ahmet & Ogul, Hasan (2019). mirLSTM: A Deep Sequential Approach to MicroRNA Target Binding Site Prediction. Communications in Computer and Information Science 1062, s 38- 44
23. C. Cortes and V. Vapnik. Support vector networks. Machine Learning, 2O:l-25, 1995.
24. Breiman, L., Friedman, J., Stone, C.J., Olshen, R.A.: Classification and Regression Trees. Belmont, California: Wadsworth (1984)
25. Devroye, L. & Wagner, T.J. (1982) "Nearest neighbor methods in discrimination, In Classification, Pattern Recognition and Reduction of Dimensionality", Handbook of Statistics, 2: 193–197. North-Holland, Amsterdam
26. Breiman, L. Random forests. Machine Learning, 45(1):5–32, 2001.
27. Douglas Montgomery, Peck, E., & Vinning, G. (2012). Introduction to Linear Regression Analysis (5th ed.). Wiley.

Recurrent Neural Networks Architectures for Accidental Fall Detection on Wearable Embedded Devices

Mirto Musci and Marco Piastra

Abstract Unintentional falls can cause severe injuries and even death, especially if no immediate assistance is given. The aim of *Fall Detection Systems* (FDSs) is to detect an occurring fall in real time therefore issuing a remote notification. An accurate FDS can drastically improve the quality of life of elderly subjects or any other person at risk. In this chapter we focus on real-time *Automatic Fall Detection* (AFD) performed onboard smart wearable devices. In particular, in this chapter we discuss the feasibility of AFDs methods based on *Deep Learning* (DL) techniques that could fit the limited computation power and memory of smaller low-power *Micro-Controller Units* (MCUs). The chapter proves that a relatively simple *Recurrent Neural Network* (RNN) architecture, based on two *Long Short-Term Memory* (LSTM) cells, could be a viable candidate for embedded AFD. Tests were performed using the *SisFall* dataset, which includes sequences of tri-axial accelerometer and gyroscope readings for simulated falls performed by volunteers. This dataset was further annotated for training the RNN architecture. The resulting AFD method is shown to outperform other methods based on statistical indicators, as reported in the literature. The embedded feasibility of such approach is validated with an implementation for the *SensorTile*® by STMicroelectronics hardware architecture.

Keywords Fall detection · Recurrent neural networks · Embedded wearable devices

---

M. Musci (✉) · M. Piastra
University of Pavia, Pavia, Italy
e-mail: mirto.musci@unipv.it; marco.piastra@unipv.it

M. Elloumi (ed.), *Deep Learning for Biomedical Data Analysis*,
https://doi.org/10.1007/978-3-030-71676-9_4

1 Introduction

Unintentional falls are the leading cause of fatal injury and the most common cause of nonfatal trauma-related hospital admissions among older adults. As stated in [23], more than 25% of people aged over 65 years old fall every year. This percentage goes up to 32–42% for those over 70. Moreover, 30–50% of people living in long-term care institutions fall each year, with almost half of them experiencing recurrent falls. Falls lead to 20–30% of mild to severe injuries and 40% of all injury-related deaths. The average cost of a single hospitalization for fall-related injuries in 65 years old people reached $17.483 in the US in 2004, with a forecast of total cost going up to $240 billion by 2040.

Elderly people are not the only group that is heavily affected by unintentional falls: any person with some sort of fragility is part of similar statistics. Examples include any kind of mild disability and post-operative patients. The situation is worsened when people live alone, so they may not receive immediate assistance in case of accident [7].

The main objective of *Fall Detection Systems* (FDSs) is to detect occurring falls automatically and in real-time, hence issuing a remote notification so that timely aid can be given. Clearly, from both a practical and psicological point-of-view, FDSs can drastically improve the quality of life of at-risk subjects.

Although, in the literature, approaches to FDS are based on either ambient-based sensors or wearable devices with on-board sensors [18], in this work we focus on the latter approach. We believe that, with proper ultra-low power, smart wearable devices, the task of fall dection can be made more ubiquitous, less intrusive and definitely less expensive.

The adoption of effective smart wearable devices entails severe constraints. In particular, due to the power consumption of wireless communication, a complete transfer of the input stream of sensor signals for a remote analysis is not feasible in general [24]. Instead, in principle, all signal processing should be performed onboard the *smart* wearable device while the wireless communication interface should be used only for alert notification and device control. Such requirement, in turn, obliges to fit any detection method within the narrow limits, in terms of both memory space and computation power, of a smaller *Micro-Controller Unit* (MCU).

For instance, the device chosen as reference for the work presented is the *SensorTile®* miniaturized board by STMicroelectronics, which includes an ARM® Cortex® M4 (i.e. STM32) MCU at 80 MHz maximum clock rate with 128 KB of RAM together with several sensors, including tri-axial accelerometers.[1]

In this work we investigate the applicability of suitable *Deep Learning* (DL) methods, namely *Recurrent Neural Network* (RNN) architectures, to the fault detection task described within the constraints imposed by the hardware architecture of choice. To do this, we selected a specific, publicly-available dataset of sequences

[1] See http://www.st.com/en/evaluation-tools/steval-stlcs01v1.html.

acquired with wearable tri-axial accelerometers worn by volunteers of different age and gender performing daily activities and simulated falls [19]. We then enhanced the original annotation in the dataset by marking specific temporal intervals in which relevant events take place thus obtaining a suitable training and test set.

We then compared two methods, both viable candidates for the implementation within the hardware constraints: one method based on statistical indicators, which were reported to perform well on that same dataset [19], and another based on a relatively simple RNN architecture, including on two stacked *Long Short-Term Memory* (LSTM) cells.

The results obtained show that the DL approach is superior to statistical indicators in the sense that its detection specificity, sensitivity and accuracy are significantly better for both young and elder subjects. The actual feasibility of the approach proposed was assessed with an embedded implementation of the RNN architecture for the *SensorTile*® device.

The rest of this chapter is organized as follows: Sect. 2 contains a review of the literature; Sect. 3 describes the design of the reference RNN architecture; Sect. 4 presents the experimental result achieved on the selected dataset; finally, Sect. 5 contains the conclusions.

2 Related Works

The analysis of the state of the art is divided into two parts: a first part describing fall detection techniques and a second part describing publicly-available datasets with sequences of tri-axial accelerometer signals from simulated falls.

2.1 Fall Detection Techniques

FDSs can be classified into two main categories: ambient-based and wearable-device-based [16]. Ambient-based systems mainly utilize video cameras, either standard or *RGB-Depth* [3, 13]. These techniques may affect privacy and can only cover areas within the range of the video cameras. *Automatic Fall Detection* (AFD) in video streams, moreover, may still represent a difficult problem to address [6].

Wearable devices, on the other hand, offer portability as they can be used regardless of the user location. The most widely adopted sensor to equip wearable devices is the 3-axis accelerometer due to its low cost and tiny size. The common availability of accelerometers in smartphones opens the possibility to use such devices as a cost-effective sensory device. Beside being widespread and economically affordable, smartphones provide a robust and powerful hardware platform (i.e., processor, screen and radio), which allows to implement fully self-contained monitoring applications. Rotational sensors (i.e. gyroscopes) are used in [4], in which very good results are reported for AFD with the use of gyroscopes measurements alone. In [14] the authors make use of both accelerometers and gyroscopes obtaining fast response with a low computational burden.

In general, AFD from wearable sensor data is considered an open problem. The common approach is to first filter the raw acceleration sensor readings and then apply a feature extraction method of some sort that detects falls from other background activities, i.e. the so-called *Activities of Daily Living* (ADL). In [19, 22], the occurrence of a fall is detected by comparing statistical indicators like acceleration magnitude and standard deviation with a predefined threshold. In [15] the authors compare two different *Machine Learning* (ML) techniques: *k-Nearest Neighbors* (kNN) and *Support Vector Machine* (SVM), while a simple feed-forward *Artificial Neural Network* (ANN) is used in [2]. Most of the above methods perform a preprocessing of the input sequence by extracting features from a *sliding window* of a predefined length. In [9], the authors combine different techniques to enhance the prediction of the classifier: they investigate the use of ANN, kNN, *Radial Basis Function* (RBF), *Probabilistic Principal Component Analysis* (PPCA) and *Linear Discriminant Analysis* (LDA). In [17] the authors propose a *Deep Convolutional Neural Network* (DCNN) followed by a RNN layer. The DCNN extracts features from sensor signals, while the RNN detects a temporal relationship among the extracted features. Detection, however, is computationally expensive and is performed on a workstation. In [20] the authors propose a *Finite State Machine* (FSM) model to extract relevant episodes from input sequences. These episodes are then subdivided into features that are fed to a kNN classifier to distinguish between falls and ADLs.

2.2 Datasets

From a general standpoint, the basic requirements for a dataset of sensor readings are: the presence of complete sequences of either falls and ADLs and the availability of the raw, unfiltered signals as read from onboard sensors. In this perspective, 6 datasets were considered for our work [5, 8, 15, 19, 21, 22]. Table 1 lists these datasets together with the main characteristics of each, highlighting the number of different subjects involved in the experiments and the number of different activities performed by volunteers.

The *SisFall* dataset was selected for the purpose of this work as it best fits our basic assumptions. Specifically, the *SisFall* dataset includes recordings from a total of 38 volunteers: 23 young subjects and 15 elderly subjects, each performing 34 different activities in a controlled scenario (19 ADLs and 15 falls) with several retries, for a total of 4510 complete sequences. In addition, the set of activities is validated by a medical staff and each activity to be performed is described in a video clip recorded by an instructor. The *SisFall* dataset was recorded with a custom board including two tri-axial accelerometers and a gyroscope, both operating at a frequency of 200 Hz.

Table 1 List of publicly-available datasets for fall detection

Dataset	Ref.	Number of subjects	Number of activities	Sensing device
DLR	[8]	16	6	Smartphone
tFall	[15]	10	8	Smartphone
Project Gravity	[22]	3	19	Smartphone
MobiFall	[21]	24	13	Smartphone
SisFall	[19]	38	34	Custom
UMAFall	[5]	17	11	Custom

3 Design

In this section, we describe the overall design of the proposed AFD method. In particular, we first describe the reasons why we enhanced the original *SisFall* dataset annotation, then we discuss the RNN architecture of choice and the training and inference procedures.

3.1 Dataset and Labeling

The original *SisFall* dataset includes annotations that are associated on entire sequences only. In other terms, each sequence is deemed to contain either a fall or just ADL and there is no indication of the temporal interval in which such fall may have occurred. Furthermore, even sequences that include a fall occurence contain other ADL sub-sequences as well, such as walking, sitting, standing, etc. It is worth mentioning that none of the datasets listed in Table 1 include a temporal annotation of this kind.

Clearly, the original *SisFall* annotation is not sufficient per se to support the training of a ML method aiming to the real time detection of falls and therefore an appropriate enhancement is in order.

In order to do so, we first defined the three classes below to be associated with sub-sequences, corresponding to temporal intervals:

- **FALL**: the time interval in which the person is experiencing an uncontrolled transition towards an unwanted, potentially catastrophic, state, i.e. a fall.
- **ALERT**: the time interval in which the person is experiencing an uncontrolled transition towards a wanted state, e.g. a stumble followed by a recovery.
- **BKG**: all time intervals in which the person is in control and in a wanted state.

Since we are interested in detecting falls, the BKG class is intended to contain all daily activities that are not related to a fall, such as walking, jumping, walking up the stairs, sitting on a chair, and so on.

The three classes above were used for marking temporal intervals in all *SisFall* sequences. More precisely, the authors have manually labeled all sequences by marking FALL and ALERT time intervals while all the remaining parts of each sequence were considered as BKG by default. The details of the annotation procedure are described in Sect. 4.

3.2 RNN Architecture

The overall RNN architecture is depicted in Fig. 1. The core of the network (Layers 4 and 6) is based on two stacked LSTM cells [11] of dimension $n = 32$.

The input preprocessing is performed by the fully-connected Layer 1, while a second fully-connected layer (Layer 8) collects the output from the LSTM cell at

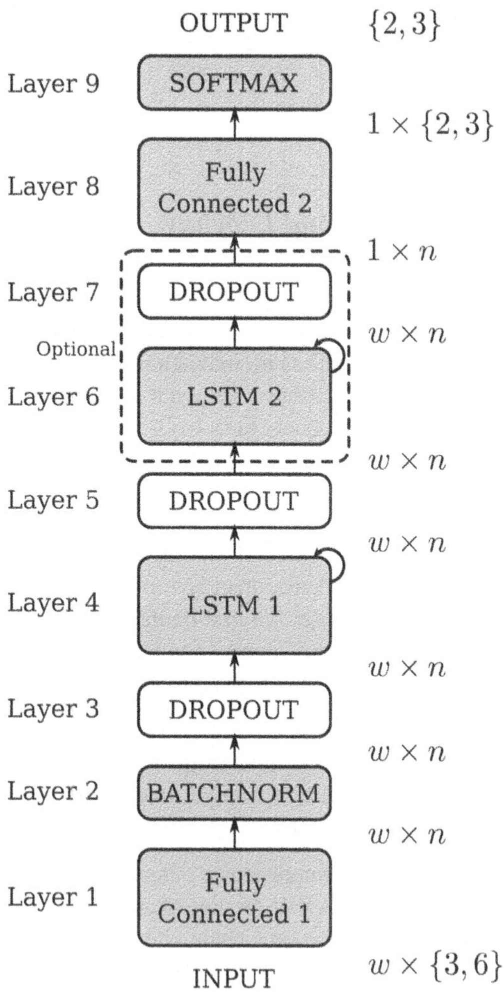

Fig. 1 The RNN architecture proposed. White blocks are active during the training phase only and are not present in the run-time module. The input size is determined by the chosen input sensors (3 if using a 3D accelerometer only, 6 if also using a 3D gyroscope) and the size w of the sliding window. The output size depends on the chosen detection classes: the size is 2 if the network discriminates between BKG and FALLs and 3 if the network also classifies ALERTs alongside BKG and FALLs. Finally, n is the internal size of each LSTM cell

Layer 6 and feeds its output to the final *SoftMax* classifier (Layer 9), which produces the classification according to the three classes described above.

The architecture includes a batch normalization layer [12] (Layer 2), to regularize input data, and three *dropout* layers [10] (Layers 3, 5 and 7). The latter layers are used during training to improve generalization, while they are removed from the run-time inference module on the embedded device.

3.3 Training and Inference

The training of LSTM cells is based on the idea of temporal unfolding [11]. Temporal unfolding entails that the input sequence is first partitioned in sub-sequences, called *windows*, having a predefined length w. Then, each LSTM cell is cascaded into exactly w copies of itself, while all such copies share the same set of numerical parameters. Each copy receives the output of the previous cell of the cascade and the input from the window at the corresponding index. Temporal unfolding allows training of a RNN, such as the LSTM, as if it was a non-recurrent deep network.

Moreover, this approach entails that input sequences in both the training set and the live sensor readings must be arranged in windows having the same fixed size w. As will be seen in Sect. 4.2, the window size w represents an hyperparameter whose tuning is critical for the effectiveness of the whole RNN architecture.

Due to its computational cost, we assume that the training process will be performed on a workstation. We also assume that a highly-optimized, detection-only run-time RNN module will be the object of the embedded implementation to be run onboard the weareable device. Of course, the run-time module will be loaded with the numerical parameters obtained from the training process.

4 Implementation and Results

This section briefly describes the implementation details of the RNN architecture and compares its results with the ones obtained with the classifiers proposed in [19], based on statistical indicators. Finally, it discusses the validation of the proposed method on the target hardware architecture.

4.1 Annotation Procedure

As already described, the annotation task consists of associating a description of the temporal intervals that correspond to the FALL and ALERT classes to each sequences in the *SisFall* dataset, while the remaining parts are considered as BKG by default.

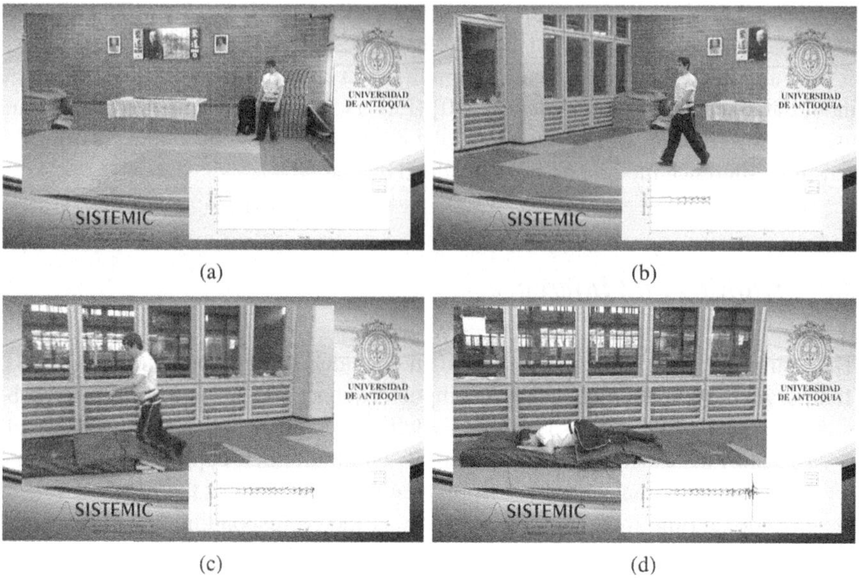

Fig. 2 An example of a *SisFall* video clip in which the activity is performed by an instructor. The sequence of sensor readings is also visible in superimposition

Each activity performed by the volounteers in the *SisFall* dataset is described by a video in which an instructor shows the type of maneuver to be performed. Each video also reports the ongoing readings of the tri-axial accelerometer, as shown in Fig. 2. In this work, the *SisFall* set of video clips were used to learn how to assign temporal intervals from the comparative analysis of the activity performed and the shape of the associated sensor readings. Such skill was used by the authors to manually annotate each individual sequence in the dataset.

The actual annotation was performed with a purpose-built software tool, developed with the *Python TkInter* library, that allows the visual analysis of a selected sequence and the marking of the temporal intervals corresponding to FALL and ALERT. An example of the visual interface of the annotation tool is shown in Fig. 3. In the figure, the upper pane shows the sequence of readings for the tri-axial accelerometers and, in this pane, the temporal intervals can be marked using the mouse pointer. The middle pane shows the resulting sequence annotation, while the lower pane shows the output produced by the C_9 statistical indicator. Figure 4 shows an example temporal annotation performed with the tool.

The extended annotation to the *SisFall* dataset is publicly available online.[2]

[2]https://bitbucket.org/unipv_cvmlab/sisfalltemporallyannotated/.

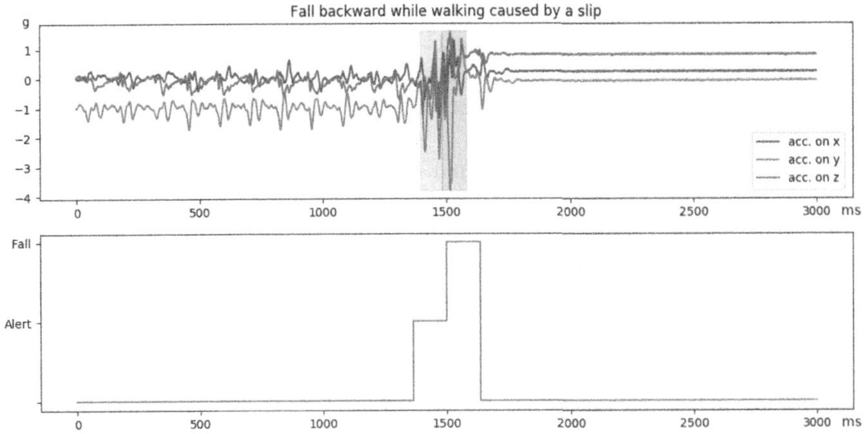

Fig. 3 The software tool developed for the annotation (see text for an explanation)

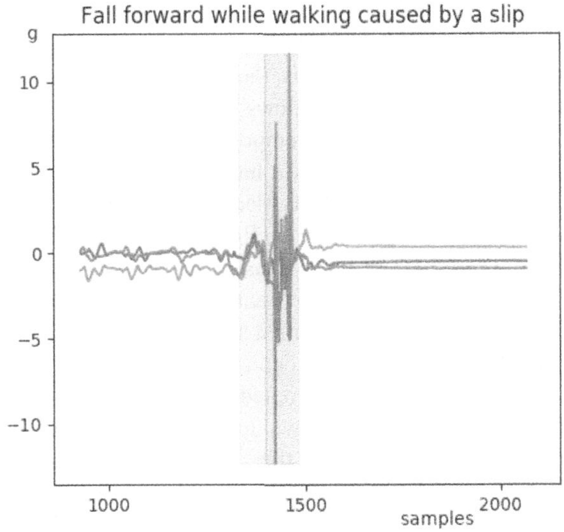

Fig. 4 An example of temporal annotation performed with the software tool. From left to right, the grayed area correspond to an ALERT and a FALL event respectively. The remaining of the sequence is classified as BKG by default

4.2 Software Implementation and Training

The RNN architecture proposed in Sect. 3 was implemented using the *TensorFlow* 1.6 [1] and the *Python* programming language. Training and testing were performed on a Dell® 5820 workstation, with a Xeon W-2133 CPU running at 3.6 GHz with 16 GB of RAM and running the latest version of the Ubuntu OS. The workstation was equipped with a *Nvidia*® *Quadro*® K5000 GPU (1536 cores, 4 GB GDDR5 RAM, 173 GB/s memory bandwidth). On average, a single experiment took circa 2 hours to complete, depending on the specifics of the particular run.

For training, we adopted a 80%/20% train/test split: given that the pool of volunteers for *SisFall* included 38 subjects, 15 elders and 23 youngsters, we decided to split by persons, to avoid the so-called identity bias. Hence, in the resulting partitioning, the training dataset included 30 subjects (12 elders), while the test dataset included 8 subjects (3 elders).

In the preprocessing step, annotated *SisFall* sequences were translated into windows having width w readings and taken at fixed intervals, i.e. *stride*, of length s.

In general, each window can span over different temporal intervals, namely it may contain readings belonging to different classes.

To assign a unique class to each window we adopted the following criteria:

- a window containing at least 10% of readings within a FALL temporal interval is tagged as FALL altogether;
- otherwise, a window containing a majority of readings within an ALERT temporal interval is tagged as ALERT;
- any other window is tagged as BKG.

The set of tagged windows resulting from the preprocessing step is definitely unbalanced. In other words, the number of windows tagged as BKG is much larger than the number of windows tagged as either FALL or ALERT. In fact, with *SisFall* sequences, BKG temporal intervals may be several seconds wide, whereas an individual FALL interval has a maximum duration of 2 s but, in many cases, it is as short as 500 ms. Thus, depending on the window size w, the number of BKG windows produced can be 50 times larger than those tagged as FALL.

Figure 6a shows the confusion matrix computed on the test set after a training with a non-weighted cross-entropy loss function and $w = 128$. As expected, BKG activities were classified accurately, whereas FALLs were poorly detected and ALERTs were almost not detected at all.

To correct this, we adopted a weighted cross-entropy loss function, in which the contribution of each window to the gradient has a weight which is inversely proportional to the size of the corresponding class in the training dataset.

More formally the weight m_i assigned to each window i is defined as:

$$m_i = \begin{cases} 1 & \text{if } i \in \text{BKG} \\ |\text{BKG}| \,/\, |\text{ALERT}| & \text{if } i \in \text{ALERT} \\ |\text{BKG}| \,/\, |\text{FALL}| & \text{if } i \in \text{FALL}. \end{cases} \quad (1)$$

Figure 6b shows the confusion matrix after a training with the weighted loss function, with a consequent drastic rise of the accuracy, in particular for the ALERT class.

As anticipated, the window width w is a critical hyperparameter, while the choice of the stride s is more dictated by practical considerations: a very low value of s entails a substantial increase in the computational burden. The choice of the values for the hyperparameters w and s was made via a sensitivity analysis, in which

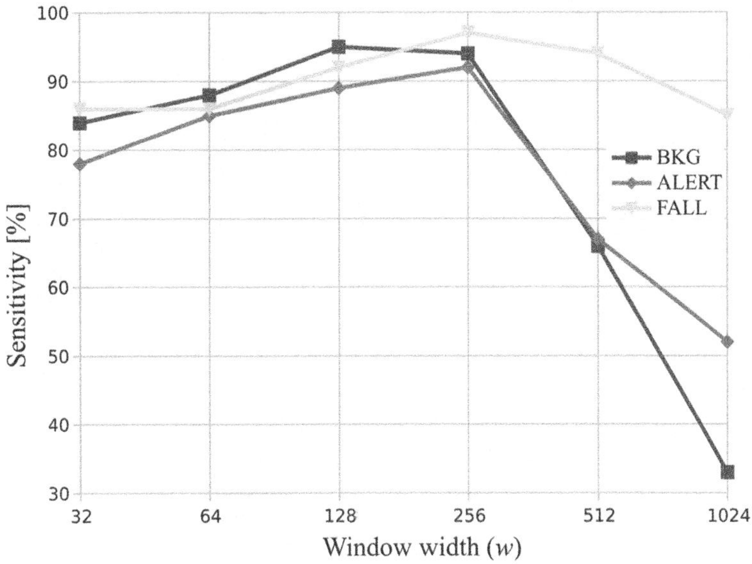

Fig. 5 Accuracy of the RNN classifier for the three classes with different window widths w and a stride of $0.5w$, when a weighted loss function is used for training

window widths varied between 32 and 1024 and stride values corresponding to 25%, 50% or 75% of w were considered.

Figure 5 describes the accuracy obtained for the three classes depending on a window width and with a fixed stride of 0.5 w. As it can be seen, a window width of 256 resulted to be the most effective for AFD.

Figure 6c shows the confusion matrix obtained with the optimal window width $w = 256$ and after an extensive grid-search optimization of the training parameters such as number of epochs, batch size and learning rate.

4.3 Comparison with Statistical Indicators

In the original paper that introduced the *SisFall* dataset [19], several statistical indicators were proposed for AFD. As we already mentioned, the original *SisFall* classifications, as either fall or ADL, were associated to entire sequences. In the same paper, each sequence was classified as representative of a fall, according to a given statistical indicator, if at any point the computed value of the same indicator was greater than a predefined threshold.

The paper reports that the following indicators were the most effective classifiers:

$$C_8 := \sqrt{\sigma^2(a_x) + \sigma^2(a_z)} \qquad (2)$$

Fig. 6 Confusion matrices obtained with $w = 128$ and with the non-weighted (**a**) and weighted (**b**) cross-entropy loss functions. The confusion matrix in (**c**) was obtained with the optimal window width $w = 256$ and after the optimization of training parameters

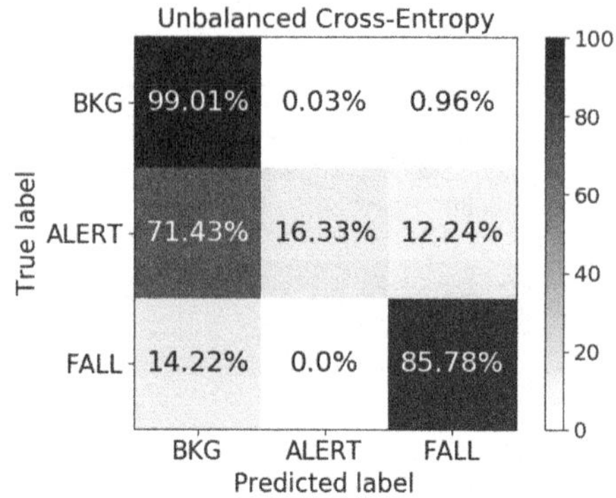

(a)

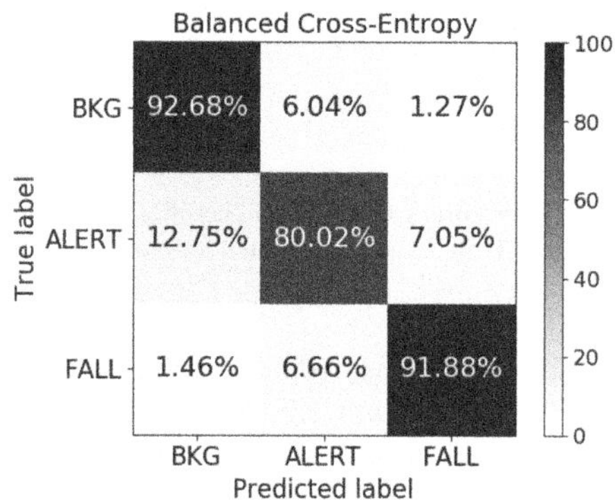

(b)

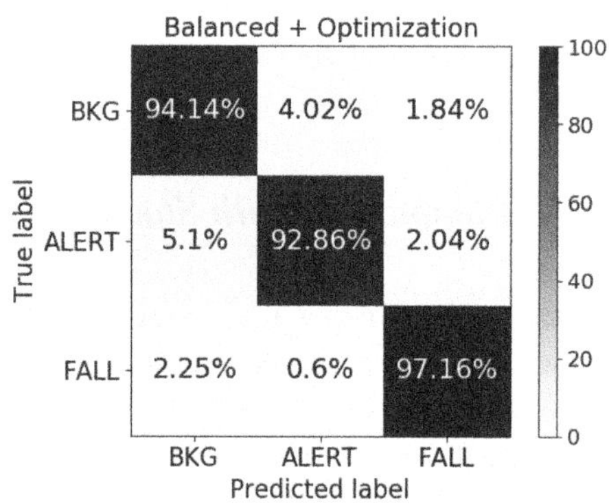

(c)

$$C_9 := \sqrt{\sigma^2(a_x) + \sigma^2(a_y) + \sigma^2(a_z)} \tag{3}$$

where σ^2 is the variance and a_x, a_y and a_z are respectively the variables representing acceleration readings along the x, y and z axes.

The difference between C_8 and C_9 is that the former relies on the standard orientation of the acquisition device used in *SisFall*, which was secured to the belt of the volunteer with the y axis lying on the sagittal plane, pointing downward, and the z axis lying on the axial plane, pointing forward. The C_9 indicator was chosen for comparison in this work because it is reported to be slightly less accurate than the C_8 counterpart while being of more general applicability.

We applied the C_9 indicator to each window obtained from the preprocessing step described above by first computing the variances σ^2 along each axis and for the same window and then comparing the value in Eq. (3) with two thresholds, a lower one for ALERT and an higher one for FALL. The window was classified as belonging to the class for which Eq. (3) was at any point greater than the higher of the two thresholds. The two thresholds in point were chosen by performing a grid search and selecting those that produced the best overall classification accuracy over the training set.

Tables 2, 3, 4 show the comparative results of the RNN classifier and the C_9 classifier on the same test set ($w = 256$, $s = 0.5w$) respectively for all volunteers and for either elder or younger ones alone. The values for *Sensitivity* (SE), *Specificity* (SP) and *Accuracy* (AC) were computed with the following definitions:

$$\mathrm{SE} = \frac{\mathrm{TP}}{\mathrm{TP} + \mathrm{FN}} \tag{4}$$

$$\mathrm{SP} = \frac{\mathrm{TN}}{\mathrm{TN} + \mathrm{FP}} \tag{5}$$

$$\mathrm{AC} = \frac{\mathrm{SE} + \mathrm{SP}}{2} \tag{6}$$

where TP, TN, FP and FN are respectively true positives, true negatives, false positives and false negatives.

From these tables, it can be seen that the RNN classifier significantly outperforms the C_9 classifier on all the three indicators and for every partitioning of the test set. In addition, the results show that the RNN classifier performs equally well on both young and elderly subjects.

For due completeness, Fig. 7 relates to a challenging activity in which the volunteers simulate a forward fall, due to a trip, that occurs while jogging. With such activity both classifiers exhibit suboptimal behavior.

The figure compares the ongoing classifications produced by the C_9 classifier (Fig. 7a) and the RNN classifier (Fig. 7b) with the ground truth of temporal annotations (Fig. 7c) on a specific sequence of sensor readings for the activity in point. The C_9 classifier is unable to distinguish the actual fall from the ADL,

Table 2 Comparison between the C_9 and the RNN classifiers on the entire test set. Bold face is used to denote the best result for each comparison

		C_9	RNN
Sensitivity	BKG	75.01	**88.39**
	ALERT	68.15	**91.08**
	FALL	75.79	**98.73**
Specificity	BKG	92.52	**97.85**
	ALERT	83.30	**90.77**
	FALL	91.57	**97.93**
Accuracy	BKG	83.77	**93.12**
	ALERT	75.73	**90.93**
	FALL	83.68	**98.33**

Table 3 Comparison between the C_9 and the RNN classifiers on elderly subjects. Bold face is used to denote the best result for each comparison

		C_9	RNN
Sensitivity	BKG	62.18	**88.83**
	ALERT	81.97	**88.52**
	FALL	54.31	**96.45**
Specificity	BKG	86.82	**94.57**
	ALERT	73.44	**90.61**
	FALL	88.71	**98.29**
Accuracy	BKG	74.50	**91.70**
	ALERT	77.70	**89.56**
	FALL	71.51	**97.37**

Table 4 Comparison between the C_9 and the RNN classifiers on young subjects. Bold face is used to denote the best result for each comparison

		C_9	RNN
Sensitivity	BKG	76.72	**87.79**
	ALERT	40.97	**92.69**
	FALL	77.57	**97.95**
Specificity	BKG	86.05	**97.57**
	ALERT	85.15	**90.94**
	FALL	91.42	**97.26**
Accuracy	BKG	81.38	**92.68**
	ALERT	63.06	**91.82**
	FALL	84.49	**97.60**

whereas the RNN classifier, although not very accurate as well, is still better in separating the two main classes.

4.4 Embedded

The actual feasibility of the embedded implementation of the RNN architecture on the target *SensorTile*® miniaturized board was assessed with a highly-optimized implementation of the runtime detection module for the ARM® Cortex® M4 MCU.

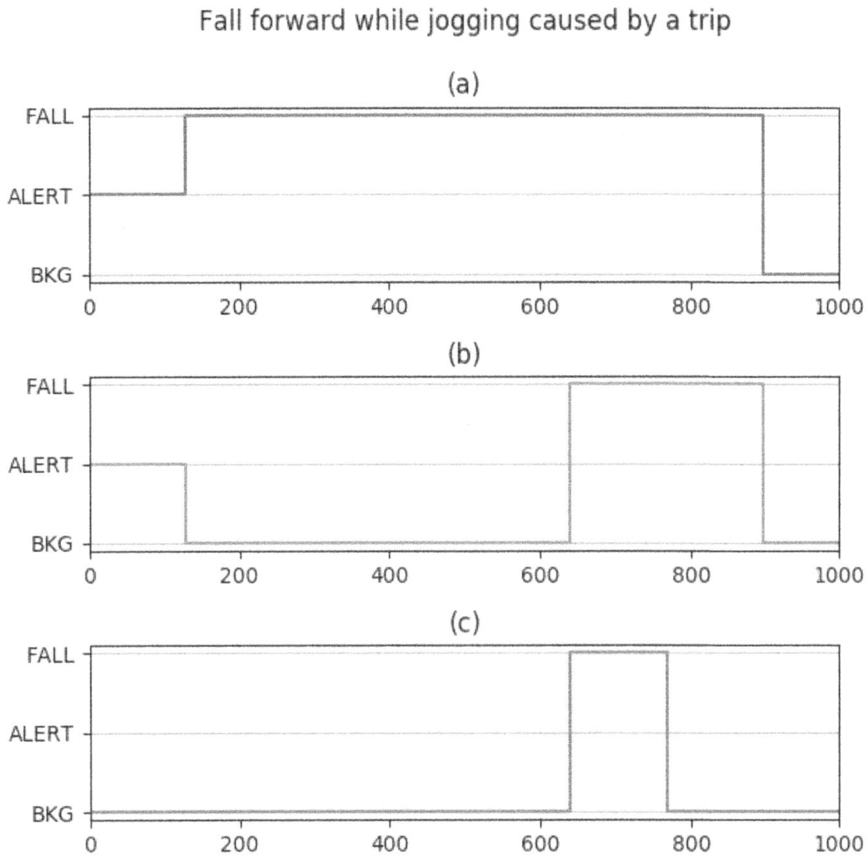

Fig. 7 Comparison of the ongoing classifications produced by the C_9 classifier (**a**) and the RNN classifier (**b**) with the ground truth, i.e. the temporal annotation (**c**). The x axis represents time expressed in sensor readings

In such implementation the original numerical representation in floating-point, 32-bit format adopted with *TensorFlow* was preserved.

The embedded implementation was validated on the test set by comparing the numerical values produced in output with those obtained with *TensorFlow* and the resulting *Mean Squared Error* (MSE) was in the order of 10^{-7}.

In terms of memory occupancy, the embedded implementation took 82 KB of the 128 KB available. The measured time-to-process ratio was about 0.3, meaning that the MCU took 0.300 s to process each second of sensor readings including window extraction and caching. This measure shows the complete suitability of the proposed RNN architecture for embedded real-time processing. Furthermore, by using the STM32CubeMX Power Consumption Calculator, we were able to estimate that a wearable device running such embedded implementation could be operative, with a battery of 100 mAh, for about 20 h without recharging.

5 Conclusions and Future Work

In this work we presented a feasibility assessment which shows that a relatively simple RNN architecture based on LSTM cells could be at the same time effective for AFD and suitable for embedded implementation on smart wearable devices using low-power MCUs.

The effectiveness of the proposed RNN architecture was evaluated in comparison with other methods by considering the publicly-available *SisFall* dataset. The dataset was extended by adding temporal annotations for three classes of relevant events.

The specificity and sensitivity and accuracy results obtained on a carefully selected test set from *SisFall* show that the proposed RNN architecture, after accurate tuning of its hyperparameters, can significantly outperform the other methods based on statistical indicators.

On the other hand, the feasibility of MCU embedding was assessed with an actual implementation of the run-time detection module of the RNN for the *SensorTile*® miniaturized board. The results obtained show the viability of the proposed approach for real-time processing and this can be obtained with a very limited power consumption.

Overall and in our opinion, the results presented suggest that, for further developments, the design strategy of DL methods for AFD will be that of finding the simplest network architecture that could accomplish the task desired, while maintaining a limited footprint in terms of computation power and memory required. In turn, such design strategy emphasizes the need to acquire complete and extensive datasets, with appropriate annotations.

In fact, one of the main directions of future work will be that of collecting a newer dataset made with a body network of several wearable sensors based on *SensorTile*® and connected to a gateway via Bluetooth. Ideally, to ease the task of annotation, in such dataset each activity performed by the volunteers should be associated to a video recording, so that temporal intervals can be identified by looking at the body posture of the volunteer instead of at the signals themselves.

In our intentions, such an extended dataset could allow a careful tuning of the design of network architectures with the objective of an even better embedded implementation for wearable devices.

Acknowledgments The authors acknowledge the financial support from Regione Lombardia, under the "Home of IoT" project (ID: 139625), co-funded by POR FESR 2014–2020. The authors would like to thank Nicola Blago, Daniele De Martini and Tullio Facchinetti for their contributions.

References

1. Martín Abadi, Paul Barham, Jianmin Chen, Zhifeng Chen, Andy Davis, Jeffrey Dean, Matthieu Devin, Sanjay Ghemawat, Geoffrey Irving, Michael Isard, et al. Tensorflow: A system for

large-scale machine learning. In *12th USENIX Symposium on Operating Systems Design and Implementation (OSDI16)*, pages 265–283, 2016.
2. Stefano Abbate, Marco Avvenuti, Francesco Bonatesta, Guglielmo Cola, Paolo Corsini, and Alessio Vecchio. A smartphone-based fall detection system. *Pervasive and Mobile Computing*, 8(6):883–899, 2012.
3. G. Baldewijns, G. Debard, G. Mertes, T. Croonenborghs, and B. Vanrumste. Improving the accuracy of existing camera based fall detection algorithms through late fusion. In *2017 39th Annual International Conference of the IEEE Engineering in Medicine and Biology Society (EMBC)*, pages 2667–2671, July 2017.
4. A. Bourke and G. Lyons. A threshold-based fall-detection algorithm using a bi-axial gyroscope sensor. *Medical Engineering & Physics*, 30(1):84–90, 2008.
5. Eduardo Casilari, Jose A. Santoyo-Ramón, and Jose M. Cano-García. Umafall: A multisensor dataset for the research on automatic fall detection. *Procedia Computer Science*, 110:32–39, 2017.
6. P. Feng, M. Yu, S. M. Naqvi, and J. A. Chambers. Deep learning for posture analysis in fall detection. In *2014 19th International Conference on Digital Signal Processing*, pages 12–17, Aug 2014.
7. Jane Fleming and Carol Brayne. Inability to get up after falling, subsequent time on floor, and summoning help: prospective cohort study in people over 90. *BMJ*, 337, 2008.
8. Korbinian Frank, Maria Josefa Vera Nadales, Patrick Robertson, and Tom Pfeifer. Bayesian recognition of motion related activities with inertial sensors. In *Proceedings of the 12th ACM International Conference - Adjunct Papers on Ubiquitous Computing*, UbiComp '10 Adjunct, pages 445–446, 2010.
9. Ryan M. Gibson, Abbes Amira, Naeem Ramzan, Pablo Casaseca de-la Higuera, and Zeeshan Pervez. Multiple comparator classifier framework for accelerometer-based fall detection and diagnostic. *Applied Soft Computing*, 39:94–103, 2016.
10. Geoffrey E. Hinton, Nitish Srivastava, Alex Krizhevsky, Ilya Sutskever, and Ruslan Salakhutdinov. Improving neural networks by preventing co-adaptation of feature detectors. *CoRR*, abs/1207.0580, 2012.
11. S. Hochreiter and J. Schmidhuber. Long short-term memory. *Neural Computation*, 9(8):1735–1780, Nov 1997.
12. Sergey Ioffe and Christian Szegedy. Batch normalization: Accelerating deep network training by reducing internal covariate shift. *CoRR*, abs/1502.03167, 2015.
13. M. Kepski and B. Kwolek. Fall detection using ceiling-mounted 3d depth camera. In *2014 International Conference on Computer Vision Theory and Applications (VISAPP)*, volume 2, pages 640–647, Jan 2014.
14. Q. Li, J. Stankovic, M. Hanson, A. Barth, J. Lach, and G. Zhou. Accurate, fast fall detection using gyroscopes and accelerometer derived posture information. In *Wearable and Implantable Body Sensor Networks*, pages 138–143, 2009.
15. Carlos Medrano, Raul Igual, Inmaculada Plaza, and Manuel Castro. Detecting falls as novelties in acceleration patterns acquired with smartphones. *PLOS ONE*, 9(4):1–9, 04 2014.
16. O. Mohamed, H. J. Choi, and Y. Iraqi. Fall detection systems for elderly care: A survey. In *2014 6th International Conference on New Technologies, Mobility and Security (NTMS)*, pages 1–4, March 2014.
17. Francisco Javier Ordóñez and Daniel Roggen. Deep convolutional and lstm recurrent neural networks for multimodal wearable activity recognition. *Sensors*, 16(1):115, 2016.
18. Natthapon Pannurat, Surapa Thiemjarus, and Ekawit Nantajeewarawat. Automatic Fall Monitoring: A Review. *Sensors*, 14(7):12900–12936, July 2014.
19. Angela Sucerquia, José López, and Jesús Vargas-Bonilla. Sisfall: A fall and movement dataset. *Sensors*, 17(1):198, 2017.
20. P. Tsinganos and A. Skodras. A smartphone-based fall detection system for the elderly. In *Proceedings of the 10th International Symposium on Image and Signal Processing and Analysis*, pages 53–58, Sept 2017.

21. George Vavoulas, Matthew Pediaditis, Charikleia Chatzaki, Emmanouil Spanakis, and Manolis Tsiknakis. The mobifall dataset: Fall detection and classification with a smartphone. *International Journal of Monitoring and Surveillance Technologies Research*, 2016.
22. T. Vilarinho, B. Farshchian, D. G. Bajer, O. H. Dahl, I. Egge, S. S. Hegdal, A. Lønes, J. N. Slettevold, and S. M. Weggersen. A combined smartphone and smartwatch fall detection system. In *2015 IEEE International Conference on Computer and Information Technology*, pages 1443–1448, Oct 2015.
23. World Health Organization. *WHO global report on falls prevention in older age*. World Health Organization Geneva, 2008.
24. Yundong Zhang, Naveen Suda, Liangzhen Lai, and Vikas Chandra. Hello edge: Keyword spotting on microcontrollers. *CoRR*, abs/1711.07128, 2017.

Part II
Deep Learning for Biomedical Image Analysis

Medical Image Retrieval System Using Deep Learning Techniques

Jitesh Pradhan, Arup Kumar Pal, and Haider Banka

Abstract *Content-based Image Retrieval* (CBIR) system uses the visual information and features present within an image, to find the most analogous images from any gigantic digital image data-set effectively and efficiently, as per the users requirements. Nowadays, the immense advancements in the field of Digital Imaging have exponentially increased the real-time applications of the CBIR techniques. Researchers around the globe are using different CBIR techniques in the field of education, defense, agriculture, remote sensing, satellite imaging, biomedical research, clinical care, and medical imaging. The Major objectives of this chapter are to provide a brief introduction to the different CBIR techniques and their applications on medical image retrieval. This chapter mainly focuses on the current *Machine Learning* (ML) and *Deep Learning* (DL) techniques to address the different issues and limitations of the traditional retrieval systems. Initially, we have discussed the different hand-crafted image features based retrieval systems to understand the perspectives of this research field. Here, we aim to congregate the weaknesses and constraints of the conventional retrieval systems and respective solutions with the help of the advanced DL algorithms. Researchers have suggested several CBIR techniques to improve the efficiency of the retrieval of medical images. In this chapter, a review of some state-of-the-art retrieval techniques and respective future research directions are provided.

Keywords Content-based image retrieval · Deep learning · Feature extraction · Machine learning · Semantic feature based image retrieval · Similarity matching · Text-based image retrieval

---

J. Pradhan (✉) · A. K. Pal · H. Banka
Indian Institute of Technology (ISM), Dhanbad, India

M. Elloumi (ed.), *Deep Learning for Biomedical Data Analysis*,
https://doi.org/10.1007/978-3-030-71676-9_5

1 Introduction

From its invention, images have been an integral part of the society for both research and recreation. In the past decades, whenever images were used in any relevant fields like medical imaging, weather forecasting, etc., it was done using printed out versions of the images. The storage and preservation of those printed images was a critical issue. Such printed images could get damaged in an accident or may get corrupted by normal wear and tear of daily use. Hence, the handling of printed images was a very challenging task. Now, with the advancement in digital technology, the use of printed versions of images has waned over the years. People have extremely portable devices capable of capturing and storing high quality images. In the meantime, with the rapid growth of Internet, sharing of digital data including images is a common practice. Hospitals and medical research centers are well equipped with all types of image capturing and scanning devices. These devices are capable of detecting even minute anomalies present in any part of the human body. On the other hand, *Closed-Circuit TeleVision* (CCTV) cameras are very portable and affordable even for the private use. Consequently, for security purposes these CCTV cameras have become very popular, especially in cities. With so many devices and usage, digital images can be found all over the Internet, exponentially increasing the size of digital repositories as well. Handling these enormous digital repositories is a very challenging task in itself. So, browsing and searching of images in these repositories become a monumental task. However, manual searching of images in these enormous digital repositories is an impractical task. So researchers have introduced automatic image retrieval systems [1–5] to solve the above issues.

Figure 1 shows the basic block diagram of the automatic image retrieval system. In this Fig. 1, we can see that the user gives the input query and the searching algorithm finds the most relevant images from the annotated image dataset. The user can provide input query in form of text as well as an image. In this automatic image retrieval system, the searching algorithm access the meta-data of the annotated data-set and uses the text/image matching approach to compare the query information and data-set information. Further, on the basis of comparison score, it will perform sorting of top matched images. Finally, from the sorted images it will retrieved the top most similar images as a final output.

After the brief introduction to the image retrieval systems, the rest of the chapter has been organized as follows: Sect. 2 shows the elaborative introductions of different kinds of image retrieval systems. Next, in Sect. 3, we have discussed the different applications of image retrieval systems. Further, in Sect. 4, we have presented the *Deep Learning* (DL) based medical image retrieval systems. Finally, in Sect. 5 we have drawn the conclusions of this book chapter.

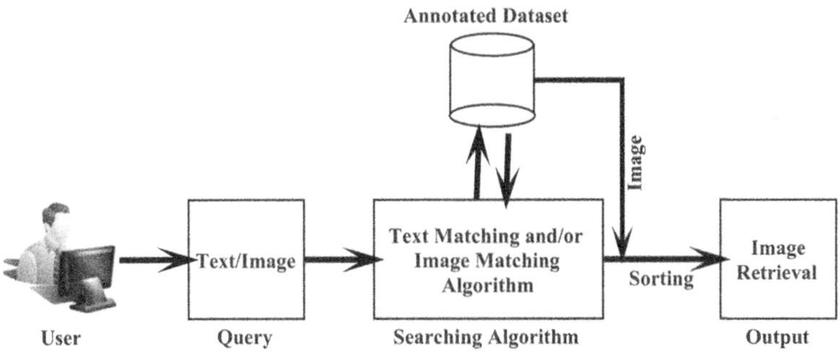

Fig. 1 Basic block diagram of an automatic image retrieval system

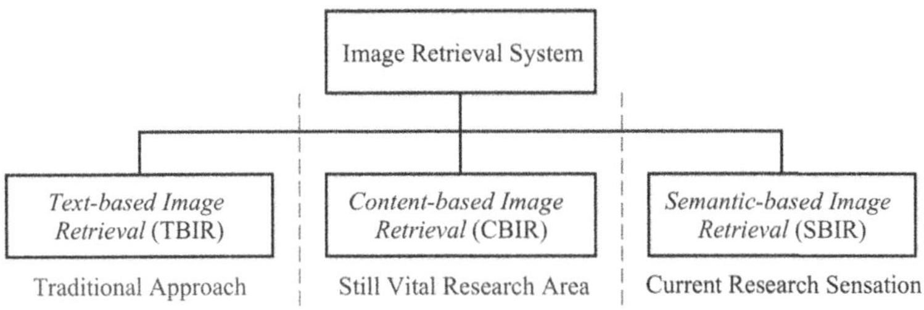

Fig. 2 Taxonomy of the image retrieval systems

2 Image Retrieval Systems

Researchers around the globe have introduced different kinds of automatic image retrieval systems which take text, meta-data, and/or image as a query input and retrieve the most similar images from the image dataset. Here, Fig. 2 shows the taxonomy of the image retrieval system.

2.1 Text-Based Image Retrieval

Traditionally, *Text-based Image Retrieval* (TBIR) systems [5–8] were used as an automatic image retrieval system. Texts were basically single/multiple keyword(s) associated with the image. These keywords can be image name, image location, file name, categories, index number, any note, and title directly or indirectly related with the image. Here, Fig. 3 shows the basic block diagram of the TBIR system. In this Fig. 3, the user gives query input in the form of a meta-data. This meta-data can be image name, image path, file name, keywords, link, etc. In result, the TBIR algorithm searches the image database for the matching meta-data description

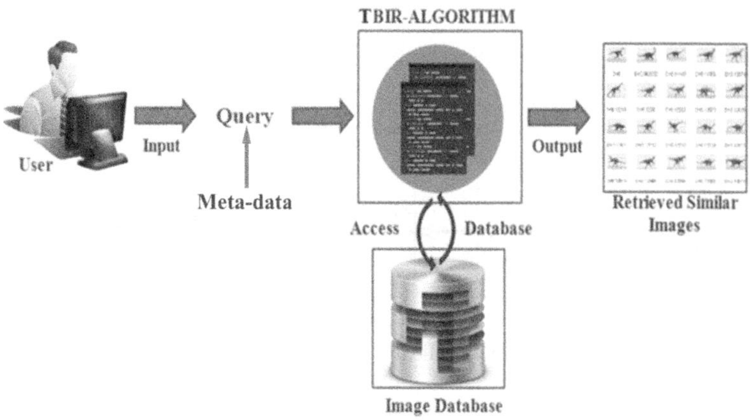

Fig. 3 Basic block diagram of the TBIR system

Algorithm 1: TBIR system

Input: Input Meta-data/Keyword(s) M_k.
Output: Top L similar images from an image dataset D_n.
Parameter: $L << n$ and n is the number of dataset images.
Select the input mata-data M_k.
Assign the suitable Meta-data/Keyword(s) M_i to the corresponding database image I_i, where $i \in \{1, 2, 3, \ldots, n\}$.
for $\forall I_i \in D_n$ **do**
  Select the corrosponding meta-data information M_i.
  Perform string comparision between M_k and M_i.
  Store the similarity distance value d_i.
end
Sort all the d_i values in non-decreasing order.
Extract image index number of top L minimum d_i values.
Finally retrieve the top L most similar images from database D_n.

and extracts the images which possess similar meta-data information. Subsequently, Algorithm 1 gives the schematic steps of a basic TBIR system.

TBIR systems have two critical flaws. First, it needs human involvement in tag assignment process in which tags were assigned to all the database images, individually. This is a very monotonous and tedious work, which creates high chance of wrong tag assignment due to incorrect human perception. Single keyword cannot be sufficient to convey the entire image information and this is the main flaw of the tag based image retrieval system. Below in Table 1, we have presented some stat-of-the-arts TBIR techniques. These techniques uses either Mata-data/Keywords or combinations of Keywords along with user's visual perception information.

Table 1 Some state-of-the-arts TBIR techniques

Authors	Year	Approach
Swain et al. [9]	1993	Proposed a tool called FINDIT in which user can locate the object and can specifies the transformation and constraints
Zhou et al. [6]	2002	Used relevance feedback and seamless query joining scheme to combine keywords and visual contents for image retrieval
Ahmad et al. [5]	2003	Proposed a *Formal Concept Analysis* (FCA) approach to perform the text-based image retrieval.
Zeng et al. [7]	2005	Proposed a relevance feedback based user interactive text-based image retrieval system.
Chen et al. [4]	2012	Proposed an adaptive classifier which uses augmented features (AFSM) in a modified *Support vector Machine* (SVM) to improve the efficiency of the TBIR system.

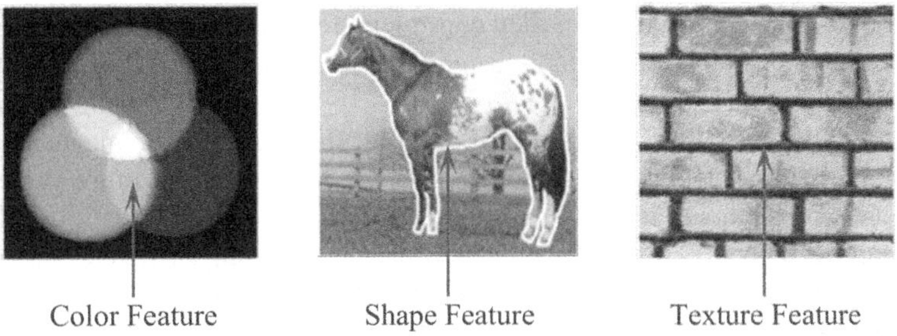

Fig. 4 Basic local features of an image (source of the images : [13, 14])

2.2 *Content-Based Image Retrieval*

Traditional TBIR systems have some major drawbacks and limitations. So researchers have solved the above issues by introducing the *Content-based Image Retrieval* (CBIR) system [1–3, 10–12]. CBIR is an automatic image searching and image retrieval system based on the actual image contents. In CBIR system actual image information were used for image searching and retrieval process. These actual image contents can be the different regular and irregular surface patterns, shapes and structures, color distribution and/or spatial relationships between the pixels present in the image. The surface patterns give the texture information about the image whereas, different shape and structures in the image give the shape information about the image. Regular and irregular color distribution in the image gives the color information about the image. All such image information is considered as local image features and these local image features are used in the CBIR process. The main local features present in an image are (1) color features, (2) shape features, and (3) texture features. Figure 4, shows the visual example of the three main local features of an image.

In CBIR system these local image information were extracted using different feature extraction techniques and stored in the form of a feature vector. It stores the actual image information as one dimensional vector, but the size of the feature vector is exceptionally smaller than the size of the actual image. Good quality feature vector must convey proper image semantics using fewer amounts of data. So performance of any CBIR system basically depends on two factors: techniques of feature extraction and formation of better feature vector for CBIR applications. Usually a CBIR system works in four steps which are as follows:

- **Feature Extraction:** In this step, the researchers use color, texture, shape, structural, co-relational feature extraction schemes or combination of these schemes to extract visual features of the query image. These extracted visual features are represented through a low dimensional feature vector.
- **Feature Space Generation:** In this step, the researchers construct feature vector for all database images. Next, they construct a feature space of database images in which they stores the feature vector of an image with corresponding to their database index value.
- **Similarity Matching:** In this step, different distance measurements schemes have been used to find the similarity between query image feature vector and the feature vectors of the feature space.
- **Image Retrieval:** This is the final step in which sorting of images have been performed based on their similarity score. After sorting, desired numbers of top images with highest similarity score have been extracted from the database as a final output.

Next, Fig. 5 shows the basic block diagram of a CBIR system. In this Fig. 5, we can observe that the user is providing the query input in form of an image. The CBIR algorithm analyzes the visual and unique contents of the input image and extracts the most similar images from the image database. Further, Algorithm 2 demonstrates the algorithmic steps of a CBIR system.

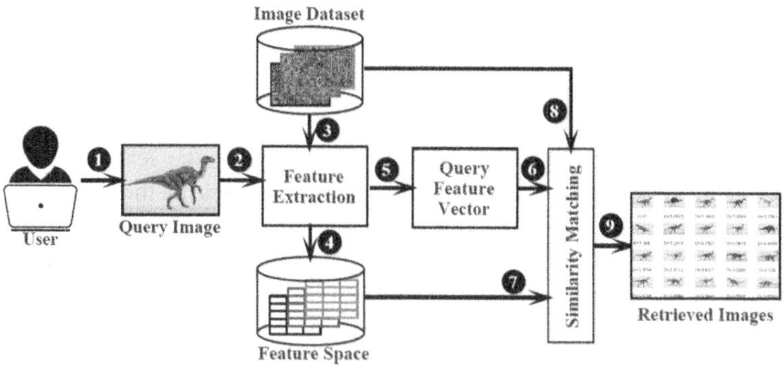

Fig. 5 Basic block diagram of a CBIR system

Algorithm 2: CBIR system

Input: Input a query image I_k.
Output: Top L similar images from an image dataset D_n.
Parameter: $L << n$ and n is the number of dataset images.
Select the input query image I_k.
Exact the unique visual features of I_k to construct the feature vector fv_k.
for $\forall I_i \in D_n$ **do**
  Extract the unique visual features for I_i.
  construct the feature vector fv_i of I_i.
  Store the fv_i at the ith index of the feature space FS_n .
end
for $\forall I_i \in D_n$ **do**
  Select the corresponding feature vector fv_i of the database image I_i.
  Perform similarity matching between fv_k and fv_i.
  Store the similarity score value d_i.
end
Sort all the d_i values in non-decreasing order.
Extract image index number of top L minimum d_i values.
Finally retrieve the top L most similar images from database D_n.

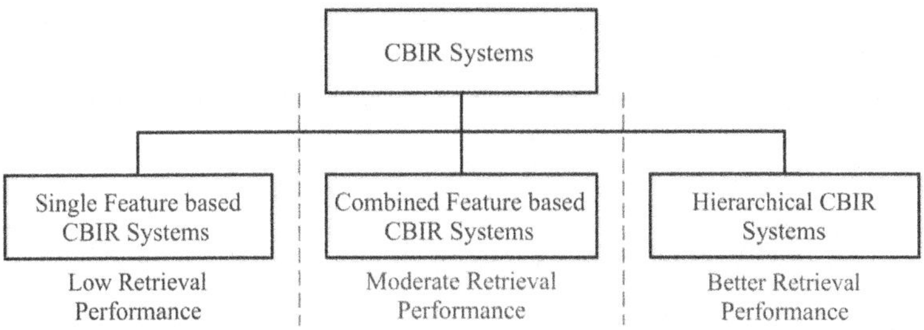

Fig. 6 Taxonomy of the CBIR systems

The performance of any CBIR system mainly depends on two factors which are suitable feature extraction and proper feature vector construction. The CBIR system works better than the TBIR system in most of the cases. At the same time CBIR system does not require any database annotation process which drastically reduces the human efforts. Subsequently, every CBIR system extracts significant amount of relevant images in the final outcomes. In literature, huge varieties of CBIR systems are present. We can broadly classify these CBIR techniques into three major categories as shown in Fig. 6.

2.2.1 Single Feature Based CBIR Systems

These are the most commonly used CBIR systems. These type of systems extract only one type of visual features from the whole input as well as database images. These visual features can be color, shape, texture, co-relational, structural of

Table 2 Some state-of-the-arts single feature based CBIR techniques

Authors	Year	Feature	Approach
Jhanwar et al. [15]	2004	Structural	Introduced a *Motif Co-occurrence Matrix* (MCM) in which they have find all structural patterns present with in a 2x2 pixel Grid.
Yahiaoui et al. [16]	2006	Shape	Proposed a new contour-based shape descriptor which improves the accuracy of real-time image search.
Kokare et al. [17]	2007	Texture	Used a 2D rotated wavelet filter in *Discrete Wavelet Transform* (DWT) to perform texture feature extraction.
Hu et al. [18]	2013	Shape	Used *Histogram of Oriented Gradients* (HOG) to perform sketch based image retrievals.
Varish et al. [19]	2015	Color	Proposed a non-uniform bin decomposition scheme to extract statistical color feature from a histogram for CBIR applications.

spatial features. Usually, natural images contain rich structural, textural and colored features. So a single type of visual feature is not sufficient to reflect the actual semantic meaning of the image. As a result, the retrieval performance of these types of CBIR systems are some what limited. Here, Table 2 shows some popular state-of-the-arts single feature based CBIR systems.

2.2.2 Combined Feature Based CBIR Systems

To address the performance issues of the single feature based CBIR systems, researchers have combined multiple features in the final feature vector. As a result, the final feature vector gives better semantic information regarding the input/database images. This also improves the retrieval efficiency of the CBIR system. Usually, the length of the final feature vector in such scheme contains more numbers of feature elements which also increases the image extraction time. Further, Table 3 demonstrates the brief description of some standard state-of-the-arts combined feature based CBIR systems.

2.2.3 Hierarchical Feature Based CBIR Systems

These are the most efficient CBIR system among all different instance-based CBIR systems. The main principal behind these types of CBIR system is to filter out irrelevant images to reduce the final search space. Thus these systems are also known as multi-level CBIR systems. In hierarchical/multi-level CBIR system, researchers use different kinds of visual features in different hierarchy/level to filter out the irrelevant images. As a result, the image search spaces get reduced after each hierarchical retrieval process. At last, the final search space contains

Table 3 Some state-of-the-arts combined features based CBIR systems

Authors	Year	Feature	Approach
Wang et al. [20]	2010	Color + Shape + Texture	Used color quantization, pseudo-zernike moments and steerable filter decomposition for color, shape, and texture feature extraction.
Bama et al. [21]	2011	Shape + Texture + Color	Used *Scale Invariant Feature Transform* (SIFT) along with Log-Gabor wavelets in HSV domain for local feature extraction.
Subramanian et al. [22]	2011	Texture + Color	Combined correlogram of Gabor wavelet with rotated and non-rotated Gabor wavelet coefficients for color and texture feature extraction.
Sajjad et al. [23]	2018	Color + Texture	Used HSV color quantization along with uniform patterns of *Rotated Local Binary Patterns* (RLBP) for color and texture feature extraction.
Pradhan et al. [1]	2019	Shape + Color + Texture	Proposed color edge map for shape and color feature extraction and used principal texture direction for texture feature extraction.

Table 4 Some state-of-the-arts hierarchical features based CBIR systems

Authors	Year	Levels	Features	Approach
Andrysiak et al. [24]	2005	3	Color-Texture-Shape	Used normalized color features, Gabor features, and Shape moments in 3 different hierarchical levels.
Shrivastava et al. [25]	2015	3	Color-Texture-Shape	Used HSV quantized histogram, Gabor features, and Fourier descriptor in 3 different hierarchical levels.
Varish et al. [26]	2017	2	Color-Texture	Used non-uniform bin decomposition based histogram and 2D *Dual-Tree Complex Wavelet Transform* (DT-CWT) in 2 different hierarchical levels.
Jian et al. [27]	2018	2	Color-Texture	Used color moments and Gabor filters in 2 different hierarchical levels.
Pradhan et al. [28]	2018	3	Shape-Color-Texture	Proposed edge join histogram and color channel co-relation histogram and also used tetrolet transform features in 3 different hierarchical levels.

the most of the relevant images which ultimately improves the retrieval efficiency of the CBIR system. Further, Table 4 summaries some popular state-of-the-arts hierarchical feature based CBIR systems.

2.3 *Semantic-Based Image Retrieval*

The normal CBIR system which is also known as instance-based CBIR system usually works better when image database contains relatively less number of images. As the size of database increases it adversely affects the retrieval efficiency and speed of the CBIR system. To address this problem researchers introduced *Semantic-based Image Retrieval* (SBIR) system [29–33]. This type of system works in two phases which are (1)semantic class predication and (2) similarity matching and retrieval. Usually, in the first phase feature extraction and feature classification process take place. Here, the main objective is to identify the class of the input image to drastically reduce the final search space. Thus SBIR system is also known as class-based CBIR system. Finally, in the second phase similarity matching is performed only for identified class of images. In the previous sub-section, we have already covered the feature extraction and similarity matching process. In SBIR also the feature extraction and similarity matching process are similar. Next, the classification works in two parts training and classification. Initially we have to select a suitable classifier like. *Random Forest*, *Support Vector Machine* (SVM), *Decision Tree*, *Multi-Layer Perception*, etc. [34–37]. Next, we have to create the training image set from a given image database. The training image set must contain only known and annotated image features. The details of training and classification process are given below:

- **Training:** Training is a process in which known annotated image features have been taken to train/create a classifier. Based on these known features the classifier learns to identify the class of any unknown image. This is an iterative process and the classifier needs to be trained through multiple set of training images.
- **Classification:** In this step, a trained classifier has been taken to identify the class of a given unknown image. The classification process is one-time process which runs on the un-annotated/unknown images.

The schematic block diagram of a SBIR system has been shown in Fig. 7. Later, Algorithm 3 demonstrates the detailed algorithmic steps for a SBIR system.

The efficiency of every SBIR system only depends on the classification accuracy. The main problem with the SBIR system is its miss-classification rate. The SBIR system works in binary form where for true classification it produces entire set of relevant images in the retrieval outcome. Simultaneously, for every miss-classification the SBIR system produces entire set of irrelevant images in the retrieval outcome. Based on the classification technique, the class-based CBIR/S-BIR systems have been divided in to two broad categories as shown in Fig. 8.

2.3.1 Machine Learning Based CBIR

Machine Learning (ML) based CBIR systems use handcrafted visual image features for feature classification. These systems works in three steps which are (1) feature

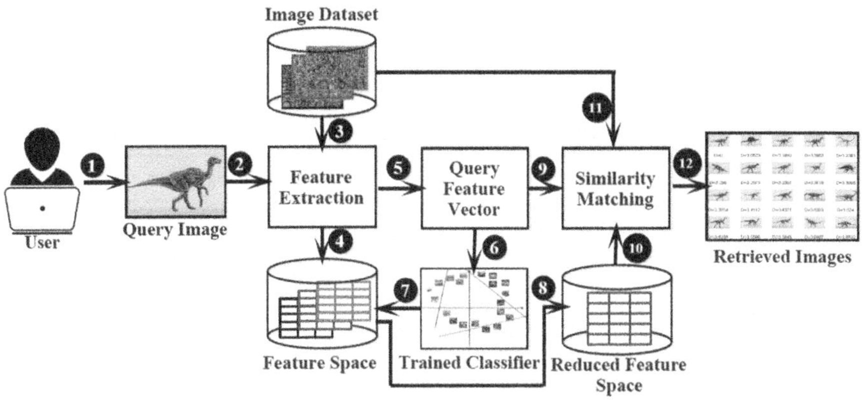

Fig. 7 Schematic block diagram of a SBIR system

Algorithm 3: SBIR system

Input: Input a query image I_k.
Output: Top L similar images from an image dataset D_n.
Parameter: $L << n$ and n is the number of dataset images.
Select the input query image I_k.
Exact the unique visual features of I_k to construct the feature vector fv_k.
for $\forall I_i \in D_n$ **do**
  Extract the unique visual features for I_i.
  construct the feature vector fv_i of I_i.
  Store the fv_i at the ith index of the feature space FS_n .
end
Select a suitable classifier CL.
Create the training feature set Tfs_m from FS_n, where $m < n$.
Train the classifier CL using Tfs_m.
Use trained CL to identify the class C_k of the input image I_k.
for $\forall I_i \in C_k$, *where* $C_k \subset D_n$ **do**
  Select the corrosponding fetaure vector fv_i of the identified class image I_i.
  Perform similarity matching between fv_k and fv_i.
  Store the similarity score value d_i.
end
Sort all the d_i values in non-decreasing order.
Extract image index number of top L minimum d_i values.
Finally retrieve the top L most similar images from database D_n.

extraction, (2) classification, and (3) retrieval. In these systems, the classification accuracy is considered as a final retrieval accuracy. ML based CBIR systems have been very useful for medical, plant leaf, fruit, natural, object, under water, and satellite image retrieval applications. Below, Table 5 shows the state-of-the-arts ML based CBIR systems.

In Table 5, SVM stands for *Support Vector Machine* [35]. MLP and MCH represent the *Multi-Layer Perception* [37] and *Moving Center Hypersphere* [32] respectively. Finally, KNN stands for *K Nearest-Neighbor* [33] classifier.

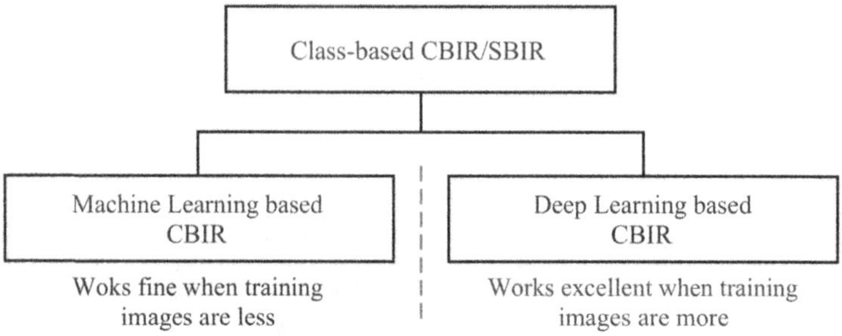

Fig. 8 Taxonomy of the class-based CBIR/SBIR systems

Table 5 Some state-of-the-arts ML based CBIR systems

Authors	Year	Classifier	Image type	Approach
Li et al. [29]	2006	SVM	Natural	Used edge histogram, *Pyramid Wavelet Transform* (PWT), and color histogram features along with SVM classifier.
Pourghassem et al. [31]	2008	MLP	X-ray	Performed preformed hierarchical medical image classification/retrieval using shape and texture features.
Wang et al. [32]	2008	MCH	Plant Leaf	Used 16 Zernike moments and 7 HU moments for plant leaf image classification/retrieval.
Haung et al. [33]	2016	SVM + KNN	Hyperspectral	Used probability maps in SVM and KNN for hyperspectral image classification/retrieval.
Guo et al. [30]	2018	SVM	Remote Sensing	Used a SVM based sequential classifier training scheme for remote sensing image classification/retrieval.

2.3.2 Deep Learning Based CBIR

Unlike ML based CBIR system, *Deep Learning* (DL) based CBIR systems do not require manual feature extraction. The DL based CBIR systems use VGG-16 [38], inceptions V3 [39], *AlexNet* [40], *ResNet* [41], etc. for direct image classification. These *Convolutional Neural Network* (CNN) models are self capable to extract different types of features in learning process. For *Artificial Neural Network* (ANN), we only need annotated image sets and the CNN model will automatically extract the important representative features to learn the semantic meaning of the image. CNN based CBIR systems have been very useful in medical, hyper-spectral, and remote sensing image retrieval applications. Next, Table 6 demonstrates some state-of-the-arts deep learning based CBIR systems.

Table 6 Some state-of-the-arts DL based CBIR systems

Authors	Year	Image type	Approach
Spanhol et al. [42]	2016	Breast Cancer	Used *LeNet* and *AlexNet* CNN model for breast cancer image clasification/retrieval.
Maggiori et al. [43]	2016	Remote Sensing	Used a feed forward CNN model for remote sensing image classification/retrieval.
Madani et al. [44]	2017	Medical	Proposed a *Generative Adversarial Network* (GAN) for medical image classification/retrieval.
Paoletti et al. [45]	2018	Hyperspectral	Performed hyperspectral image classification/retrieval using a 3D CNN model.
Zeng et al. [46]	2019	Medical	Performed medical image classification/retrieval using *Synergic Deep Learning* (SDL) model.

3 Applications of CBIR/SBIR

The CBIR/SBIR technologies have been employed in divers application like medical imaging, agriculture, education, historical research, remote sensing, crime prevention, hyper-spectral imaging, biodiversity information system, fingerprint identification, etc. [43, 45, 46]. The CBIR systems are especially very useful for students to search images for their educational projects. CBIR systems are also useful for medical image retrieval. Doctors and practitioners can use medical images to find similar images from the medical database. It helps them to understand the medical condition of the patient by analyzing different retrieved medical images. Next, Fig. 9, shows the basic and popular applications of the CBIR/SBIR systems. Later, in Sect. 4, we have explained the application of DL based CBIR/SBIR for medical image retrieval in detail.

4 Deep Learning Based Medical Image Retrieval System

Nowadays, medical imaging is playing crucial role on identifying the grade of a cancer. So, huge amount of digital images have been recorded in the process of cancer treatment and research. These captured images are highly heterogeneous in nature. Manual management of such huge volume heterogeneous data is the real-time scenario is practically not feasible. Thus, there is a dire need of an efficient and effective medical image retrieval system for medical practice, cancer practice and medical research.

The image retrieval research area is well studied and tremendous variety of CBIR and SBIR systems are available in the literature. Visual features and unique characteristics of medical images are very different from the features of natural

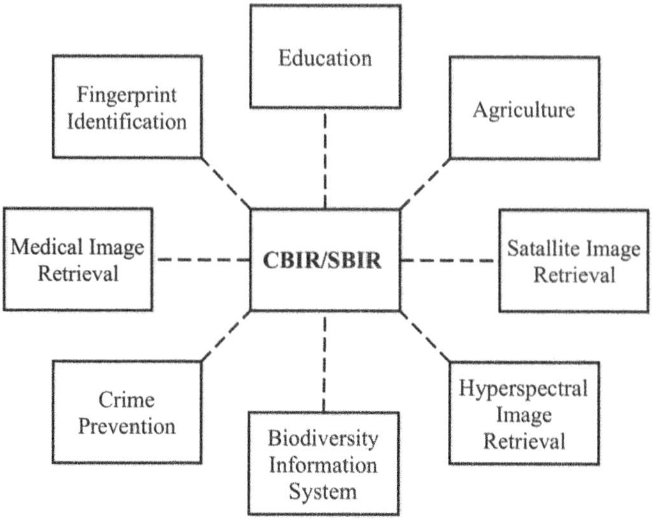

Fig. 9 Different applications of CBIR/SBIR system

images. Thus direct use of all these CBIR and SBIR systems for medical image retrieval applications usually gives unsatisfactory retrieval outcomes. To address this problem, researchers around the globe are widely using DL models (CNN models) [38–41] for medical image retrieval. In the following sub-sections, we discuss the CNN models and their use in medical image retrieval applications.

4.1 Deep Learning Models

The most widely used CNN models for medical image retrieval are inceptions V3 [39], *ResNet* V2 [47], and VGG-16 [38].

- ***Inceptions V3:*** Inceptions V3 is the third version of the *Google's* inception CNN architecture. It was trained using *ImageNet* dataset [48] containing 1000 general classes and 1 million images. The architecture of inceptions V3 has been shown in Fig. 10.
- ***ResNet V2:*** *ResNet V2* model is another variation of inceptions V3 model. This model is also trained on the *ImageNet* images [48] just like inception V3 which consists of 1 million images representing 1000 class. The architecture of inception *ResNet* V2 has been shown in Fig. 11.
- ***VGG-16:*** As the name suggests VGG-16 consists of 16 weight layers. This ANN is known for its simplicity but still for giving good accuracies on *ImageNet* [48]. It uses only 3 × 3 convolution layers which are placed at the top of each other in order of increasing depth. Reduction in the volume size in the ANN is handled

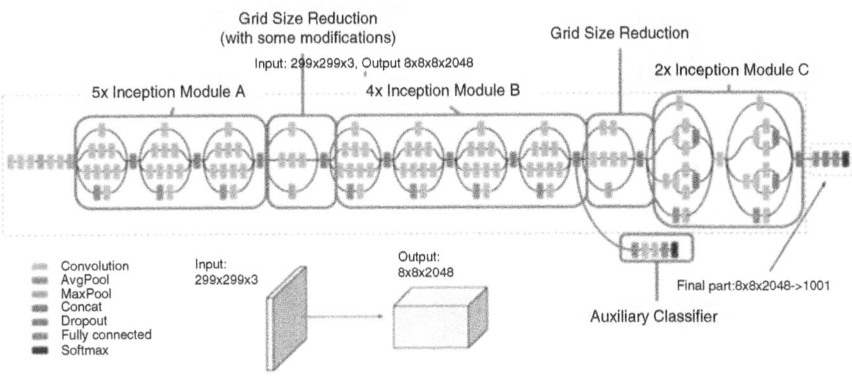

Fig. 10 Architecture of inceptions V3 model [49]

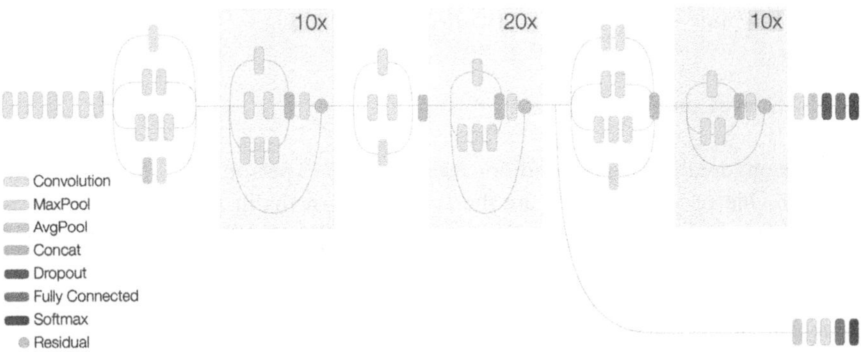

Fig. 11 Architecture of inceptions *ResNet* V2 model [49]

by the max pooling layer. In this model, two fully connected layers also known as dense layers are present which contains 4096 nodes. At last, a softmax layer has been used in the VGG-16 ANN. The schematic architecture of the VGG-16 CNN has been shown in Fig. 12.

4.2 *Deep Learning Based Skin Cancer Image Retrieval*

The ANNs discussed in the above sub-section were trained on the skin cancer image data that were taken from a competition that was hosted by the *International Skin Imaging Collaboration* (ISIC) [50, 51]. The goal of the competition was to support research and development of algorithms for automated detection of melanoma, which is the most lethal form of the skin cancer. A broad overview of relevant information about the skin cancer and the methods used for their classification and retrieval has been described as follows:

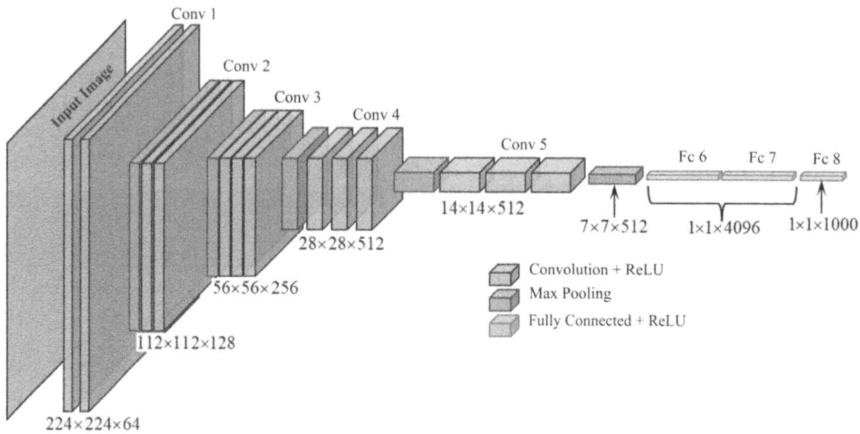

Fig. 12 Schematic architecture of the VGG-16 CNN model

4.2.1 Survey on Skin Lesions

Skin lesions are the medical conditions that appear on a person due to many different reasons. One of these reasons are the abnormal growths in skin tissues, which is defined as a cancer. This disease is spread over more than 14.1 million people and had caused more than 8.2 million deaths, worldwide. If we analyze the last 30 years records for skin cancer, it is troublesome to know that the number of cases which are diagnosed with skin cancer have increased tremendously. It is even more troublesome when money comes into picture, seeing that the large amount of money is spent in the analysis of the patient. Out of many steps starting from diagnosis to medicines if any of the steps are minimized, it would mean a decrease in whole dermatology sectors expenditure.

There are mainly three types of cancers are present which are basal cell skin cancer, melanoma and squamous cell. Basal cell and squamous cell in company with some other less common skin cancers, are known as non-melanoma skin cancer (NMSC). Basal cancer grows slowly and damages the tissue around it but it never spread to the distant areas or results in the death. It generally appears as a painless raised area on skin. Squamous cancer generally spreads and it is usually present in form of hard lump and can also form an ulcer. *Melanoma* are the most dangerous form of the skin cancer. The basic signs of *melanoma* are changed in color, shape, and size of a mole. Some times it has irregular edges and more than one color. In the affected area the patient can also feel itchiness and bleedings.

Between 20% to 30% of the melanoma cancers develop from the moles. People with the lighter skin are at more risk as compared to people with darker skin. Non-melanoma skin cancers are generally curable. Skin cancer accounts to 40% cases out of total cases of cancer so is the most common form of a cancer. The most common form of skin cancer is non-melanoma which occurs to at least 2–3 millions people per year. Basal cell and squamous cell skin cancer rarely causes death of the person.

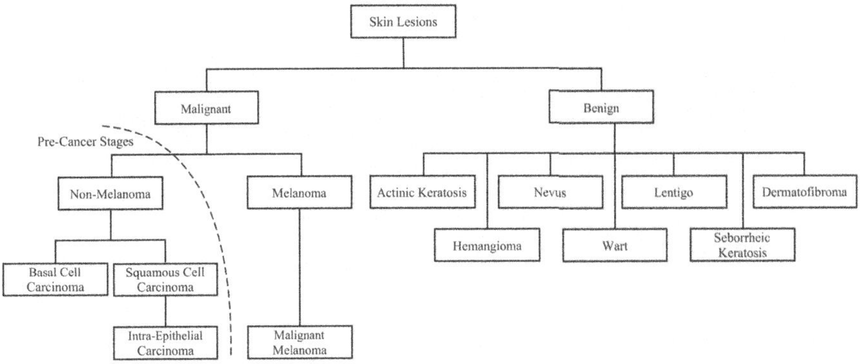

Fig. 13 Different categories of the skin cancer

Table 7 Distribution of the images for training and retrieval

Images for	Total images	*Melanoma*	*Sebborhic Keratosis*	*Nevus*
Training	2236	391	274	1571
Validation	150	30	42	78
Testing	364	100	70	194

Figure 13 represents the different categories of the skin cancers. Further, in this chapter we have mainly focused on the classification and retrieval of the malignant and other two Benign class forms of the cancer.

4.2.2 Dataset

The ISIC dataset [50, 51] has been chosen because it is publicly available. This dataset contains 2000 training images in total, in which, 374 images belong to *Melanocytic Melanoma*, 254 images belong to *Sebborhic Keratosis* and 1372 images belong to *Nevus* category of skin lesion. Along with, training images there were 150 validation images and 600 test images are also present. Here, we have also merged some of the test images into the training images to increase the quantity of the training images. So finally, we have used 421 training images and 100 testing images for *Melanoma*. Total 316 training images and 70 testing images for *Sebborhic Keratosis*. For *Nevus*, we have used 1649 training and 194 testing images. The distribution of the dataset has been shown in Table 7. Further, sample image of *Melanoma*, *Sebborhic Keratosis*, and *Nevus* have been displayed in Fig. 14.

4.2.3 Data Augmentation

This technique is used when quantity training images are not sufficient train the CNN model. This techniques is used to increase the data by employing different

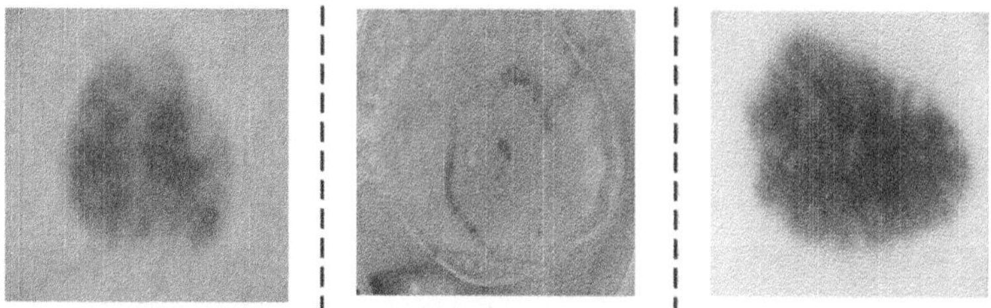

Fig. 14 Sample image of different grades of skin cancer

Table 8 Different transformations used for data augmentation

Transformation technique	Probability
Rotation	0.5
Random noise	0.5
Flip horizontally	0.7
Flip vertically	0.5
Random blur	0.8
Lightning variance	0.5

transformations on the data. As the data set used here is highly biased for *nevus* image class. At the same time, the total amount of data is also very small in number. So data augmentation techniques has been used for up-sampling the dominated class. It also increases the total number of images. Some of the transformations which have been used are random rotation, vertical flip, horizontal flip, random zoom, blurring of the images, etc. The different transformations used to increases the number of images have been given in Table 8. Further, Fig. 15 shows the sample outcome of the different transformations.

4.2.4 Retrieval Using Inception V3 Model

The above discussed inception V3 model has been trained with the training dataset. Here, we have trained this model for multiple (5 times) times for 200 epochs. We have used *Adam* [52] as optimizer with batch size of 32 and learning rate is 1. After training the selected model, we have checked the classification accuracy for test dataset. Here, the classification accuracy and retrieval accuracy will be the same because for every true classification this model will give 100% retrieval precision. Similarly, for every miss-classification this model will give 0% retrieval precision. Further, Table 9 shows the test accuracies of inception V3 for multiple trainings. From this Table 9 we can infer that the overall text accuracy of the inception V3 model is 54%. Later, Fig. 16 shows the top 10 retrieved images for a *nevus* query image when true classification occurs. In this Fig. 16 we can clearly observe that all the retrieved image are also from *nevus* category. Next, Fig. 17 shows the top 10

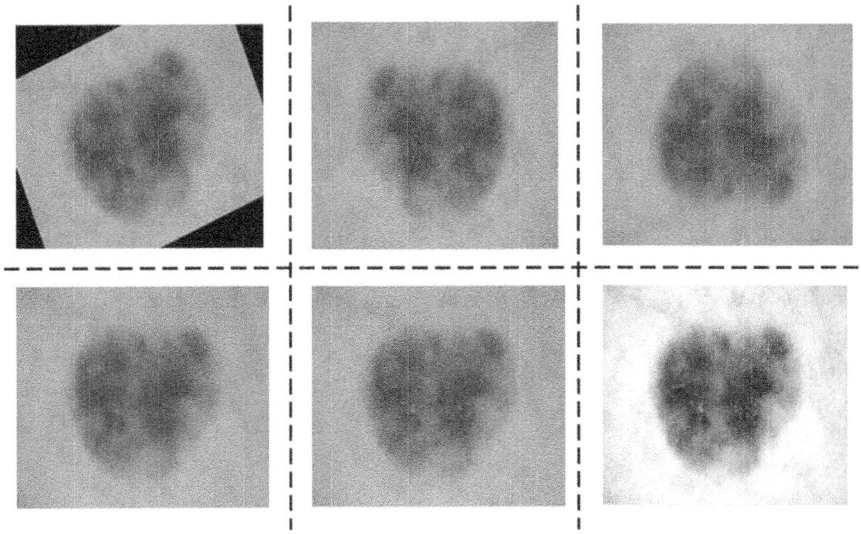

Fig. 15 Sample outcomes of different transformation used for data augmentation

Table 9 Test accuracies of inception V3 model for multiple trainings

Training instance	Category wise test accuracies		
	Melanoma	*Sebborhic Keratosis*	*Nevus*
1	41%	31%	89%
2	42%	30%	88%
3	40%	33%	88%
4	42%	30%	89%
5	43%	33%	90%
Average	42%	31%	89%

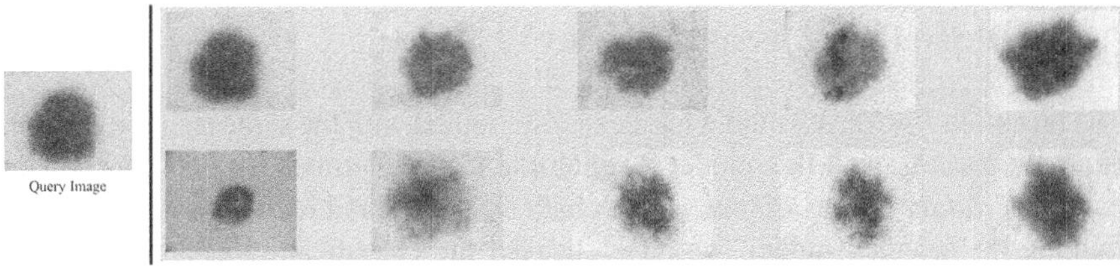

Fig. 16 Top 10 outcomes for a *nevus* class query image (for true classification)

retrieved outcomes in case of miss-classification for a *melanoma* query image. The inception V3 model has classified the query image in to *nevus* category. Thus all the retrieved images shown in Fig. 17 belong to *nevus* category.

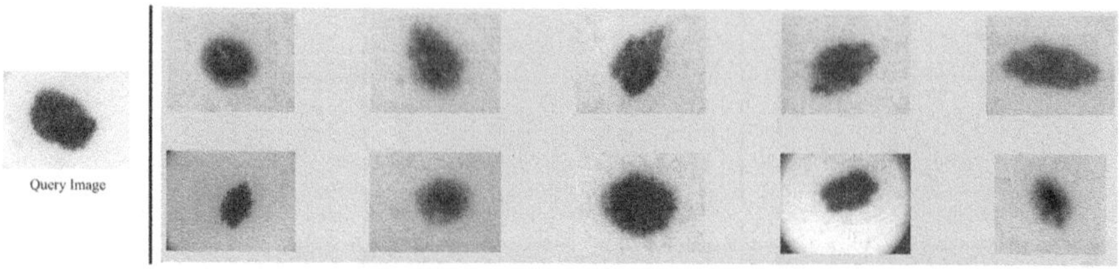

Fig. 17 Top 10 outcomes for a *melanoma* class query image (for miss-classification)

Table 10 Test accuracies of inception *ResNet V2* model for multiple trainings

Training instance	Category wise test accuracies		
	Melanoma	*Sebborhic Keratosis*	*Nevus*
1	13%	61%	84%
2	15%	64%	85%
3	14%	64%	84%
4	15%	64%	84%
5	13%	61%	85%
Average	14%	63%	84%

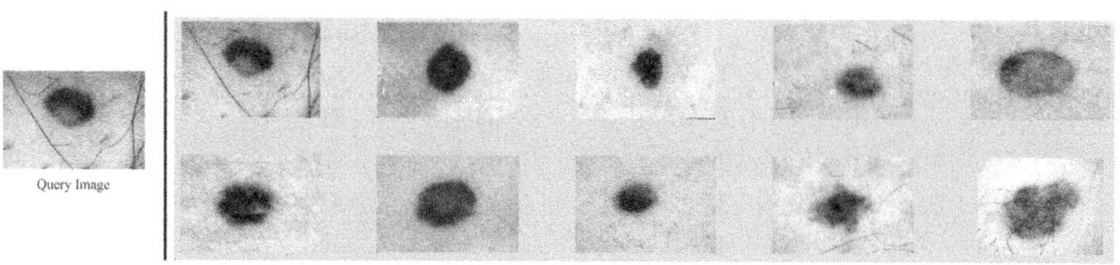

Fig. 18 Top 10 outcomes for a *nevus* class query image (for true classification)

4.2.5 Retrieval Using Inception ResNet V2 Model

The inception *ResNet V2* model has been also trained with the same training dataset. Here, we have trained this model for multiple (5 times) times for 200 epochs. We have used *Adam* [52] as optimizer with bath size of 32 and learning rate is 1. After training the selected model, we have checked the classification accuracy for test dataset. Below, Table 10 demonstrates the test accuracies of inception *ResNet V2* for multiple trainings. From this Table 10 we can depict that the overall text accuracy of the inception *ResNet V2* model is 53.6%. Later, Figs. 18 and 19 shows the retrieval outcomes of *nevus* and *sebborhic keratosis* query image. For *nevus* query image, this model has shown true classification whereas for *sebborhic keratosis* it has shown miss classification. Thus the retrieved images shown in Fig. 19 are from *nevus* image category.

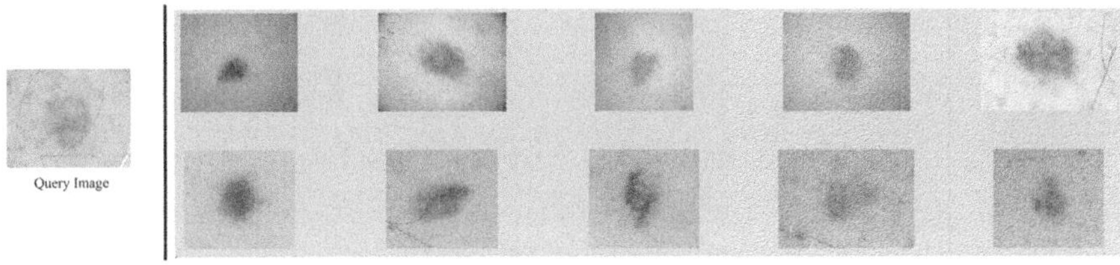

Fig. 19 Top 10 outcomes for a *sebborhic keratosis* class query image (for miss-classification)

Table 11 Test accuracies of VGG-16 model for multiple trainings

Training instance	Category wise test accuracies		
	Melanoma	*Sebborhic Keratosis*	*Nevus*
1	70%	69%	27%
2	71%	70%	29%
3	70%	69%	27%
4	71%	69%	28%
5	71%	67%	27%
Average	70.6%	68.8%	27.6%

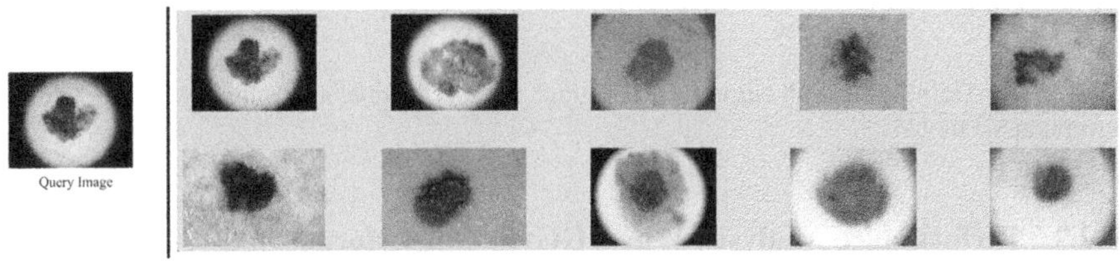

Fig. 20 Top 10 outcomes for a *melanoma* class query image (for true classification)

4.2.6 Retrieval Using VGG-16 Model

The VGG-16 model has been also trained with the same training image dataset. In this case also, we have trained the model for multiple (5 times) times with 200 epochs. We have used *Adam* [52] as optimizer with bath size of 32 and learning rate is 1. After training of VGG-16 model, we have find the classification accuracy for the test dataset. Next, Table 11 shows the test accuracies of the VGG-16 model for multiple trainings. From Table 11 we can calculate the overall text accuracy of the VGG-16 model as 56%. Later, Figs. 20 and 21 demonstrate the extracted images for *melanoma* and *nevus* class query images. This model accurately identifies the class of *melanoma* query image. But for *nevus* query images it has shown miss classification. Thus the retrieved images shown in Fig. 21 are from *sebborhic keratosis* image category.

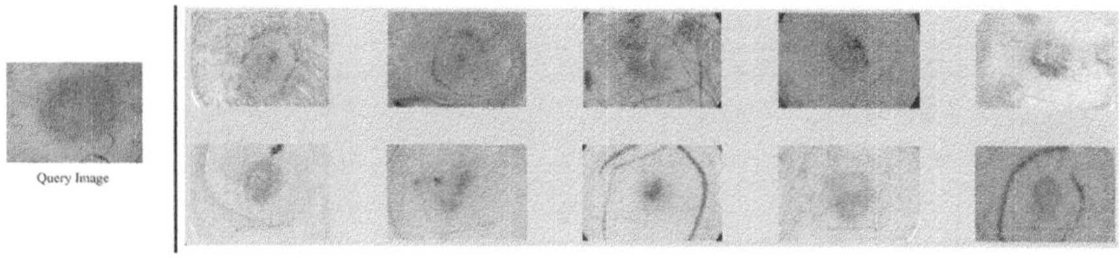

Fig. 21 Top 10 outcomes for a *nevus* class query image (for miss-classification)

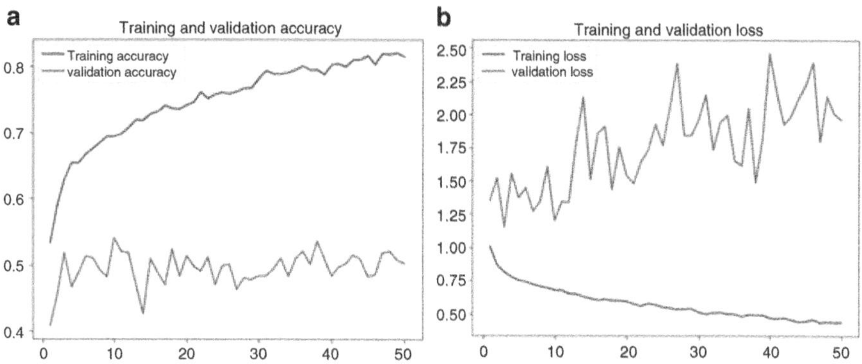

Fig. 22 (**a**) Training and validation accuracy graph and (**b**) training and validation loss graph for inception V3 model

4.2.7 Performance Analysis: Why VGG-16 Outperforms

Although a significant difference in the performance of the VGG-16 from other two models can not be seen from above results, but the reasons for superior performance of VGG-16 model is not only the test accuracies but other factors too, which will be explained in this section in detail.

To understand the reason behind inferior performances of the inception V3 and inception *Resnet* V2, we have further analyzed two graphs of these models. Now, all the three models were again trained on the training dataset. But this time we have used 50 epochs only and rest of all the parameters are same. After the training process, we have generated the (1) training and validation accuracy graph and (2) training and validation loss graph of all three models. Figures 22, 23, and 24 demonstrate the training and validation accuracy graph along with training and validation loss graph of all three models. In these three figures, the x-$axis$ represents the number of epochs. The Y-$axis$ of the training and validation accuracy graphs represent the accuracy values. Similarly, the Y-$axis$ of the training and validation loss graphs represent the loss values.

On analyzing the graphs of inception V3 and *Resnet* V2 model, we can clearly see that both the models are over-fitting for the work in hand. As a result, training

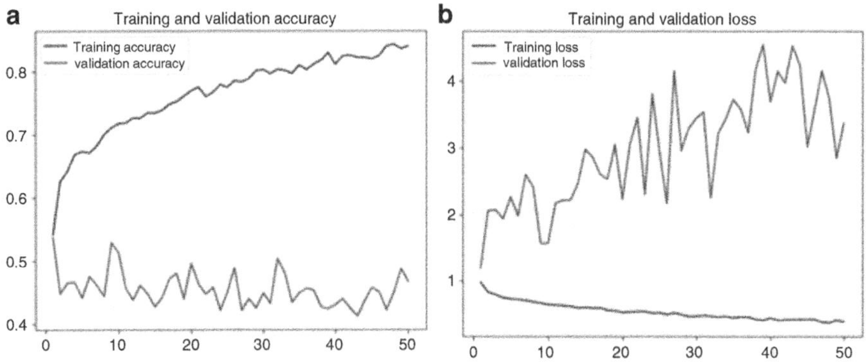

Fig. 23 (**a**) Training and validation accuracy graph and (**b**) training and validation loss graph for inception *ResNet* V2 model

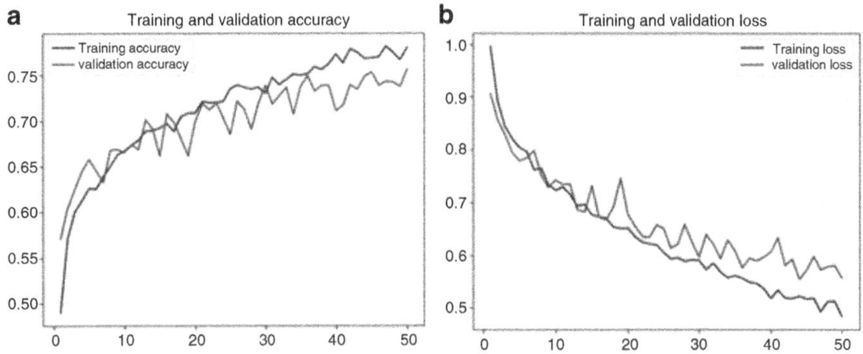

Fig. 24 (**a**) Training and validation accuracy graph and (**b**) training and validation loss graph for VGG-16 model

accuracy is increasing and training loss is decreasing. At the same time, these two models are not showing any improvement in the validation loss and accuracy. The validation loss and accuracy values are constantly fluctuating and are unstable throughout the entire training process. On the other hand, VGG-16 model is showing more stable behavior for both validation loss and validation accuracy improvement. The smaller size of training dataset is the main reason behind the unstable behavior of inception V3 and *Resnet* V2 models. Inception V3 and inception *Resnet* V2 both are highly complex models and it is a proven fact that highly complex model will over-fit the data set easily as compared to the simpler model. The complex structure of these two models are adversely affecting the performance.

Table 12 Category wise test accuracies of modified VGG-16 model

CNN model	Category wise test accuracies			Average test accuracy
	Melanoma	*Sebborhic Keratosis*	*Nevus*	
Modified VGG-16	83%	87%	86%	85%

4.2.8 Retrieval Using Modified VGG-16 Model

The VGG-16 model has shown best results among all the three discussed models. So in this sub-section we have further improved the performance of the VGG-16 model by utilizing two loss functions during the training process. In this experiment, the training of the VGG-16 model was completed in two different phases which are as follows:

- **Phase I :** In this phase, we have loaded the *ImageNet* weights on the initial layers of the VGG-16 ANN (i.e. Transfer Learning). Next, we have used *RMSProp* optimizer [53] and performed the training of the model for 100 epochs. After the training process, we have stored the weights of the model. These weights will be used in the second phase of the training.
- **Phase II :** In this phase, we have loaded the saved weights obtained from *Phase I* into the initial layers of the VGG-16 model instead of *ImageNet* weights. Next, we have performed training on the last convolutional block and classifier of the VGG-16 model. Here, we have used *SGD* optimizer [54] with a learning rate of 0.0001. This training has been carried out for 200 epochs. After the training process, the weights were saved and these weights represent the final weight which will be used for prediction of skin lesions.

The retrieval results achieved through this methodology are the best results amongst all the discussed CNN models. This model has shown significant improvements as compared to the above three CNN based models. The test accuracies for this approach have been shown in Table 12. Later, Figs. 25 and 26 demonstrate the (1) training and validation accuracy graph and (2) training and validation loss graph of Phase I and II training process. In these graphs, we can observe a constant and steady improvements on the validation loss and accuracy.

5 Conclusions

In this chapter, the basic concepts of different automatic image retrieval systems have been discussed. In the literature, three kinds of automatic retrieval systems are present. These systems are *Text-based Image Retrieval* (TBIR), *Content-based Image Retrieval* (CBIR), and *Semantic-based Image Retrieval* (SBIR) system. This chapter elaborated the algorithmic steps, merits, and demerits of these systems. These image retrieval systems have been widely used by the

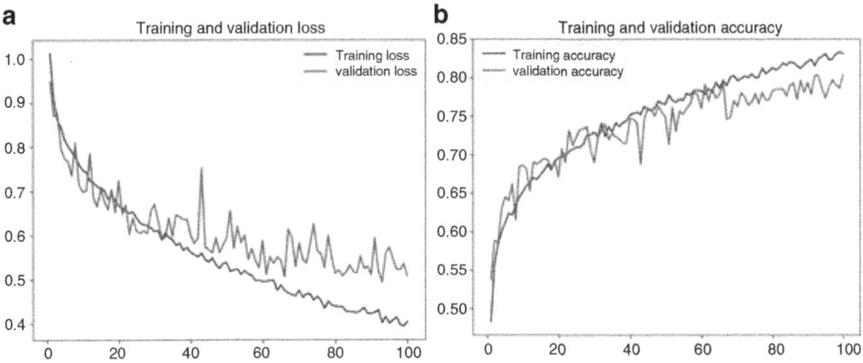

Fig. 25 (**a**) Training and validation accuracy graph and (**b**) training and validation loss graph of *Phase I* training process

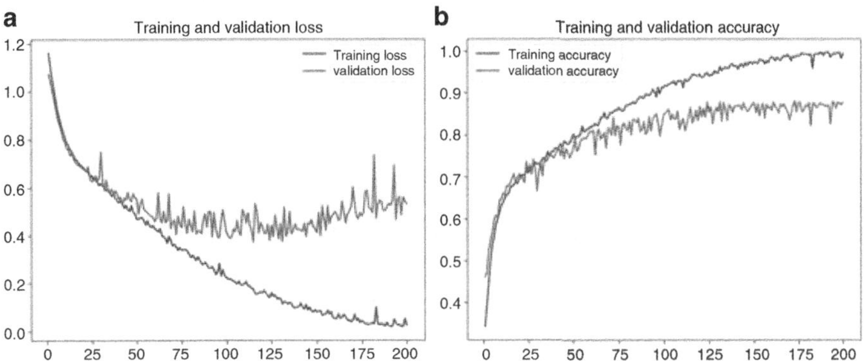

Fig. 26 (**a**) Training and validation accuracy graph and (**b**) training and validation loss graph of *Phase II* training process

researchers for different kinds of applications such as medical imaging, agriculture, education, historical research, remote sensing, crime prevention, hyper-spectral imaging, biodiversity information system, fingerprint identification, etc. Further, different deep learning models and their application on medical image retrieval have been discussed in detail. In this chapter, inception V3, inception *ResNet* V2, and VGG-16 based skin lesion retrieval systems have been explained in detail. The VGG-16 based medical image retrieval system have shown better performance than the other two CNN based system. The reason for better retrieval performance of VGG-16 model has been also explained. A modified VGG-16 based skin lesion retrieval system has been also explained in this chapter. This modified VGG-16 model has shown significant improvements as compared to the rest of CNN based systems.

Nowadays, medical imaging are playing critical role in diagnosis and treatment of several diseases. As a result, a huge amount of digital images have been captured during course of the treatment. Manual searching and management of all these high volume medical images are not a practical task. Thus, there is dire need to develop a CNN based model which can classify and retrieve all kinds of medical images. Till now, no such a effective and efficient medical image retrieval system exists in the literature which can handle all kinds of medical images. So, development of such a medical retrieval system is still an active research problem for the research community.

References

1. Jitesh Pradhan, Arup Kumar Pal, and Haider Banka. Principal texture direction based block level image reordering and use of color edge features for application of object based image retrieval. *Multimedia Tools and Applications*, 78(2):1685–1717, 2019.
2. Nishant Shrivastava and Vipin Tyagi. Content based image retrieval based on relative locations of multiple regions of interest using selective regions matching. *Information Sciences*, 259:212–224, 2014.
3. Jitesh Pradhan, Sumit Kumar, Arup Kumar Pal, and Haider Banka. Texture and color visual features based cbir using 2d dt-cwt and histograms. In *International Conference on Mathematics and Computing*, pages 84–96. Springer, 2018.
4. Lin Chen, Dong Xu, Ivor W Tsang, and Jiebo Luo. Tag-based image retrieval improved by augmented features and group-based refinement. *IEEE Transactions on Multimedia*, 14(4):1057–1067, 2012.
5. I Ahamd and Taek-Sueng Jang. Old fashion text-based image retrieval using FCA. In *Proceedings 2003 International Conference on Image Processing (Cat. No. 03CH37429)*, volume 3, pages III–33. IEEE, 2003.
6. Xiang Sean Zhou and Thomas S Huang. Unifying keywords and visual contents in image retrieval. *Ieee Multimedia*, 9(2):23–33, 2002.
7. Chen Zhang, Joyce Y Chai, and Rong Jin. User term feedback in interactive text-based image retrieval. In *Proceedings of the 28th annual international ACM SIGIR conference on Research and development in information retrieval*, pages 51–58. ACM, 2005.
8. Wen Li, Lixin Duan, Dong Xu, and Ivor Wai-Hung Tsang. Text-based image retrieval using progressive multi-instance learning. In *2011 International Conference on Computer Vision*, pages 2049–2055. IEEE, 2011.
9. Michael J Swain. Interactive indexing into image databases. In *Storage and Retrieval for Image and Video Databases*, volume 1908, pages 95–103. International Society for Optics and Photonics, 1993.
10. Jun Yue, Zhenbo Li, Lu Liu, and Zetian Fu. Content-based image retrieval using color and texture fused features. *Mathematical and Computer Modelling*, 54(3–4):1121–1127, 2011.
11. Jitesh Pradhan, Arup Kumar Pal, and Haider Banka. A prominent object region detection based approach for CBIR application. In *2016 Fourth International Conference on Parallel, Distributed and Grid Computing (PDGC)*, pages 447–452. IEEE, 2016.
12. Jitesh Pradhan, Ashok Ajad, Arup Kumar Pal, and Haider Banka. Multi-level colored directional motif histograms for content-based image retrieval. *The Visual Computer*, pages 1–22, 2019.
13. Safia Abdelmounaime and He Dong-Chen. New brodatz-based image databases for grayscale color and multiband texture analysis. *ISRN Machine Vision*, 2013, 2013.
14. Jia Li and James Z Wang. Real-time computerized annotation of pictures. *IEEE transactions on pattern analysis and machine intelligence*, 30(6):985–1002, 2008.

15. N Jhanwar, Subhasis Chaudhuri, Guna Seetharaman, and Bertrand Zavidovique. Content based image retrieval using motif cooccurrence matrix. *Image and Vision Computing*, 22(14):1211–1220, 2004.
16. Itheri Yahiaoui, Nicolas Hervé, and Nozha Boujemaa. Shape-based image retrieval in botanical collections. In *Pacific-Rim Conference on Multimedia*, pages 357–364. Springer, 2006.
17. Manesh Kokare, Prabir K Biswas, and Biswanath N Chatterji. Texture image retrieval using rotated wavelet filters. *Pattern recognition letters*, 28(10):1240–1249, 2007.
18. Rui Hu and John Collomosse. A performance evaluation of gradient field hog descriptor for sketch based image retrieval. *Computer Vision and Image Understanding*, 117(7):790–806, 2013.
19. Naushad Varish and Arup Kumar Pal. Content based image retrieval using statistical features of color histogram. In *2015 3rd international conference on signal processing, communication and networking (ICSCN)*, pages 1–6. IEEE, 2015.
20. Xiang-Yang Wang, Yong-Jian Yu, and Hong-Ying Yang. An effective image retrieval scheme using color, texture and shape features. *Computer Standards & Interfaces*, 33(1):59–68, 2011.
21. B Sathya Bama, S Mohana Valli, S Raju, and V Abhai Kumar. Content based leaf image retrieval (CBLIR) using shape, color and texture features. *Indian Journal of Computer Science and Engineering*, 2(2):202–211, 2011.
22. Murala Subrahmanyam, RP Maheshwari, and R Balasubramanian. A correlogram algorithm for image indexing and retrieval using wavelet and rotated wavelet filters. *International Journal of Signal and Imaging Systems Engineering*, 4(1):27–34, 2011.
23. Muhammad Sajjad, Amin Ullah, Jamil Ahmad, Naveed Abbas, Seungmin Rho, and Sung Wook Baik. Integrating salient colors with rotational invariant texture features for image representation in retrieval systems. *Multimedia Tools and Applications*, 77(4):4769–4789, 2018.
24. Tomasz Andrysiak and Michal Choraś. Image retrieval based on hierarchical gabor filters. *International Journal of Applied Mathematics and Computer Science*, 15:471–480, 2005.
25. Nishant Shrivastava and Vipin Tyagi. An efficient technique for retrieval of color images in large databases. *Computers & Electrical Engineering*, 46:314–327, 2015.
26. Naushad Varish, Jitesh Pradhan, and Arup Kumar Pal. Image retrieval based on non-uniform bins of color histogram and dual tree complex wavelet transform. *Multimedia Tools and Applications*, 76(14):15885–15921, 2017.
27. Muwei Jian, Yilong Yin, Junyu Dong, and Kin-Man Lam. Content-based image retrieval via a hierarchical-local-feature extraction scheme. *Multimedia Tools and Applications*, 77(21):29099–29117, 2018.
28. Jitesh Pradhan, Sumit Kumar, Arup Kumar Pal, and Haider Banka. A hierarchical cbir framework using adaptive tetrolet transform and novel histograms from color and shape features. *Digital Signal Processing*, 82:258–281, 2018.
29. Jing Li, Nigel Allinson, Dacheng Tao, and Xuelong Li. Multitraining support vector machine for image retrieval. *IEEE Transactions on Image Processing*, 15(11):3597–3601, 2006.
30. Yiqing Guo, Xiuping Jia, and David Paull. Effective sequential classifier training for svm-based multitemporal remote sensing image classification. *IEEE Transactions on Image Processing*, 27(6):3036–3048, 2018.
31. Hossein Pourghassem and Hassan Ghassemian. Content-based medical image classification using a new hierarchical merging scheme. *Computerized Medical Imaging and Graphics*, 32(8):651–661, 2008.
32. Xiao-Feng Wang, De-Shuang Huang, Ji-Xiang Du, Huan Xu, and Laurent Heutte. Classification of plant leaf images with complicated background. *Applied mathematics and computation*, 205(2):916–926, 2008.
33. Kunshan Huang, Shutao Li, Xudong Kang, and Leyuan Fang. Spectral–spatial hyperspectral image classification based on knn. *Sensing and Imaging*, 17(1):1, 2016.
34. Andy Liaw, Matthew Wiener, et al. Classification and regression by randomforest. *R news*, 2(3):18–22, 2002.

35. Johan AK Suykens and Joos Vandewalle. Least squares support vector machine classifiers. *Neural processing letters*, 9(3):293–300, 1999.
36. S Rasoul Safavian and David Landgrebe. A survey of decision tree classifier methodology. *IEEE transactions on systems, man, and cybernetics*, 21(3):660–674, 1991.
37. Sankar K Pal and Sushmita Mitra. Multilayer perceptron, fuzzy sets, and classification. *IEEE Transactions on neural networks*, 3(5):683–697, 1992.
38. Karen Simonyan and Andrew Zisserman. Very deep convolutional networks for large-scale image recognition. *arXiv preprint arXiv:1409.1556*, 2014.
39. Christian Szegedy, Vincent Vanhoucke, Sergey Ioffe, Jon Shlens, and Zbigniew Wojna. Rethinking the inception architecture for computer vision. In *Proceedings of the IEEE conference on computer vision and pattern recognition*, pages 2818–2826, 2016.
40. Alex Krizhevsky, Ilya Sutskever, and Geoffrey E Hinton. Imagenet classification with deep convolutional neural networks. In *Advances in neural information processing systems*, pages 1097–1105, 2012.
41. Kaiming He, Xiangyu Zhang, Shaoqing Ren, and Jian Sun. Deep residual learning for image recognition. In *Proceedings of the IEEE conference on computer vision and pattern recognition*, pages 770–778, 2016.
42. Fabio Alexandre Spanhol, Luiz S Oliveira, Caroline Petitjean, and Laurent Heutte. Breast cancer histopathological image classification using convolutional neural networks. In *2016 international joint conference on neural networks (IJCNN)*, pages 2560–2567. IEEE, 2016.
43. Emmanuel Maggiori, Yuliya Tarabalka, Guillaume Charpiat, and Pierre Alliez. Convolutional neural networks for large-scale remote-sensing image classification. *IEEE Transactions on Geoscience and Remote Sensing*, 55(2):645–657, 2016.
44. Ali Madani, Mehdi Moradi, and Tanveer F Syeda-Mahmood. Medical image classification based on a generative adversarial network trained discriminator, June 27 2019. US Patent App. 15/850,116.
45. ME Paoletti, JM Haut, J Plaza, and A Plaza. A new deep convolutional neural network for fast hyperspectral image classification. *ISPRS journal of photogrammetry and remote sensing*, 145:120–147, 2018.
46. Jianpeng Zhang, Yutong Xie, Qi Wu, and Yong Xia. Medical image classification using synergic deep learning. *Medical image analysis*, 54:10–19, 2019.
47. Christian Szegedy, Sergey Ioffe, Vincent Vanhoucke, and Alexander A Alemi. Inception-v4, inception-resnet and the impact of residual connections on learning. In *Thirty-First AAAI Conference on Artificial Intelligence*, 2017.
48. Jia Deng, Wei Dong, Richard Socher, Li-Jia Li, Kai Li, and Li Fei-Fei. Imagenet: A large-scale hierarchical image database. In *2009 IEEE conference on computer vision and pattern recognition*, pages 248–255. Ieee, 2009.
49. Alex Alemi. Improving inception and image classification in tensorflow. *Google Research Blog*, 2016.
50. Noel Codella, Veronica Rotemberg, Philipp Tschandl, M Emre Celebi, Stephen Dusza, David Gutman, Brian Helba, Aadi Kalloo, Konstantinos Liopyris, Michael Marchetti, et al. Skin lesion analysis toward melanoma detection 2018: A challenge hosted by the international skin imaging collaboration (isic). *arXiv preprint arXiv:1902.03368*, 2019.
51. Philipp Tschandl, Cliff Rosendahl, and Harald Kittler. The ham10000 dataset, a large collection of multi-source dermatoscopic images of common pigmented skin lesions. *Scientific data*, 5:180161, 2018.
52. Diederik P Kingma and Jimmy Ba. Adam: A method for stochastic optimization. *arXiv preprint arXiv:1412.6980*, 2014.
53. Christian Igel and Michael Hüsken. Improving the rprop learning algorithm. In *Proceedings of the second international ICSC symposium on neural computation (NC 2000)*, volume 2000, pages 115–121. Citeseer, 2000.
54. Léon Bottou. Large-scale machine learning with stochastic gradient descent. In *Proceedings of COMPSTAT'2010*, pages 177–186. Springer, 2010.

Medical Image Fusion Using Deep Learning

Ashif Sheikh, Jitesh Pradhan, Arpit Dhuriya, and Arup Kumar Pal

Abstract Usually, digital devices and cameras suffer from the limited depth-of-focus problem so they cannot focus on every prominent object present within a single frame of reference. As a result, in a captured picture most of the objects become background information while only a few objects acquire maximum attention. To address this issue, researchers have combined all the partially focused distinct images of the same frame of references into a single fully focused image. This approach is known as an image fusion approach wherein the final fused image all the prominent objects get equal attention. Nowadays, with the rapid advancements in the field of digital technology has escalated the use of image fusion techniques in a variety of real-life applications such as advanced photography, medical diagnosis, surveillance, object detection, remote sensing, etc. Recently, image fusion in the field of medical imaging has emerged as a grand success for medical diagnostics and analysis. So, medical image fusion has become a suitable subsidiary for medical experts. This chapter aims to focus on different aspects of image fusion techniques and their applications in the area of medical imaging. This chapter also explicitly explains the different *Deep Learning* (DL) techniques to address the various issues related to the field of medical image fusion. Specifically, we first condense the drawbacks of the conventional image fusion techniques and recapitulate the advantage of DL techniques. Besides, this chapter also presents a description of the significant medical image fusion techniques along with various relevant imaging modalities and their application domains.

Keywords Deep learning · Dense block · Image fusion · Local binary pattern · Medical imaging modalities

A. Sheikh (✉) · J. Pradhan · A. Dhuriya · A. K. Pal
Indian Institute of Technology (ISM), Dhanbad, India

M. Elloumi (ed.), *Deep Learning for Biomedical Data Analysis*,
https://doi.org/10.1007/978-3-030-71676-9_6

1 Introduction

Over the last few decades, image fusion has been evolved as a significant area of research. Image fusion is an approach of combining salient information from various source images of the identical scenario captured at different times and different positions. The fused image contains all the relevant, complementary and salient information of the source images by providing a suitable representation of the scenario. The prerequisite for image fusion is to avoid the mis-registration and noise for all the input images. The input images to be combined are termed to as source images and the final resultant image as a fused image. Generally, image fusion techniques can be divided into two major domains, spatial and transform domain. The techniques of spatial domain follow a linear or non-linear approach to combine the pixels of the input images. Whereas in transform domain, initially, the source images are transformed into frequency domain using suitable transform tools. Later, the final fused image has been achieved by employing an appropriate fusion approach on the transformed images. The mathematical expression [1] for the general image fusion technique is denoted as $I_F = \phi(I_1, I_2, \ldots, I_M) = \gamma_1 I_1 + \gamma_2 I_2 + \ldots + \gamma_M I_M$, where the image fusion function ϕ takes the different source images i.e. $I_1, I_2, \ldots, I_M$ and they are fused with different γ values (γ_i). The benefits of image fusion are mobilized representation of information, to reduce uncertainty, to remove artifacts and to improve the reliability in the final fused image. So researchers around the globe have used image fusion techniques in different aspects like the medical image fusion has achieved enormous success in medical image diagnostics and analysis. Medical experts use the medical images as a suitable and helpful assistant in the course of treatment for different diseases. Besides, medical image fusion deals with different types of imaging modalities such as *Computed Tomography* (CT), *Magnetic Resonance Imaging* (MRI), *Single Photon Emission Computed Tomography* (SPECT), *Positron Emission Tomography* (PET), visible and infrared images, etc. Among these images, visible and infrared images are fused using a generative adversarial *Artificial Neural Network* (ANN) termed as *FusionGAN* [2]. Radiation differences in infrared images help in distinguishing targets from their backgrounds and visible images provide texture details with high spatial resolution. To obtain a hyperspectral image of high spatial resolution, a multispectral image of high spatial resolution is fused [3] with a hyperspectral image of low spatial resolution by a novel *Spatial-Spectral Sparse Representation* (SSSR).

After the explicit presentation of introduction section, the chapter is organized as follows: in Sect. 2, the different applications of image fusion technique has been presented. Later, in Sect. 3, the basic image fusion approaches have been given. The step-by-step process involved in the general medical image fusion procedure is shown in Sect. 4. Next, in Sect. 5, a brief introduction regarding various relevant methods have been presented. The *Deep Learning* (DL) based medical image fusion has been given in Sect. 6. In Sect. 7, the experimental results and the related discussions have been presented. Finally, Sect. 8 concludes the chapter with some future scope.

2 Applications of Image Fusion Techniques

Image fusion provides more information and better description of different input images in the form of a final fused image. As discussed earlier, image fusion is very effective and widely adopted in different real-life applications such as medical imaging [4–6], object detection [7–10], recognition [11–16], tracking[17–19], color vision [20–22], remote sensing [23, 24] , surveillance [25–29], etc. Figure 1 demonstrates the different applications of image fusion techniques. Further, a brief discussion on the various applications of image fusion have been presented in the following sub-sections.

2.1 Fusion in Medical Imaging

Medical diagnostics using medical images confront severe scientific challenges such as brain cancer detection, tissue classification, malignant tumor detection, and etc. Generally, doctors and medical experts cannot get enough information about any disease from a single imaging modality. Thus, different types of images are combined to get more information regarding patients medical health condition. Several fusion algorithms are widely used and embedded in the medical diagnostic process for the proper assessment of medical problems. For example, Guan et al. [4] have designed an approximate sparse representation based on multiresolution analysis for medical image fusion. The approximate sparse representation coefficients help

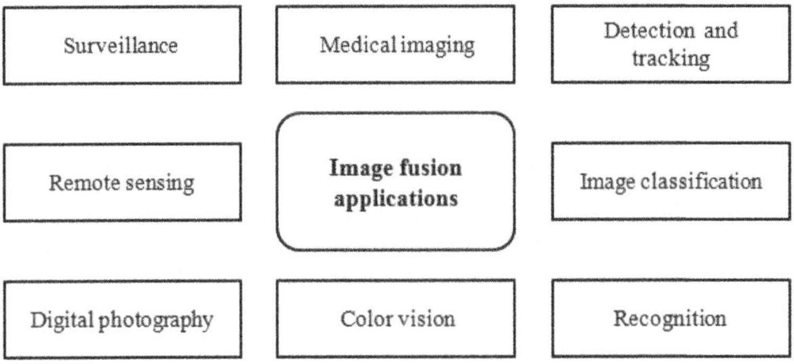

Fig. 1 Image fusion applications

to express the high-frequency and low-frequency information of the images. The algorithm has a better visual effect and successfully highlights the target information. Jiang et al. [6] have proposed similar sparse representation method which takes the benefit of edge-preserving in multi-scale decomposition for sustaining structural information of the source image. Recently, the wavelet transform based analysis has been extended by fractional wavelet transform to analyze the different fused images in the time-frequency domain. Similarly, the discrete fractional wavelet helps to analyze the multimodality images more accurately using multimodal medical image fusion method [30]. Various other widely used medical image fusion applications can be found in the field of breast cancer detection, image segmentation, multimodal image fusion, etc. For example, several medical imaging modalities are fused to get significant information of the human body [6]. Principal application domains of various medical imaging modalities are:

- **Lungs:** The lung is an essential organ of the human body which is responsible for the respiratory system and normally, happens to damage from air pollution. It is very difficult to distinguish between a healthy tissue and an infected tissue of the lungs. In this case, the multi-modality medical image fusion approaches are seen to be ideal for checking the condition of the lungs. Therefore, the imaging modalities such as CT, PET-CT, and SPECT are combined to identify the infected and healthy tissues.
- **Brain:** The brain is a vital human organ and known as the CPU of the human body. The medical imaging modalities such as SPECT, PET, CT, endoscopy, and EEG are commonly used to diagnosis the condition of human brain. The fusion of modalities has been introduced to many applications such as segmentation of the brain tissues, abnormal brain tissue classification, brain tumor detection, and monitoring.
- **Cardiac:** Cardiac is also an important human organ. To maintain a healthy condition of human cardiac, several medical imaging modalities are used. Examples of such modalities are MRI, CT, and Electrocardiography. Most of the cases, the fusion of these modalities helps in cardiac deformation recovery.
- **Prostate:** The prostate is significantly benefited from multi-modality medical image fusion and the medical imaging modalities that are concerned with the prostate includes MRI, CT, PET, SPECT, and uroscopic. The fusion of different modalities is done for applications such as prostate cancer detection, prostate localization, prostate gland motion evaluation, etc.
- **Breast:** Over the last few decades, the number of breast cancer patients specifically women has been rapidly increased. The individual modalities that are used to analyze the condition of breasts include mammogram, CT, MRI, SPECT, and PET. Also, the medical imaging modalities that are most frequently used for multi-modality image fusion include mammogram- MRI, mammogram-CT, and mammogram-MRI-CT. These modalities have been applied in several applications such as breast cancer detection, and malignant breast tumor prediction.

2.2 *Fusion in Detection and Tracking*

Researchers around the globe have used different image fusion techniques in real-world applications such as road, object, fault, pedestrian, region, and fire detection. Object detection aims to trace the most significant region of interest of an image. Whereas, tracking shares the exact position of the object(s) in real-time scenario. Target tracking algorithms are expected to be more efficient since the tracking is performed in real-time video or images. Gao et al. [8] have proposed a suitable method for moving object detection which provides a specific framework based on low rank and sparse decomposition. Also, Luo et al. [31] have introduced a hybrid tracking framework for fusion tracking with the help of an adaptive weighting scheme. A robust fusion tracking approach acquires complementary information through thermal infrared and visible image sequence using this framework. Later, Leang et al. [18] have designed a framework for combining a set of trackers for single-object tracking which shows good performance for the on-line fusion of trackers.

2.3 *Fusion in Recognition*

Recognition aims to identify the class of an individual target object in a given scene which can have single or multiple different objects. Face recognition has been considered as one of the most popular recognition applications. Hariharan et al. [13] have used *Empirical Mode Decomposition* (EMD) on visible and infrared images to perform the image fusion which ultimately improves the accuracy of face recognition. The EMD decomposes the input images into *Intrinsic Mode Functions* (IMF) to perform image fusion. As a result, the EMD fused images provide a significantly higher face recognition rate as compared to the raw visible and infrared fused images. Over the few years, it is observed that face recognition suffers from several difficult problems such as light changes, expressions, and wearing of stuffs (including hats, sunglass, and scarf). Recently, Umer et al. [16] have shown a fusion approach of different feature learning techniques for face recognition. A standard face recognition system has three components: (a) preprocessing, (b) feature extraction and (c) classification. In the pre-processing phase, identification of face region is performed on the input image. Thereafter, the features are extracted from the identified region to distinguish salient features. Then classification tasks are performed on the extracted features to identify the actual class of different faces. Action recognition is also a challenging problem in real-life scenario. However, an effective multistage local fusion ANNs with residual connection improves the performance of video action recognition. For example, He et al. [15] have designed a local fusion block that can employ residual connections for video action recognition.

2.4 Fusion in Color Vision

Color plays an essential role in the human visual system to discriminate the different objects. The human eyes grasp only the visible light. Hence, it is required to convert the information of the target scene into a pseudo-color image. Colorization methods consist of two classes, in which first-class deals with the color propagation and the second class deals with the color transfer. The color propagation process is accomplished through the amplification of the color data into the grayscale image where color data is initiated manually. Mainly this method is used in diffusion and some other techniques. Wherein, the color transfer method transfers the colors of a source color image into a target gray-scale image. The major drawback of the color transfer methods is the spatial color coherency. To address this problem, Jin et al. [32] have proposed a fused color transfer method with the help of *Discrete Fourier Transform* (DFT) and variance features for image colorization. In most of the cases, color contrast has a significant role in color image fusion. Normally, color contrast enhances the human visual capabilities to discriminate the salient features from other background features. So, Yin et al. [21] have introduced a color contrast enhancement scheme that increases the color contrast of targets in the final fused infrared and visible images.

2.5 Fusion in Remote Sensing

Remote sensing is a popular application of image fusion technique which captures the detailed information about a scene through a remote sensing system. The applications of remote sensing include surveillance, oceanography, military services, agriculture, Earth surface and atmosphere monitoring, etc. Remote sensing system provides important coverage and mapping of soil, water, and forest. Typically, the fusion in remote sensing applications is accomplished with the help of infrared and visual images that deal with the radiation differences. For example, Li et al. [23] have designed a remote sensing system of *Unmanned Aerial Vehicles* (UAVs) in medium-altitude. The system is installed on the integrated cameras that are mounted on UAVs. These medium-altitude UAVs are significant information acquisition platform. The multicontourlet transform [24] based infrared and visible remote sensing image fusion scheme outperforms the wavelet-transform-based and contourlet-transform-based image fusion schemes.

2.6 Fusion in Surveillance

Nowadays, criminal activities, accidents and vandalisms are enormously increasing everywhere. Therefore, the smart surveillance systems are direly needed in our day-to-day life activities. A smart video surveillance system ensures citizens' well-being by constantly monitoring public and private environments. For this, surveillance system requires clear, lively, robust and explicit images or videos. Recently, the data comes from several surveillance applications that are of different types and complex in nature. Hence, it is highly required to analyze and standardize the visual data to increase its robustness, reliability, and accuracy. Infrared images mainly captures the thermal radiation of the objects. So these images do not suffer from the adverse effects of the fluctuating weather and luminance conditions. Normally, visible images contain high spatial resolution and diverse colors that helps to identify objects easily. Therefore, the fusion of infrared and visible image improves the quality of final fused image. Sometimes, a fusion of infrared and visible images based on swarm intelligence using discrete cosine transform [19] generates a single informative fused image. Similarly, the information collected from a visible spectral video is fused with the thermal infrared video to enhance the effectiveness of the surveillance system. In this process, Kumar et al. [28] have introduced an intelligent fusion method using fuzzy logic and Kalman filtering technique to perform the object tracking.

3 Image Fusion Approaches

The algorithms of the image fusion approach are performed at different stages: (1) pixel level, (2) feature level, and (3) decision level [1] as shown in Fig. 2. At the pixel-level approach, fusion is done by combining or aggregating each input image pixel by pixel. At the feature-level approach, image fusion is achieved on the extracted features of the source images. The feature-level image fusion deals with regional based fusion techniques. Decision-level image fusion approach is performed on probabilistic decision information that is derived from extracted features. Generally, the input to the decision-level is an initial decision that is taken from the feature level approach of image fusion.

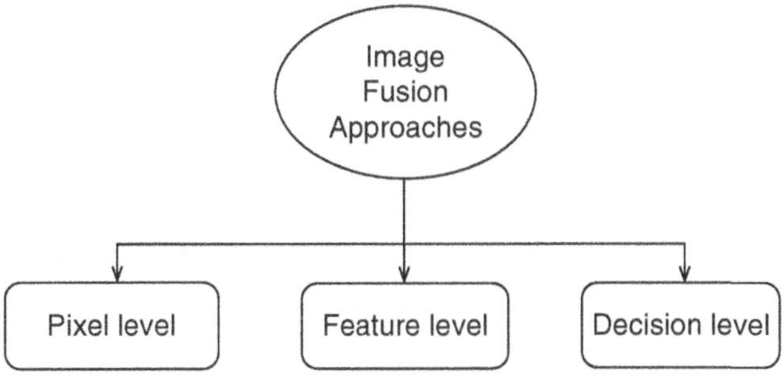

Fig. 2 Image fusion approaches

3.1 Pixel Level

Image fusion approaches are generally summarized by three major stages as shown in Fig. 2. Indeed, pixel-level image fusion methods get preference for fusion compared to other level methods because of their simplest implementation. In the pixel-level fusion approach, individual pixels are fused to get a single gray or chromatic image. The fused image needs to follow some requirements: (a) the fusion process should preserve the complimentary and salient information of the input images; (b) It must avoid artifact and deteriorative states; (c) the fused image should be effective for human vision. A general pixel-level image fusion system consists of a few stages: image acquisition, image registration, pixel/block level statistical parameters calculation, fusion strategy, and final fused image as shown in Fig. 3.

The fusion methods which are frequently used in pixel-level approach are as follows: (1) the *Sparse Representation* (SR) based methods such as group sparse representation, gradient constrained sparse representation, joint sparsity model; (2) the multi-scale based decomposition methods such as pyramid, wavelet, complex wavelet, log-Gabor transform; (3) different other methods such as PCA, ICA, fuzzy theory, *Intensity-Hue-Saturation* (HIS) transform; (4) the combination of different transforms such as hybrid wavelet-contourlet, multi-scale transform-SR, IHS-wavelet. The SR based method uses different fusion strategies such as choose-max, coefficients combining, weighted average. In a multi-scale based decomposition method, the fusion strategies such as optimization, choose-max, weighted-average and coefficient based measurements are used. Besides, different *Machine Learning* (ML)-based fusion strategies are also used in the pixel-level approach. Over the last few years, pixel-level image fusion approach has been used in different applications such as digital photography, surveillance and navigation, Medical diagnosis, and remote sensing.

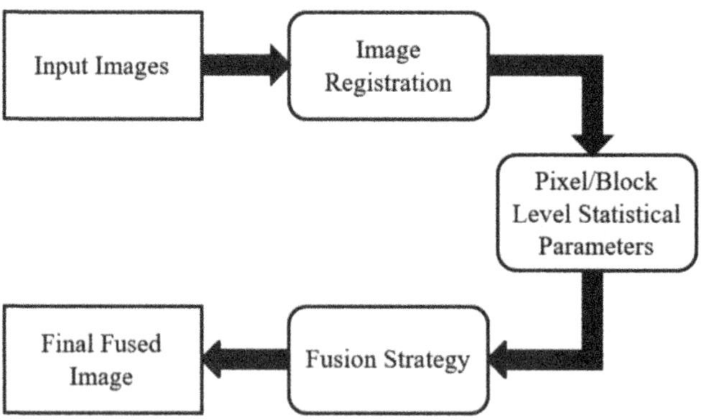

Fig. 3 Stages of a generic pixel/block level image fusion approach

3.2 Feature Level

Due to several limitations of pixel-level based methods such as misregistration, high sensitivity to the artifacts or noise and blurring effect, the feature-level image fusion approach is used. The feature-level image fusion techniques are more complex than the pixel-level image fusion techniques. As a result, it represents more details in the fused image. A general feature-level image fusion system consists of a few stages: image acquisition, image registration, image transformation, feature extraction, feature combination, inverse transformation, and the final fused image as shown in Fig. 4.

The feature-level image fusion approaches can be broadly classified [1] into three major categories which are region-based, ML, and similarity matching to content-based as shown in Fig. 5. In region-based image fusion approach, initially, several suitable segmentation techniques are used to divide the input images into different regions. Next, some effective fusion strategies are employed on the extracted features from different regions to compute the final fused image. An appropriate classifier on the extracted features is used for fusion in ML method. Content-based

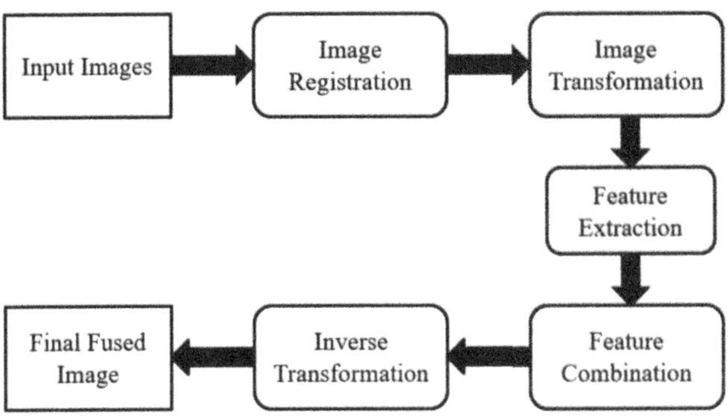

Fig. 4 Stages of a generic feature level image fusion approach

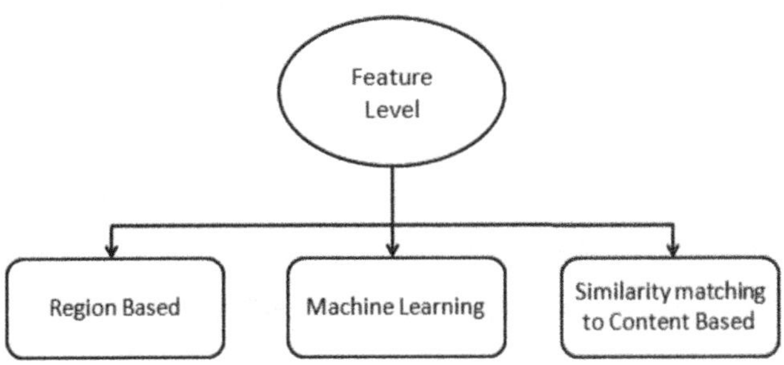

Fig. 5 Classification of feature level image fusion methods

image retrieval deals with the features such as texture, shape, and color that helps in similarity matching by aggregating relevant features.

The frequently used fusion methods in feature-level image fusion technique are *Non-Subsample Contourlet Transform* (NSCT), *Dual-Tree Complex Wavelet Transform* (DT-CWT), *Discrete Wavelet Frame Transform* (DWFT), *Shift-Invariant Shearlet Transform* (SIST), *Artificial Neural Network* (ANN), *Independent Component Analysis* (ICA), ICA-SVM and *Bi-directional Empirical Mode Decomposition* (BEMD). Major application domain of feature-level fusion involves medical imaging, remote sensing, surveillance, and photography applications.

3.3 Decision Level

Decision-level fusion is considered as the highest processing level of information. At decision-level fusion, information from several individual data sources is combined. Hence, the results from different classifiers are fused to conclude the final decision. Normally, the information is extracted individually from the input images and then different decision strategies are used to fuse the information to get better quality of the final fused image. The decisions which are taken as input are always labeled based on some considered criteria. A general feature-level image fusion system consists of a few stages: image acquisition, image registration, image transformation, feature extraction, classification, decision fusion strategy, and the final fused image as shown in Fig. 6.

The most frequently used decision methods are Bayesian inference, fuzzy decision rule, voting, rank-based, hybrid methods based on consensus, joint measures method, Dempster-Shafer methods, and adaptive decision fusion. Although, the decision level fusion is very challenging specifically in case of low-quality images but in the recent years, the application of decision-level image fusion has been

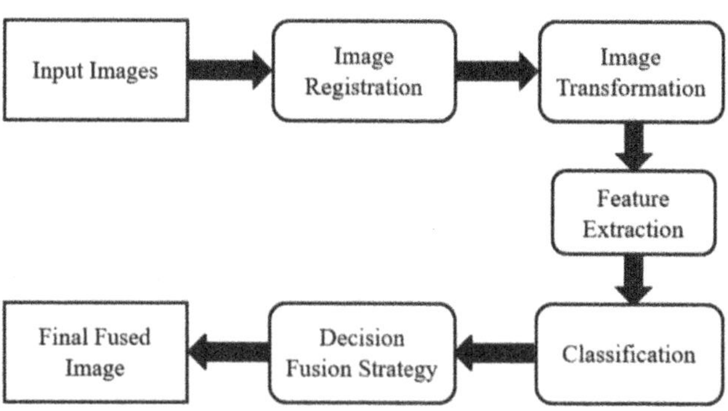

Fig. 6 Stages of a generic decision level image fusion approach

widely used. The various applications of the decision-level fusion methods are medical imaging, fingerprint verification, image classification, remote sensing, etc.

4 The General Medical Image Fusion Procedure

Normally, in the medical image fusion procedure, multiple medical images collected from single or multiple imaging modalities are combined to reduce the redundancy and improve the quality of imaging for proper validation of medical problems. The general medical image fusion procedure consists of few stages as shown in Fig. 7.

- **Image Registration:** The step-by-step procedure of medical image fusion starts with image registration. Image registration aligns images of the same scene concerning each other according to a coordinate system. In this process, the original image (also known as reference image or fixed image) is considered to be the source image. The remaining source images (often called *sensed images*) use geometric transformation for aligning with the reference image. Therefore, the process of image registration is initiated with the mapping of the input (target) images using the reference image. Consequently, the mapping provides the correct match of the similar images using definite features.
- **Feature Extraction:** The feature extraction step provides several representatives and salient features that are extracted from the registered input images. As a result, each input image generates several feature maps.
- **Decision Labeling:** In this stage, the feature maps for the registered input images are labeled depending upon some condition. Therefore, various decision maps that rely on labeling are executed as the output.
- **Semantic Equivalence:** Sometimes, the feature maps or decision maps obtained from input images might not indicate the same object or phenomena. As a result,

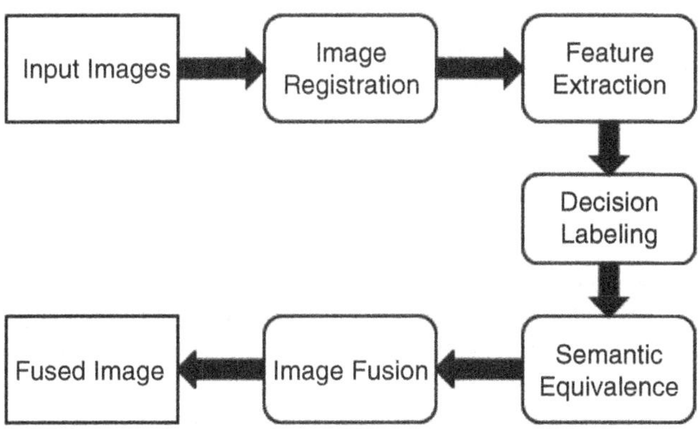

Fig. 7 General medical image fusion procedure

fusion might lead to inconsistency. Therefore, the different inputs to common objects or phenomena are made semantically equivalent.

- **Image Fusion:** In the ultimate stage, the resulting images are combined into an integral fused image that contains a more helpful and excellent representation of the scene. Finally, the fused image is expressed compactly than any of the individual input images.

5 Major Medical Image Fusion Methods

Various medical image fusion methods were introduced to achieve medical image fusion goals.
Major medical image fusion methods are:

- **Pyramid-based Methods:** The idea of the pyramid method was introduced in the 1980s with a binocular fusion using a low-pass Laplacian pyramid [33]. The aims of pyramid-based methods start with the generation of sub-images of spatial frequency band with different scales using the decomposition approach. Therefore, the decomposition approach provides a pyramid data structure. The pyramidal decomposition is suitable in several applications of image processing such as multi-scale data representation using faster methods. The Gaussian pyramid is an example of the most frequently used pyramid method that is formed with the help of a Gaussian kernel and a 2-D downsampler. The Gaussian pyramid faces information lost of high frequencies in between adjacent levels. The Laplacian pyramid helps to rectify the information loss and represent the high frequencies. Some other methods of pyramidal decomposition include gradient, steerable and ratio-of-low-pass pyramid.
- **Morphological Methods:** Image processing deals with several morphology operators and the relevant information contained in medical images is identified using morphological operators. Generally, morphology signifies the structural components of an image using different morphological operations. Several filtering techniques of morphological methods have been extensively used in medical image fusion. In some cases, fusion based on morphological techniques is done for more than one modalities such as CT, PET, MRI, Mammogram and Ultrasound images. The applications of morphological methods include brain diagnosis, tissue classification, segmentation, and medical image retrieval.
- **Wavelet-based Methods:** Grossman and Morlet [34] introduced the idea of wavelet transform at the beginning of the 1980s. In the case of wavelet-based fusion, the explicit information of images extracted using wavelet transform methods are combined. The explicit information is expressed in terms of low frequency. There exist several methods based on wavelet transform including *Discrete Wavelet Transform* (DWT), stationary wavelet transforms, lifting wavelet transform, dual-tree-DWT and Multi-wavelet transform coupled.

The discrete wavelet transform faces some elementary problems such as shift variance, lack of directionality, and aliasing. The dual-tree-DWT rectifies the complications of shift variance and lack of directionality. Sometimes, wavelets are combined with other approaches to get improved quality of the fused image. A few examples of such combinations are wavelet-SVM, ICA-wavelet, wavelet-MRA, and wavelet-texture measure. The wavelets have significant applications in image fusion including medical diagnostic, segmentation, pseudo coloring, 3D conformal radiotherapy, and super-resolution treatment planning.

- **Sparse Representation-based Methods:** Sparse representation is prevalent as an ideal approach to characterize the human visual system. There are several applications of sparse representation such as image processing, pattern recognition, automation and computer vision. In the very beginning of the sparse representation technique, separate pixel vectors are generated from every individual image and thereby produces sparse coefficients vector by applying a sparse coding technique with the help of over-complete dictionary. Thereafter, fusion rule is employed to fuse the sparse coefficient vectors for generating a meaningful representation of the source images. At the end, reconstruction is done on the fused image with the help of the learned dictionary. Here, sparse representation indicates few non-zero elements or coefficients which show the salient information of the source image. As a result, the process increases the suitability of the represented source images.
- **Subspace-based Methods:** The methods based on Subspace are concerned with the dimensions. The source images of higher dimensions are projected to subspaces of lower dimension spaces. Normal natural images of higher dimension consist of redundant information. Therefore, few methods of dimensionality reduction such as *Principal Component Analysis* (PCA), *Independent Component Analysis* (ICA) are used to remove the redundancies by maintaining the original data. Moreover, the lower dimension spaces are easy to process as it utilizes a small memory and thereby consumes less time. PCA derives the similarity and differences of correlated variables by transforming them into uncorrelated variables. Again, *Robust Principal Component Analysis* (RPCA) helps to rectify the problem of gross errors faced by PCA. The RPCA has a significant role in image fusion for retaining significant information by removing sparse noise. Image fusion methods using region-based ICA divides the input images into several regions and thereby obtains the ICA coefficients. Sometimes, a combination of ICA and other approach is executed for better representation of fused image in image fusion.
- **Neural Network-based Methods:** The idea of ANN is motivated by the biological neural systems of the human brain that is consists of millions of neurons. The neurons of one layer are strongly connected to neurons of the next layer through a weight vector and they transmit the processed information to each other. The ANN models take the input data and learn from the data to predict and analyze the information without any complex mathematical operations. Therefore, the

ANN model is referred to as a decision-making tool. The applications of ANN in medical image fusion include cancer detection, feature generation, classification and medical diagnostic. Sometimes, the ANN model[1] is sequentially processed with other techniques for improvement of the quality of the extracted features. There exists some hybrid model of ANNs such as SVM-ANN, wavelet-ANN, ANN-fuzzy and SVM-ANN-GMM. Recently, DL has been evolved as a popular state-of-the-art tool in place of the ANN model. The application of DL has spread over a wide range of fields including image processing, pattern recognition, remote sensing, image fusion and computer vision.

- **Knowledge-based Methods:** To consider any decision related to medical imaging such as segmentation designing, labeling and image registration, the knowledge of any medical practitioner is highly acknowledged. To some extent, the knowledge is needed to be domain-specific. The applications of knowledge-based methods in image fusion include brain diagnostic, classification of tissue, segmentation, breast cancer tumor detection and color visualization. The hybrid combination of knowledge-based methods with some other methods improves the significance of feature-level data. Example of such other hybrid combination includes wavelet-SVM, wavelet-MRA, neuro-fuzzy-wavelet and ICA-wavelet.

6 Deep Learning Based Medical Image Fusion

In this chapter, a DL based approach for the fusion of medical images is shown explicitly. For the last few years, CNN architecture has achieved huge success in the field of image processing. In our experiment, the grayscale medical images are considered in the fusion process for the desired results that are shown in the 'Experimental results and discussion section'.

In the experiment, $I_1, \ldots, I_n$ indicates the grayscale input images and $n \geq 2$. It is to be noted that the preregistration for all the input images is done accordingly. The ANN architecture used for accomplishing the fusion task consists of three parts: *Encoder*, *Fusion layer* and *Decoder*. The detailed ANN architecture of the fusion process is shown in Fig. 8.

The deep features from the input images are extracted in the encoder stage which itself consists of two sub-stages (*C1* and *DenseBlock*). In the initial sub-stage, the rough features are extracted in the *C1* layer (i.e. first layer) that deals with 3×3 filters. The encoder part incorporates the *DenseBlock* as the next sub-stage, which is the origin of the term *DenseFuse*. This method helps the ANN to preserve useful information in the middle layer.The *DenseBlock* that contains 3×3 filters consists of three convolutional layers in the next sub-stage where the output of each layer cascades as the next layer input. Normally, the input images are padded in the reflection mode within the ANN.

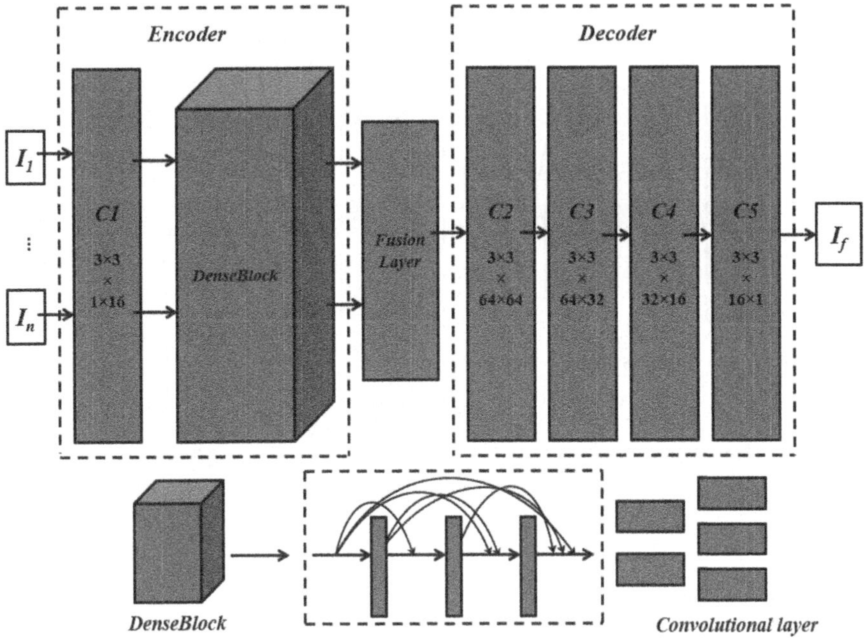

Fig. 8 The architecture of *DenseFuse* method [35]

6.1 Training

To reconstruct the input images through the training phase, the fusion layer is excluded by considering only encoder and decoder ANNs to be trained. Thereafter, an adaptive fusion strategy is used for the fusion of the deep features generated through encoder. Note that it is done, once the weights of encoder and decoder are fixed. The explicit framework for the training phase of the ANN is shown in Fig. 9, and the detailed architecture is described in Table 1.

Figure 9 shows that the encoder ANN consists of convolutional layer *C1* which deals with 3 × 3 filters and the *DenseBlock* consists of convolutional layers *DC1*, *DC2*, and *DC3*. In the ANN, each layer is cascaded to other layers. The decoder ANN contains four layers: *C2*, *C3*, *C4*, *C5*; and all these layers are used for the reconstruction of the input images.

For the appropriate reconstructions of the input images through the training of the encoder and decoder ANN, the loss function L is minimized as,

$$L = \lambda L_{ssim} + L_p \tag{1}$$

where Eq. (1) signifies a weighted combination of structural similarity (SSIM) loss (L_{ssim}) with weight (λ) and pixel loss (L_p) . Calculation of pixel loss is done as,

$$L = \|O - I\|_2 \tag{2}$$

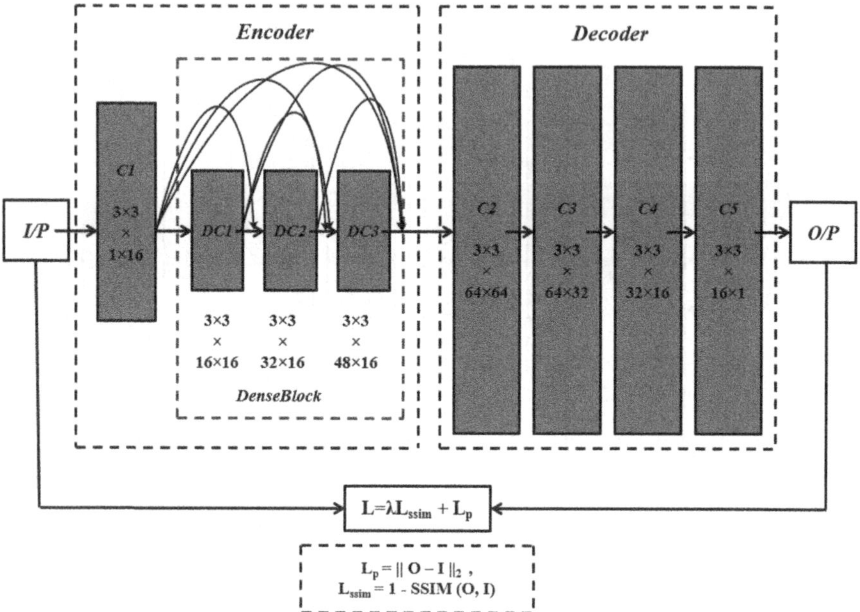

Fig. 9 The training process block diagram

Table 1 The training process architecture. Conv indicates the convolutional block (convolutional layer + activation)

	Layer	Size	Stride	Channel (input)	Channel (output)	Activation
Encoder	Conv(C1) DenseBlock	3	1	1	16	ReLu
Decoder	Conv(C2)	3	1	64	64	ReLu
	Conv(C3)	3	1	64	32	ReLu
	Conv(C4)	3	1	32	16	ReLu
	Conv(C5)	3	1	16	1	ReLu
DenseBlock	Conv(DC1)	3	1	16	16	ReLu
	Conv(DC2)	3	1	32	16	ReLu
	Conv(DC3)	3	1	48	16	ReLu

where input and output images are denoted by I and O, respectively. The Euclidean distance is calculated between the input (I) and output (O) from the above Eq. (2). Again, the L_{ssim} (SSIM loss) is calculated by Eq. (3),

$$L_{ssim} = 1 - SSIM(O, I) \tag{3}$$

where SSIM () indicates the structural similarity between two images and the value of λ is set to 10. The grayscale input images of the brain, breast, and prostate are collected from TCIA public access [36] for the fusion task accomplishing. Due to insufficient data of grayscale images in TCIA public access, MS-COCO [37] is used as input images for training the ANN and each image is resized to 224 × 224. The task is executed with NVIDIA Tesla K80 (Google colab) and using Keras (Tensorflow backend).

6.2 Fusion Strategy

Mean Strategy After the training phase, two or more input images are passed through the encoder as shown in Fig. 8. The mean fusion strategy combines the salient feature maps that are generated by encoder. The procedure of mean fusion strategy is shown in Fig. 10. In the ANN, the number of feature maps are represented by M where M=64 and $k \in \{1, 2, \ldots, M\}$. The feature map index obtained from the input image is indicated by $n \geq 2$. The mean strategy equation is represented by Eq.(4),

$$f^k(x, y) = \sum_{i=1}^{n} \phi_i^k(x, y)/2 \tag{4}$$

where f^k represents the feature maps after fusion, ϕ_i^k $(i = 1, \ldots, n)$ represents the feature maps acquired by the encoder and (x, y) indicated the corresponding feature map position. The decoder takes f^k as input for the reconstruction of the final fused image.

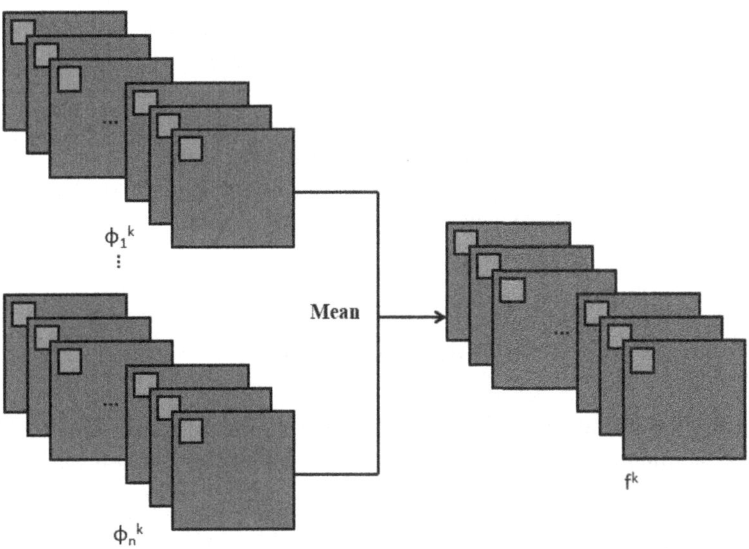

Fig. 10 The procedure of mean strategy

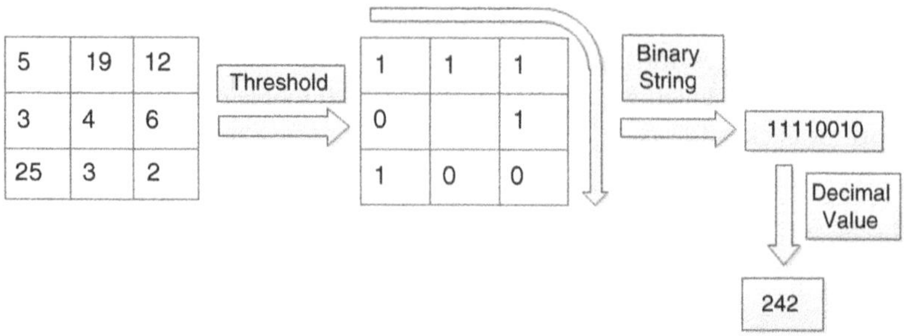

Fig. 11 LBP conversion from gray scale image to binary string and the corresponding decimal value

Average LBP Strategy *Local Binary Pattern* (LBP) is a standard feature descriptor for texture features. The LBP conversion from grayscale image to binary string and corresponding decimal value is shown in Fig. 11. The average LBP descriptor can be represented by,

$$LBP_{P,R} = \sum_{i=0}^{P-1} l\,(g_i - g_c)\,2^i/2 \tag{5}$$

where,

$$l(x) = \begin{cases} 1, & if\ x \geq 0 \\ 0, & otherwise \end{cases} \tag{6}$$

Here radius is denoted by R and the number of pixels in the neighborhood is denoted by P. The central pixel intensity is the threshold value that is to be compared with the neighboring pixels. The comparison starts with the top-left pixel and continues clockwise or anti-clockwise. If the neighborhood pixel intensity is larger than or equal to the central pixel intensity, the value is set to '1' or '0'. Therefore, a binary pattern is produced for all neighborhoods and the decimal value corresponding to the pattern is marked with the central pixel.

7 Experimental Results and Discussion

Training Phase Analysis The input images have been collected from MS-COCO [37] for the training of the DL architecture. We have used 79000 images as input image and 1000 images to validate the ability of reconstruction for the images in every iteration.

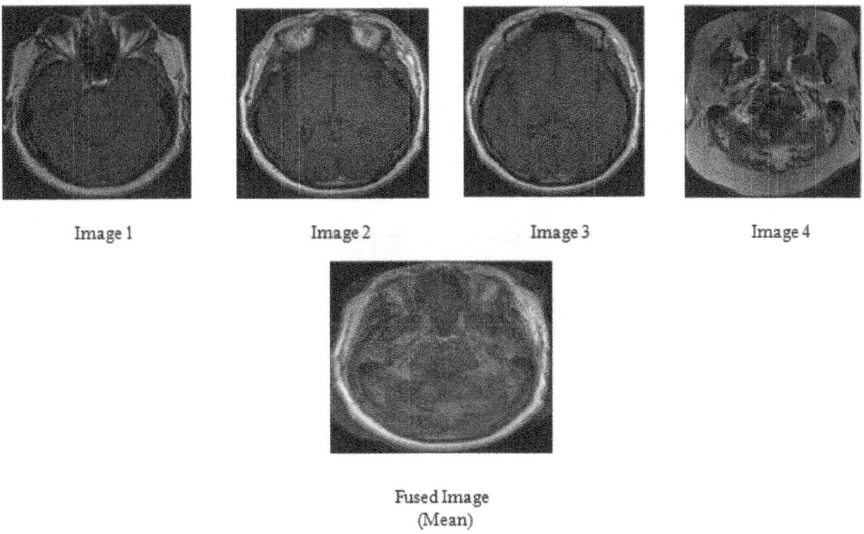

Fig. 12 Four source images of the human brain and the corresponding fused image (Mean)

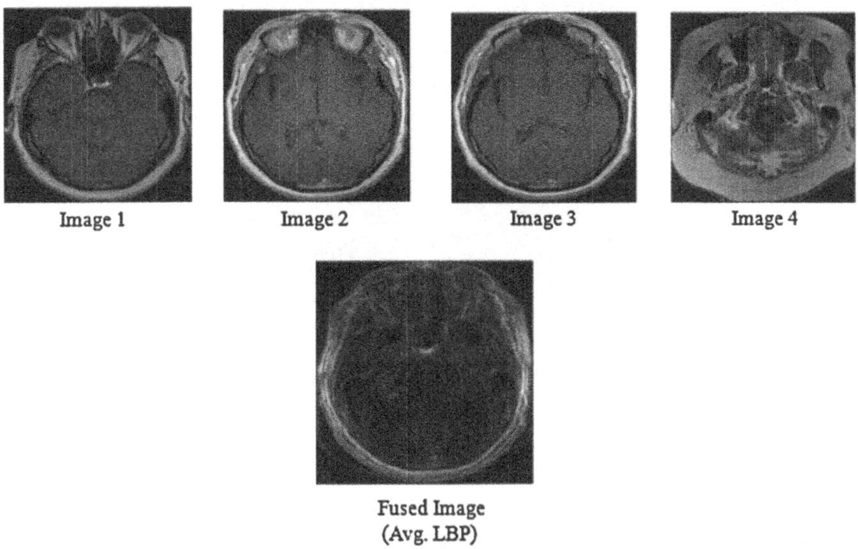

Fig. 13 Four source images of the human brain and the corresponding fused image (Avg. LBP)

Results of the Fusion Approach The experiment is carried out to validate the fusion technique for the gray-scale medical imaging. The result of the fusion approach for (1) brain; (2) breast and (3) prostate images are shown in Figs. 12, 13, 14, 15, 16, and 17.

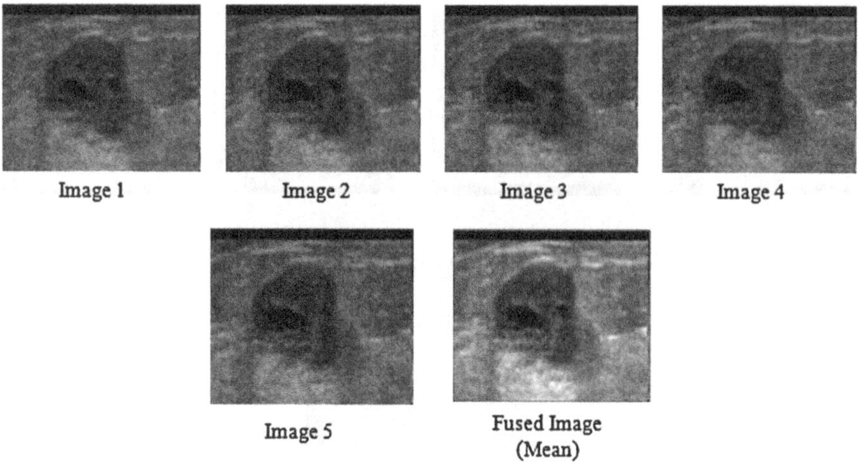

Fig. 14 Five source images of the human breast and the corresponding fused image (mean)

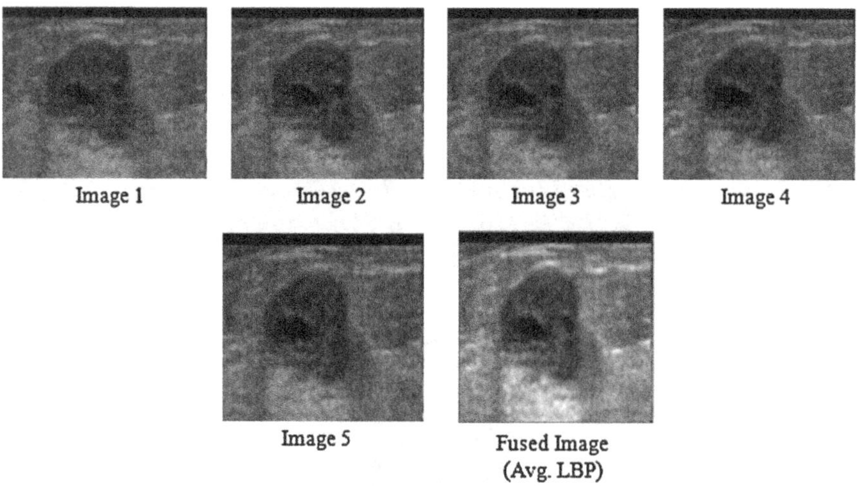

Fig. 15 Five source images of the human breast and the corresponding fused image (Avg. LBP)

Evaluation of the Results The entropy of all the source input images and their corresponding fused images are shown in Tables 2, 3, and 4. The fused images are statistically evaluated by a quality metrics i.e. entropy for the fusion strategy such as mean and average LBP. The entropy of the fused images obtained by the average LBP fusion strategy is less than the mean fusion strategy for all the human organs i.e. brain, breast, and prostate. The resultant fused images are noise-free and quite clear than every individual image as shown in Figs. 12, 13, 14, 15, 16, and 17.

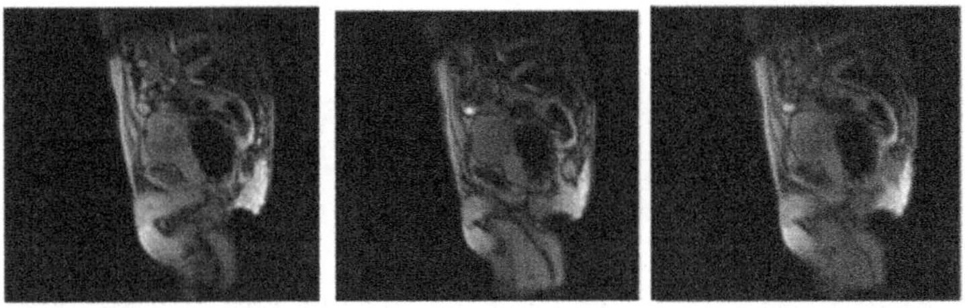

Fig. 16 Two source images of the prostate and the corresponding fused image (mean)

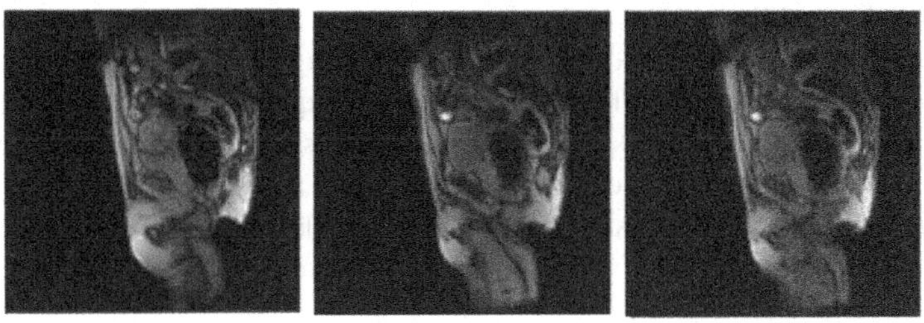

Fig. 17 Two source images of the prostate and the corresponding fused image (Avg. LBP)

Table 2 The entropy of the brain images and the fused image for the fusion strategy: mean; avg. LBP

	Brain			
	Image 1	Image 2	Image 3	Image 4
Strategy	3.64074	3.78477	3.67800	3.90555
Mean	3.90515			
Avg. LBP	3.55992			

Table 3 The entropy of the breast images and the fused image for the fusion strategy: mean; avg. LBP

	Breast				
	Image 1	Image 2	Image 3	Image 4	Image 5
Strategy	3.98946	4.00226	4.00487	4.01268	4.00116
Mean	3.99156				
Avg. LBP	3.97891				

8 Conclusion

The area of medical image analysis and problem diagnosis confronts various scientific and technical challenges such as blurring effects, artifact or noise, etc. With the advancement of technologies, the image representation quality and the imaging accuracies of medical imaging have been improved remarkably. Also, new technologies in medical imaging have explored and shown the advantages of

Table 4 The entropy of the prostate images and the fused image for the fusion strategy: mean; avg. LBP

	Prostate	
	Image 1	Image 2
Strategy	3.12712	3.10732
Mean	3.13902	
Avg. LBP	3.05960	

multimodal medical images. Although for a specific medical problem, choosing a proper combination of medical imaging modalities is very difficult and challenging. The frequently used approaches like wavelet transform and morphological methods are very successful for the analysis of medical images. Also, several new techniques like fuzzy logic, ML, and ANNs have shown improved image representation. This chapter has shown a detail description of the image fusion procedure and the medical image fusion explicitly. Also, the DL technique has come up with a global opportunity to the researchers for medical image fusion problems. Since different challenges are still open in medical image diagnosis and the medical image fusion techniques have shown to be successful, it is highly desired that more advancement of clinical problem diagnosis would appear in the coming days.

References

1. Bikash Meher, Sanjay Agrawal, Rutuparna Panda, and Ajith Abraham. A survey on region based image fusion methods. *Information Fusion*, 48:119–132, 2019.
2. Jiayi Ma, Wei Yu, Pengwei Liang, Chang Li, and Junjun Jiang. Fusiongan: A generative adversarial network for infrared and visible image fusion. *Information Fusion*, 48:11–26, 2019.
3. Renwei Dian, Shutao Li, Leyuan Fang, and Qi Wei. Multispectral and hyperspectral image fusion with spatial-spectral sparse representation. *Information Fusion*, 49:262–270, 2019.
4. Jian-sheng Guan, Shao-bo Kang, and Yuan Sun. Medical image fusion algorithm based on multi-resolution analysis coupling approximate spare representation. *Future Generation Computer Systems*, 98:201–207, 2019.
5. Huafeng Li, Xiaoge He, Dapeng Tao, Yuanyan Tang, and Ruxin Wang. Joint medical image fusion, denoising and enhancement via discriminative low-rank sparse dictionaries learning. *Pattern Recognition*, 79:130–146, 2018.
6. Wei Jiang, Xiaomin Yang, Wei Wu, Kai Liu, Awais Ahmad, Arun Kumar Sangaiah, and Gwanggil Jeon. Medical images fusion by using weighted least squares filter and sparse representation. *Computers & Electrical Engineering*, 67:252–266, 2018.
7. Pingping Zhang, Wei Liu, Yinjie Lei, and Huchuan Lu. Hyperfusion-net: Hyper-densely reflective feature fusion for salient object detection. *Pattern Recognition*, 93:521–533, 2019.
8. Shibo Gao, Yongmei Cheng, and Yongqiang Zhao. Method of visual and infrared fusion for moving object detection. *Optics letters*, 38(11):1981–1983, 2013.
9. Ju Han and Bir Bhanu. Fusion of color and infrared video for moving human detection. *Pattern Recognition*, 40(6):1771–1784, 2007.
10. I Ulusoy and H Yuruk. New method for the fusion of complementary information from infrared and visual images for object detection. *IET Image Processing*, 5(1):36–48, 2011.
11. Ning Sun, Qi Li, Ruizhi Huan, Jixin Liu, and Guang Han. Deep spatial-temporal feature fusion for facial expression recognition in static images. *Pattern Recognition Letters*, 2017.

12. Jingu Heo, Seong G Kong, Besma R Abidi, and Mongi A Abidi. Fusion of visual and thermal signatures with eyeglass removal for robust face recognition. In *2004 Conference on Computer Vision and Pattern Recognition Workshop*, pages 122–122. IEEE, 2004.
13. Harishwaran Hariharan, Andreas Koschan, Besma Abidi, Andrei Gribok, and Mongi Abidi. Fusion of visible and infrared images using empirical mode decomposition to improve face recognition. In *2006 International Conference on Image Processing*, pages 2049–2052. IEEE, 2006.
14. Seong G Kong, Jingu Heo, Besma R Abidi, Joonki Paik, and Mongi A Abidi. Recent advances in visual and infrared face recognition—a review. *Computer Vision and Image Understanding*, 97(1):103–135, 2005.
15. Feixiang He, Fayao Liu, Rui Yao, and Guosheng Lin. Local fusion networks with chained residual pooling for video action recognition. *Image and Vision Computing*, 81:34–41, 2019.
16. Saiyed Umer, Bibhas Chandra Dhara, and Bhabatosh Chanda. Face recognition using fusion of feature learning techniques. *Measurement*, 2019.
17. Hao Zhu, Henry Leung, and Ka-Veng Yuen. A joint data association, registration, and fusion approach for distributed tracking. *Information Sciences*, 324:186–196, 2015.
18. Isabelle Leang, Stéphane Herbin, Benoît Girard, and Jacques Droulez. On-line fusion of trackers for single-object tracking. *Pattern Recognition*, 74:459–473, 2018.
19. Mengjie Hu, Zhen Liu, Jingyu Zhang, and Guangjun Zhang. Robust object tracking via multi-cue fusion. *Signal Processing*, 139:86–95, 2017.
20. Marko Barjaktarović, Milica M Janković, Marija Jeremić, and Milovan Matović. Hybrid vision-fusion system for whole-body scintigraphy. *Computers in biology and medicine*, 96:69–78, 2018.
21. Songfeng Yin, Liangcai Cao, Yongshun Ling, and Guofan Jin. One color contrast enhanced infrared and visible image fusion method. *Infrared Physics & Technology*, 53(2):146–150, 2010.
22. Daniela Oppermann, Juergen Schramme, and Christa Neumeyer. Rod-cone based color vision in seals under photopic conditions. *Vision research*, 125:30–40, 2016.
23. Hongguang Li, Wenrui Ding, Xianbin Cao, and Chunlei Liu. Image registration and fusion of visible and infrared integrated camera for medium-altitude unmanned aerial vehicle remote sensing. *Remote Sensing*, 9(5):441, 2017.
24. Xia Chang, Licheng Jiao, Fang Liu, and Fangfang Xin. Multicontourlet-based adaptive fusion of infrared and visible remote sensing images. *IEEE Geoscience and Remote Sensing Letters*, 7(3):549–553, 2010.
25. Juan Gómez-Romero, Miguel A Serrano, Jesús García, José M Molina, and Galina Rogova. Context-based multi-level information fusion for harbor surveillance. *Information Fusion*, 21:173–186, 2015.
26. Kejiang Xiao, Rui Wang, Hua Deng, Lei Zhang, and Chunhua Yang. Energy-aware scheduling for information fusion in wireless sensor network surveillance. *Information Fusion*, 48:95–106, 2019.
27. Ciarán O'Conaire, Noel E O'Connor, Eddie Cooke, and Alan F Smeaton. Comparison of fusion methods for thermo-visual surveillance tracking. In *2006 9th International Conference on Information Fusion*, pages 1–7. IEEE, 2006.
28. Praveen Kumar, Ankush Mittal, and Padam Kumar. Fusion of thermal infrared and visible spectrum video for robust surveillance. In *Computer Vision, Graphics and Image Processing*, pages 528–539. Springer, 2006.
29. Giovanni Simone, Alfonso Farina, Francesco Carlo Morabito, Sebastiano B Serpico, and Lorenzo Bruzzone. Image fusion techniques for remote sensing applications. *Information fusion*, 3(1):3–15, 2002.
30. Xiaojun Xu, Youren Wang, and Shuai Chen. Medical image fusion using discrete fractional wavelet transform. *Biomedical signal processing and control*, 27:103–111, 2016.
31. Chengwei Luo, Bin Sun, Ke Yang, Taoran Lu, and Wei-Chang Yeh. Thermal infrared and visible sequences fusion tracking based on a hybrid tracking framework with adaptive weighting scheme. *Infrared Physics & Technology*, 99:265–276, 2019.

32. Zhengmeng Jin, Lihua Min, Michael K Ng, and Minling Zheng. Image colorization by fusion of color transfers based on dft and variance features. *Computers & Mathematics with Applications*, 77(9):2553–2567, 2019.
33. Peter Burt and Edward Adelson. The laplacian pyramid as a compact image code. *IEEE Transactions on communications*, 31(4):532–540, 1983.
34. Jetal Morlet, G Arens, E Fourgeau, and D Giard. Wave propagation and sampling theory—part II: Sampling theory and complex waves. *Geophysics*, 47(2):222–236, 1982.
35. Hui Li and Xiao-Jun Wu. Densefuse: A fusion approach to infrared and visible images. *IEEE Transactions on Image Processing*, 28(5):2614–2623, 2018.
36. Kenneth Clark, Bruce Vendt, Kirk Smith, John Freymann, Justin Kirby, Paul Koppel, Stephen Moore, Stanley Phillips, David Maffitt, Michael Pringle, et al. The cancer imaging archive (tcia): maintaining and operating a public information repository. *Journal of digital imaging*, 26(6):1045–1057, 2013.
37. Tsung-Yi Lin, Michael Maire, Serge Belongie, James Hays, Pietro Perona, Deva Ramanan, Piotr Dollár, and C Lawrence Zitnick. Microsoft coco: Common objects in context. In *European conference on computer vision*, pages 740–755. Springer, 2014.

Deep Learning for Histopathological Image Analysis

Cédric Wemmert, Jonathan Weber, Friedrich Feuerhake, and Germain Forestier

Abstract Anatomical Pathology dates back to the nineteenth century when Rudolf Virchow introduced his concept of cellular pathology and when the technical improvements of light microscopy enabled wide-spread use of structural criteria to define diseases. Since then, the quality of optical instruments has been constantly evolving. However the central element of the diagnostic process remains the knowledge and experience of pathologists visually classifying observations according to internationally agreed guidelines (e.g., *World Health Organisation* (WHO) classification), and much of the pre-analytical steps of specimen preparation (e.g., fixation, embedding, sectioning, staining) is only partially automated and still requires many manual steps. Thanks to the recent advent and cost-effectiveness of digital scanners, tissue histopathology slides can now be fully digitized and stored as *Whole Slide Images* (WSI). With the availability and analysis of a much larger set of variables combined with sophisticated imaging and analytic techniques, the traditional paradigm of pathology based on visually descriptive microscopy can be complemented and substantially improved by digital pathology, utilizing screen-based visualization of digital tissue sections and novel analysis tools potentially combining the conventional evaluation by pathologists with a computer-based diagnostic aid system. A central element of such evolving medical utilities and decision support systems will be image analysis, a field in which Deep Learning (DL) has recently made immense progress and especially the development of very large Artificial Neural Networks (ANNs) that are revolutionizing the field. Indeed, they have surpassed all existing image processing methods in most fields (segmentation, object detection, classification, etc.). All current methods

C. Wemmert (✉)
ICube, University of Strasbourg, Strasbourg, France
e-mail: wemmert@unistra.fr

J. Weber · G. Forestier
IRIMAS, University of Haute-Alsace, Mulhouse, France

F. Feuerhake
Hannover Medical School, Hannover, Germany

University Clinic of Freiburg, Freiburg, Germany

M. Elloumi (ed.), *Deep Learning for Biomedical Data Analysis*,
https://doi.org/10.1007/978-3-030-71676-9_7

applied to histopathological image analysis will be presented as well as the future technological issues and challenges of this discipline.

Keywords Machine learning · Digital pathology · Whole slide images · Deep learning

1 Introduction

Pathology is the gold standard for the diagnostic evaluation and the understanding of many of the underlying biological and pathophysiological mechanisms. This typically involves visual evaluation by pathologists of a sample of cells under a microscope, to identify structural tissue properties. Currently, visual evaluation of microscopic specimens is largely an unassisted process, and pathologist's accuracy is established through long training, standardization and benchmarking, quality control by peer reviews, and personal experience. But this field has experienced several technological revolutions in recent years with the advent of virtual microscopy (conversion of glass slides into high-resolution images i.e. digital slides), often referred to as *Digital Pathology* (DP). In this domain, important efforts have been made to design image analysis tools, to identify for example basic biological structures (e.g. stroma, immune cells), in order to facilitate the task of biologists for (semi-) automated digital slide interpretation. Digital Pathology is currently regarded as one of the most promising avenues of diagnostic medicine in order to achieve better and faster diagnosis, prognosis and prediction of important diseases. With the recent advent of *Whole-Slide Imaging* (WSI) i.e. the scanning of entire slides, the field of digital pathology produces daily a massive amount of images with related metadata (e.g. patient information, diagnosis, treatment).

At the same time, automatic image analysis algorithms have recently made extraordinary progress, particularly with the advent of the *Deep Learning* (DL) methods introduced by Lecun et al. [32]. Indeed, the performance of these methods has exploded in recent years, in particular allowing the detection, classification and segmentation of objects of interest in images with very high precision [21]. Although the technical progress holds great promise for digitization and improvements regarding standardization and increasing efficiency, the majority of institutes for pathology still prefer the conventional microscopy approach and only few hospitals are moving towards a totally digital pathology service [59]. Apart from the huge investment cost for the hospital and the difficulty of changing the daily practices of an entire department, there are also many challenges to overcome in order to integrate WSI in routine diagnostic workflows. Indeed, these images contain billion of pixels and are highly heterogeneous (Fig. 1), in terms of signal (acquisition devices, inter-centric variability [49]), and of semantics (e.g. multiple types of cells), causing practical difficulties for the processing/visualization by conventional algorithms. As pointed out in [28], the main difficulties to analyze histopathological images are their huge size, the lack of sufficient labeled data, the different levels of

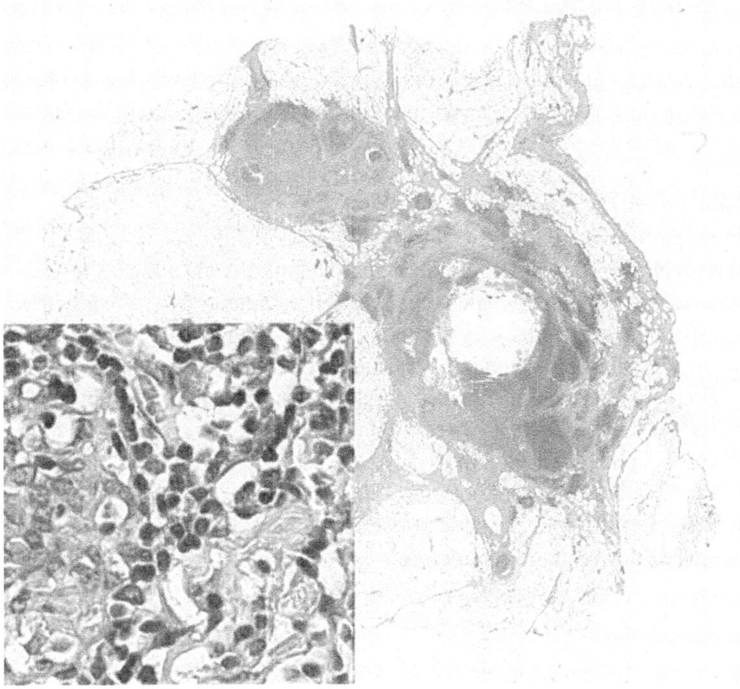

Fig. 1 Example of digitized WSI of a breast cancer surgical resection after previous diagnostic biopsy and subsequent neoadjuvant chemotherapy stained with Haematoxylin&Eosin ($18,000 \times 15,000$ pixels)

magnification resulting in different levels of details, the nature itself of WSI (which are more like texture images), the color variations and presence of artifacts.

The rest of the chapter is organized as follows: in Sect. 2, we first present current existing DL models for WSI analysis and explain their strength and limits.In Sect. 3, all the challenges and opportunities, induced by this emerging domain, are discussed. Finally in the last section, we conclude that even if DL has proven its efficiency in the field of histological images analysis, its broad use and acceptance in clinical routine remains challenging.

2 Current Deep Learning Models for Digital Pathology

Numerous DL approaches have been proposed in the last years [23, 36]. In this section, we present the most popular ones and analyze their strengths and weaknesses after having introduced what is DL and how it can be applied to solve two major tasks in histopathological image analysis: classification and segmentation.

2.1 What Is Deep Learning?

DL methods are a family of *Machine Learning* (ML) methods that are based on a representation of a model to be estimated in the form of a so-called *deep Artificial Neural Networks* (ANN). The concept of learning from an ANN is not new and dates back to the 1960s with Rosenblatt's definition [48] of the perceptron (Fig. 2). The learned model is a function of the form: $y = f(\sum_{i=1}^{n} \Phi_i x_i)$.

Subsequently, the 1980s saw the emergence of multilayer ANNs [38]. Each *neuron* is an elementary processing unit and the entire ANN allows a representation of a non-linear model. The learning is based on the *gradient backpropagation* method [50], which remains very computationally intensive (Fig. 3).

It is in the 2010s, following the democratization of *Graphics Processing Units* (GPU) cards allowing to have very important computing powers, that this technology got popular, with ANNs containing many layers and offering exceptional performances on many problems. In particular, in the field of image analysis (object detection, classification, segmentation, etc.), the so-called *Convolutional Neural Networks* (CNN) make it possible to achieve unprecedented levels of efficiency. An example of this type of architecture is given in Fig. 4.

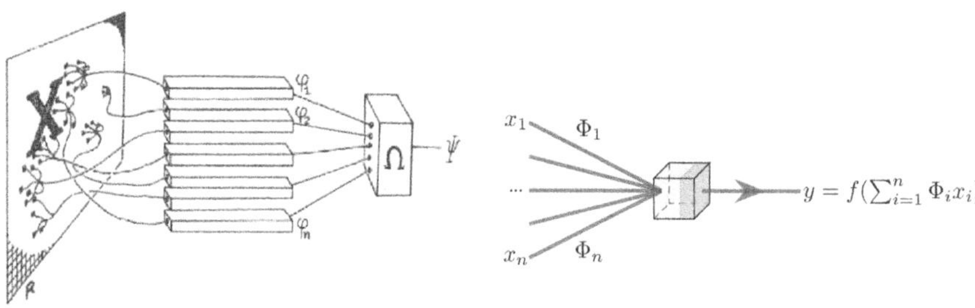

(a) Original schema given by Rosenblatt. (b) Perceptron definition

Fig. 2 The perceptron proposed by Rosenblatt in 1958. (**a**) Original schema given by Rosenblatt. (**b**) Perceptron definition

Fig. 3 A multilayer perceptron with multiple layers of neurons, p neurons in the input layer and K neurons in the output layer

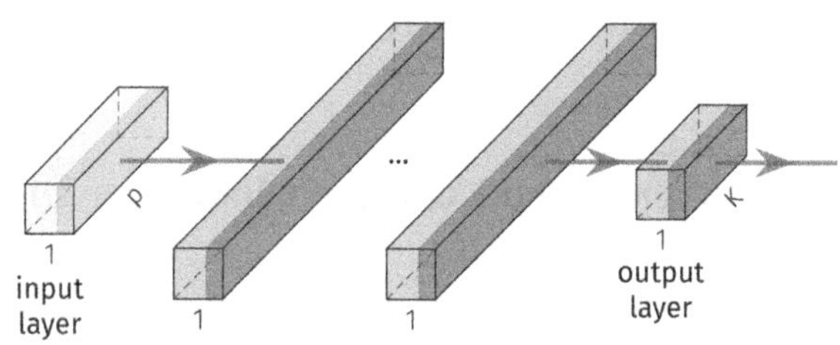

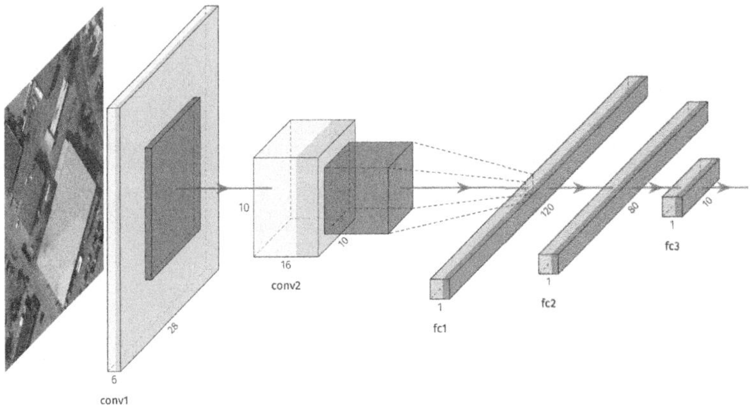

Fig. 4 Example of CNN, the *LeNet* architecture[33]

To train the model $f_\Theta(x) \rightarrow y$, the algorithm needs a training set composed of labeled samples: $(x_1, y_1), \ldots, (x_n, y_n)$. Then, the training of the ANN consists in an optimization problem of a loss function $\Delta(\hat{y}, y) \in \mathbb{R}^+$ as follows:

$$\Theta^* = \arg\min_{\Theta} \frac{1}{n} \sum \Delta(f_\Theta(x_i), y_i)$$

2.2 *Deep Learning for Classification*

The classification task is the classical task for ANNs, in digital pathology application it aims to label patches extracted from WSIs. For this purpose, CNNs are the most effective and most widely used ANNs today. They have a similar methodology to traditional supervised image classification methods: they receive input images, extract features of each of them, and then train a classifier (fully connected multilayer perceptron like in Fig. 3) on them. However, features are automatically learned by the ANN. Indeed, during the training phase, the classification error is minimized in order to optimize the classifier parameters and the features to be extracted from the image (in the form of convolutions applied at different levels of image resolution). The first architecture developed was the *LeNet* ANN (Fig. 4) by LeCun et al. [33]. Soon, more convolution layers were added to achieve deeper architectures, allowing better results like VGG16 proposed in [55] (Fig. 5).

This type of architecture has been widely applied to histopathological image analysis for different tasks mainly in cancer: detection of particular cells or regions of interest, tissue classification, scoring.

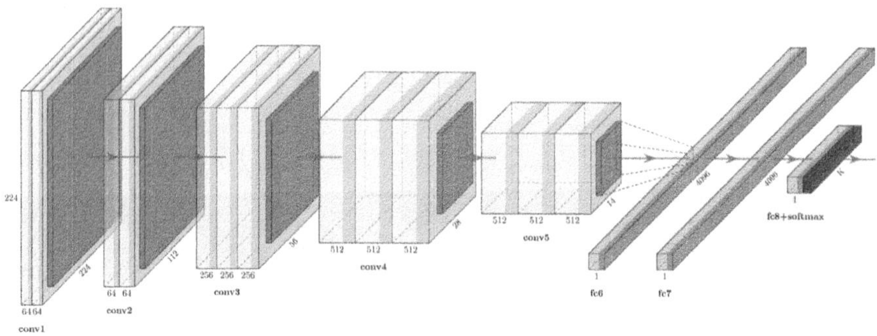

Fig. 5 Example of deep CNN for classification, the VGG16 architecture [55]

2.2.1 Detection

Mitosis detection is an important topic in cancer diagnostic, in [52] the authors proposed an interesting approach which uses both DL and hand-crafted features. The idea is to train a CNN, composed of five convolutional layers and two fully connected layers to classify mitosis image patches and non mitosis image patches and to combine it with 55 hand-crafted features plugged to the first fully connected layer. Adding hand-crafted features greatly improves the DL approach results.

In [12], the authors compared the efficiency of classical pre-defined features and features learned from a DL architecture, more precisely an autoencoder CNN [6], for *basal-cell carcinoma cancer* detection. The learned representations performs better than the the pre-defined and allows better cancer prediction.

2.2.2 Scoring

Tumor Proportional Scoring (TPS) plays an important role in the identification of *non-small-cell-lung-cancer*, as it represents the level of the *Programmed Death-Ligand 1* (PD-L1) expression [67]. In [26], the authors propose to use an *Auxiliary Classifier - Generative Adversarial Network* (AC-GAN) [42] which works as a classical *Generative Adversarial Network* (GAN) [19] but instead of just using noise as input for the generator also use hot-encoded desired class information. To be able to produce TPS score, discriminator not only indicates if the input patch is fake or real but also predict it class (positive tumor cell region or negative tumor cell region) which is further used to compute the TPS score. The performance of this network are good and allows to quickly obtain the TPS score in seconds.

In [53] used DL classification to estimate the spatial organization of immune cells in the tumor microenvironment. For this, they generate maps of *Tumor-Infiltrating Lymphocytes* (TILs) using CNNs enabling to evaluate a prognostic factors, like the Immunoscore [17], that quantify such spatial TILs densities in different tumor regions.

2.2.3 Tissue Classification

Gecer et al. [18] proposes a cascade architecture, first four fully CNNs aims to detect salient regions of interest which are then classified by a CNN into five different type of diagnostics.

Alternatively, in [11], the authors aims to classify tissue between *adenocarcinoma*, squamous cell *carcinoma* and normal tissues using *inception-v3* [62] ANN. Moreover, their ANN is able to predict six of the ten most commonly mutated genes in *adenocarcinoma* which could have a great impact on treatments.

In [70], the authors deal with the problem of not having enough annotated histological data for a specific tissue type. To tackle this problem, they propose to train a *GoogLeNet* architecture [63] on a different annotated dataset (with other tissue type) and then fine-tuning it on their dataset with less annotations.

Deep Multiple Instance Learning [45] is a rising topic in biomedical ML. The basic idea is to use multiple labels to annotate an area, with multiple examples, the network will learn which structure belongs to which labels. It is useful for WSI classification. It has been used for classification of breast histopathology [13] and precursor lesions of *esophageal adenocarcinoma* [66].

2.3 *Deep Learning for Segmentation*

While CNNs have proven their efficiency for classification tasks, by splitting images into small patches to be classified, specific architectures have quickly emerged to obtain more accurate detection of objects of interest with their outlines [4, 47].

These architectures consist of two parts (Fig. 6), one for encoding the information contained in the image given to the input layer (composed of a succession of encoder layers) and the other for decoding (composed of a succession of decoder layers). Each encoder layer applies convolution, batch normalization and non-linearity, then applies a max pooling on the result. Decoders are similar to encoders, but they

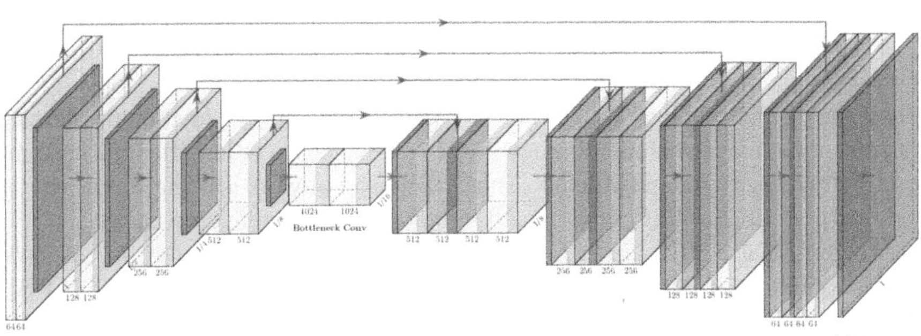

Fig. 6 Example of CNN for segmentation, the UNet architecture[47]

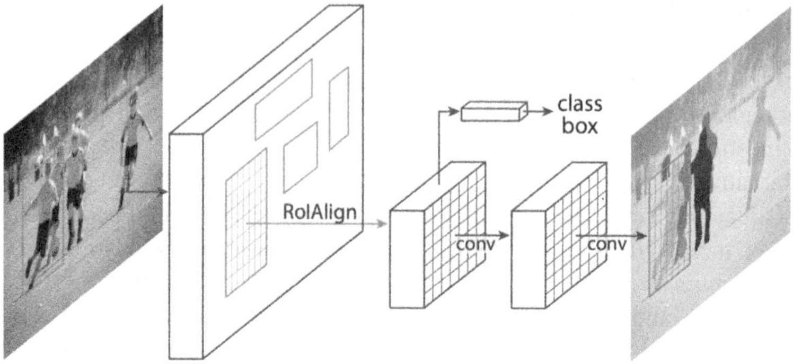

Fig. 7 The Mask-R-CNN architecture for instance segmentation [21]

oversample their input, using indices stored from the encoding step. After the final decoder, the output is sent to a classifier which gives the final prediction corresponding to the segmentation of the image given as input (each channel of the prediction corresponds to a class of objects to segment).

More recently, He et al. [21] proposed a novel approach (Mask-R-CNN) to detect objects in an image while simultaneously generating a high-quality segmentation mask for each instance. The architecture of this ANN is presented on Fig. 7. It is based on the Faster-R-CNN network proposed in [46]. The Mask-R-CNN network is composed of the classical R-CNN performing the classification and bounding box regression and of a branch for predicting segmentation masks on each instance extracted.

All these methods have been applied to histopathological data mainly for two types of applications: the segmentation of cell nuclei to go beyond an estimation of cell density by colorimetric deconvolution of images, and the segmentation of larger or composite objects of interest (tumor, glomeruli, lobules, etc.).

2.3.1 Cell Segmentation

The analysis of histopathological images and the resulting diagnosis are mainly done by quantifying the immune or cancerous cells present in the biopsy or by observing the morphology of the cells. To go beyond simply estimating the density of cells or to analyze the morphology of the cell nucleus, it is important to segment each cell precisely and individually. This is why a lot of work is being conducted on cell segmentation in WSI of any type (brightfield or immunofluorescence).

Naylor et al. [39] studied multiple DL segmentation methods on a same dataset (publicly available[1]). Three different approaches were compared: FCN [35], *PangNet* [43] and *DeconvNet* [41]. The authors proposed a post-processing of the

[1] https://peterjacknaylor.github.io/.

probability map which resulted in a F-score of around 0.8 for the individual cell segmentation.

Many recent work have been proposed based on new architectures dedicated to this specific problem [40, 51, 57], using adversarial ANN [3] or Mask-R-CNN [71].

2.3.2 Large Regions of Interest or Composite Objects Segmentation

The ANNs presented in the previous section have also been applied to larger objects segmentation. Indeed, they are very efficient at capturing the texture and shape of composite or complex objects, since they are based on convolutional layers at several levels of resolution (encoder part of the ANN).

For example, they were successfully applied many times in breast cancer for region of interest segmentation (stroma, tumor area, etc.) [29, 60, 72] or for objects segmentation like lobules [2] or glands [10]. Other organs or pathologies were also investigated like, for example, colon cancer [9, 24, 56, 64, 68], brain cancer [61, 73] or glomeruli segmentation in renal pathology [14, 25, 65].

3 Challenges and Opportunities

As seen before, artificial learning methods, notably ANNs, have demonstrated their interest in the field of computer vision and more specifically in the analysis of histopathological images. However, there remains many challenges to be overcome before seeing the emergence of tools in clinical routine that are sufficiently reliable, generic and whose diagnostic proposals are explicable. The main challenge, far more important than the definition of new learning or image analysis algorithms, concerns the data themselves. In fact, the majority of new methods that are emerging in computer vision and that could be applied to medical images come up against a problem of access to data, the creation of a sufficiently large learning set (annotations) and/or data quality (heterogeneity of data, etc.) in the majority of projects.

3.1 Annotations

Acquiring annotated data is difficult and time consuming in all ML project. While large databases of annotated simple objects (e.g. car, dog, car, bridge, etc.) are now available through public datasets like *ImageNet* [15] or COCO [34], large repositories of annotated biomedical images are still rare. This is probably due to the high degree of qualification required to discriminate between pathology from images compared to annotating a dog or a cat in an image.

This problem is particularly present in DL where ANN models generally require thousands of images to be trained efficiently. Furthermore, medical images are sensitive data, and they also requires a carefully data privacy policy to be able to share and distribute the images. In addition, expensive hardware is often required to capture the images compared to everyday objects that can be obtained using a simple smart-phone. However, despite these limitations, some current projects are now available and contain medical images, like *The Cancer Genome Atlas Program* (TCGA) [69].

However, the data are rarely annotated and only the raw images are available. Another point concerns the high degree of class unbalancing that can appear in medical image applications. For example, for many tasks, (e.g. mitosis detection), a negative label ("no mitosis") is much more frequent than the positive label ("mitosis"). This can disturb the training of an ANN that will naturally over-fit the most present class.

Multiple solutions are currently developed by the ML community to try to cope with these limitations. For example, data augmentation [44] which consists in creating synthetic data from real data is most of the time used in current DL model. Performing data augmentation allows to perform an explicit regularization of the ANN. An alternative to obtain more annotated data is to use crowdsourcing [20], which consists in using a population of annotators to perform annotations manually. It has been shown that the level of expertise of the annotators can be limited [20] and reliable annotation can be obtained by merging of combining multiple "weak" annotations of the same image (i.e. the power of the crowd [1]). Finally, domain adaptation or transfer learning [54] consists in using a pretrained ANN instead of using a brand new, randomly initialized model. This is particularly promising as it allows to avoid starting "from scratch" every time a new task has to be performed. It has been shown in [27] that results obtained using pretrained ANNs are quite competitive compared to newly trained ones.

Finally, recent results [8, 58] have demonstrated that models can be transferred and adapted to solve different, but related, tasks.

3.2 *Multiple Stainings*

An important part of digital pathology is the analysis of multiple digitized WSI from differently stained tissue sections. It is common practice to mount consecutive sections containing corresponding microscopic structures on glass slides, and to stain them differently to highlight specific tissue components. These multiple staining modalities result in very different images but include a significant amount of consistent image information. DL approaches have recently been proposed to analyze these images in order to automatically identify objects of interest for pathologists.

The analysis and integration of information from different stainings is usually performed with reference to a specific organ, structure, or pattern observed in the tissue. For example, to diagnose pathologies such as breast cancer or kidney allograft rejection it is necessary to study the inflammatory microenvironment of the organ. In these cases, the relevant information is the distribution of immune cells (e.g. macrophages or lymphocytes) in relation to important structures of the organ, such as glomeruli for the kidney, or lobules for the breast.

To automatically perform such an analysis, the structure of interest (glomeruli, lobules, etc.) should be detected in each section irrespective of the individual staining modality. In order to avoid having to annotate each staining, strategies have been proposed in [31] to build an ANN that is robust to color variations between the WSI.

Another source of heterogeneity comes from slides originating from different hospital or centers. Indeed, depending of the operator preparing the slide, section thickness, chemical formulations, lab protocols etc. two images, even using the same staining, can look very different [37]. In order to address this problem of inter-centric heterogeneity, several teams proposed approaches based on domain adaptation [16, 30].

3.3 Generative Adversarial Network

A *Generative Adversarial Network* (GAN) is a class of ML systems invented by Ian Goodfellow [19]. These models are composed of a generative ANN that generates candidates and a discriminate network that evaluates them. The generative ANN learns to map from a latent space to a data distribution, while the discriminative ANN distinguishes candidates produced by the generator from the true data. GAN is use extensively in computer vision and are starting to be used in DP too. For example, Neslihan et al. [5] proposed a method that uses dimension reduction and conditional adversarial generative ANNs to transform unstained hyperspectral tissue image to their *Haematoxylin&Eosin* (H&E) equivalent. The goal is to create a virtual digital H&E staining that could automate some of the tasks in the diagnostic pathology workflow.

Zanjani et al. [74] also explored how GAN could be used to normalize stainings. By replacing the latent representation of a source image with those extracted from a template image in the trained model, the proposed model can generate a new color copy of the source image while preserving the important tissue structures.

Alternatively, Burlingame et al. [7] proposed the SHIFT method that uses GAN to translate histopathological images to immunofluorescent images. This method has the potential to improve our understanding of the mapping of histological and morphological profiles into protein expression profiles. decision-making.

GAN can also be used to cope with the lack of annotated data (see Sect. 3.1). For example, Hou et al. [22] used GAN to synthesize histopathological images, in order to train supervised CNN with the generated data. The synthetic images are generated with the mask corresponding to the expected segmentation and adapted to the reference style. This enables to boost the performance of the trained CNN by using on-the-fly generated adversarial examples.

4 Conclusion

Thanks to the recent advent and increasing cost-effectiveness of digital scanners, tissue histopathology slides can now be completely digitized and stored as digital images. With the availability and analysis of a much larger set of variables combined with sophisticated imaging and analytic techniques, the traditional paradigm of pathology and microscopy could quickly be complemented and potentially also partially replaced by digital pathology, based on a screen-based visualization of digital tissue sections and an analysis combining a pathologist and a computer-based diagnostic aid system. In this chapter, we presented challenges and opportunities of using DL techniques to process these data. We discussed current DL models for digital pathology and discussed the problems related to the acquisition of annotations and the heterogeneity of the data (e.g. slides coming from multiples centers, different types of stainings, etc.). With the broad access to state-of-art of DL models and the availability of efficient computation systems, DL for digital pathology will continue to make progress in the coming years.

Acknowledgments This work was performed in the framework of SYSIMIT (FKZ:01ZX1308A) funded by German Ministry for Education and Research (BMBF) and the ERACoSysMed project SysMIFTA under the grant agreement No. 031L0085 (BMBF, Projektträger Jülich, Germany) and No. ANR-15-CMED-0004-03 (ANR, France). This work was also supported under the framework of the IdEx Unistra and benefits from a funding from the state managed by the French National Research Agency as part of the Investments for the future program.

References

1. Shadi Albarqouni, Christoph Baur, Felix Achilles, Vasileios Belagiannis, Stefanie Demirci, and Nassir Navab. AggNet: deep learning from crowds for mitosis detection in breast cancer histology images. *IEEE transactions on medical imaging*, 35(5):1313–1321, 2016.
2. Grégory Apou, Nadine S Schaadt, Benoît Naegel, Germain Forestier, Ralf Schönmeyer, Friedrich Feuerhake, Cédric Wemmert, and Anne Grote. Detection of lobular structures in normal breast tissue. *Computers in biology and medicine*, 74:91–102, 2016.
3. Assaf Arbelle and Tammy Riklin Raviv. Microscopy cell segmentation via adversarial neural networks. In *2018 IEEE 15th International Symposium on Biomedical Imaging (ISBI 2018)*, pages 645–648. IEEE, 2018.
4. Vijay Badrinarayanan, Alex Kendall, and Roberto Cipolla. Segnet: A deep convolutional encoder-decoder architecture for image segmentation. *IEEE transactions on pattern analysis and machine intelligence*, 39(12):2481–2495, 2017.
5. Neslihan Bayramoglu, Mika Kaakinen, Lauri Eklund, and Janne Heikkila. Towards virtual h&e staining of hyperspectral lung histology images using conditional generative adversarial networks. In *Proceedings of the IEEE International Conference on Computer Vision*, pages 64–71, 2017.

6. Y. Bengio, A. Courville, and P. Vincent. Representation learning: A review and new perspectives. *IEEE Transactions on Pattern Analysis and Machine Intelligence*, 35(8):1798–1828, Aug 2013.
7. Erik A Burlingame, Adam A Margolin, Joe W Gray, and Young Hwan Chang. Shift: speedy histopathological-to-immunofluorescent translation of whole slide images using conditional generative adversarial networks. In *Medical Imaging 2018: Digital Pathology*, volume 10581, page 1058105. International Society for Optics and Photonics, 2018.
8. Jongwon Chang, Jisang Yu, Taehwa Han, Hyuk-jae Chang, and Eunjeong Park. A method for classifying medical images using transfer learning: a pilot study on histopathology of breast cancer. In *2017 IEEE 19th International Conference on e-Health Networking, Applications and Services (Healthcom)*, pages 1–4. IEEE, 2017.
9. Hao Chen, Xiaojuan Qi, Lequan Yu, Qi Dou, Jing Qin, and Pheng-Ann Heng. Dcan: Deep contour-aware networks for object instance segmentation from histology images. *Medical image analysis*, 36:135–146, 2017.
10. Hao Chen, Xiaojuan Qi, Lequan Yu, and Pheng-Ann Heng. Dcan: deep contour-aware networks for accurate gland segmentation. In *Proceedings of the IEEE conference on Computer Vision and Pattern Recognition*, pages 2487–2496, 2016.
11. Nicolas Coudray, Paolo Santiago Ocampo, Theodore Sakellaropoulos, Navneet Narula, Matija Snuderl, David Fenyö, Andre L Moreira, Narges Razavian, and Aristotelis Tsirigos. Classification and mutation prediction from non–small cell lung cancer histopathology images using deep learning. *Nature medicine*, 24(10):1559, 2018.
12. Angel Alfonso Cruz-Roa, John Edison Arevalo Ovalle, Anant Madabhushi, and Fabio Augusto González Osorio. A deep learning architecture for image representation, visual interpretability and automated basal-cell carcinoma cancer detection. In Kensaku Mori, Ichiro Sakuma, Yoshinobu Sato, Christian Barillot, and Nassir Navab, editors, *Medical Image Computing and Computer-Assisted Intervention – MICCAI 2013*, pages 403–410, Berlin, Heidelberg, 2013. Springer Berlin Heidelberg.
13. Kausik Das, Sailesh Conjeti, Abhijit Guha Roy, Jyotirmoy Chatterjee, and Debdoot Sheet. Multiple instance learning of deep convolutional neural networks for breast histopathology whole slide classification. In *2018 IEEE 15th International Symposium on Biomedical Imaging (ISBI 2018)*, pages 578–581. IEEE, 2018.
14. Thomas de Bel, Meyke Hermsen, Bart Smeets, Luuk Hilbrands, Jeroen van der Laak, and Geert Litjens. Automatic segmentation of histopathological slides of renal tissue using deep learning. In *Medical Imaging 2018: Digital Pathology*, volume 10581, page 1058112. International Society for Optics and Photonics, 2018.
15. Jia Deng, Wei Dong, Richard Socher, Li-Jia Li, Kai Li, and Li Fei-Fei. Imagenet: A large-scale hierarchical image database. In *2009 IEEE conference on computer vision and pattern recognition*, pages 248–255. Ieee, 2009.
16. Michael Gadermayr, Martin Strauch, Barbara Mara Klinkhammer, Sonja Djudjaj, Peter Boor, and Dorit Merhof. Domain adaptive classification for compensating variability in histopathological whole slide images. In *International Conference on Image Analysis and Recognition*, pages 616–622. Springer, 2016.
17. Jérôme Galon, Anne Costes, Fatima Sanchez-Cabo, Amos Kirilovsky, Bernhard Mlecnik, Christine Lagorce-Pagès, Marie Tosolini, Matthieu Camus, Anne Berger, Philippe Wind, et al. Type, density, and location of immune cells within human colorectal tumors predict clinical outcome. *Science*, 313(5795):1960–1964, 2006.
18. Baris Gecer, Selim Aksoy, Ezgi Mercan, Linda G. Shapiro, Donald L. Weaver, and Joann G. Elmore. Detection and classification of cancer in whole slide breast histopathology images using deep convolutional networks. *Pattern Recognition*, 84:345–356, 2018.
19. Ian Goodfellow, Jean Pouget-Abadie, Mehdi Mirza, Bing Xu, David Warde-Farley, Sherjil Ozair, Aaron Courville, and Yoshua Bengio. Generative adversarial nets. In *Advances in neural information processing systems*, pages 2672–2680, 2014.

20. Anne Grote, Nadine S Schaadt, Germain Forestier, Cédric Wemmert, and Friedrich Feuerhake. Crowdsourcing of histological image labeling and object delineation by medical students. *IEEE transactions on medical imaging*, 2018.
21. Kaiming He, Georgia Gkioxari, Piotr Dollár, and Ross Girshick. Mask r-cnn. In *Proceedings of the IEEE international conference on computer vision*, pages 2961–2969, 2017.
22. Le Hou, Ayush Agarwal, Dimitris Samaras, Tahsin M Kurc, Rajarsi R Gupta, and Joel H Saltz. Unsupervised histopathology image synthesis. *arXiv preprint arXiv:1712.05021*, 2017.
23. Andrew Janowczyk and Anant Madabhushi. Deep learning for digital pathology image analysis: A comprehensive tutorial with selected use cases. *Journal of pathology informatics*, 7, 2016.
24. Philipp Kainz, Michael Pfeiffer, and Martin Urschler. Segmentation and classification of colon glands with deep convolutional neural networks and total variation regularization. *PeerJ*, 5:e3874, 2017.
25. Shruti Kannan, Laura A Morgan, Benjamin Liang, McKenzie G Cheung, Christopher Q Lin, Dan Mun, Ralph G Nader, Mostafa E Belghasem, Joel M Henderson, Jean M Francis, et al. Segmentation of glomeruli within trichrome images using deep learning. *Kidney International Reports*, 2019.
26. Ansh Kapil, Armin Meier, Aleksandra Zuraw, Keith E Steele, Marlon C Rebelatto, Günter Schmidt, and Nicolas Brieu. Deep semi supervised generative learning for automated tumor proportion scoring on nsclc tissue needle biopsies. *Scientific reports*, 8(1):17343, 2018.
27. Brady Kieffer, Morteza Babaie, Shivam Kalra, and Hamid R Tizhoosh. Convolutional neural networks for histopathology image classification: Training vs. using pre-trained networks. In *2017 Seventh International Conference on Image Processing Theory, Tools and Applications (IPTA)*, pages 1–6. IEEE, 2017.
28. Daisuke Komura and Shumpei Ishikawa. Machine learning methods for histopathological image analysis. *Computational and structural biotechnology journal*, 16:34–42, 2018.
29. V Kovalev, A Kalinovsky, and V Liauchuk. Deep learning in big image data: Histology image classification for breast cancer diagnosis. In *Big Data and Advanced Analytics, Proc. 2nd International Conference, BSUIR, Minsk*, pages 44–53, 2016.
30. Maxime W Lafarge, Josien PW Pluim, Koen AJ Eppenhof, Pim Moeskops, and Mitko Veta. Domain-adversarial neural networks to address the appearance variability of histopathology images. In *Deep Learning in Medical Image Analysis and Multimodal Learning for Clinical Decision Support*, pages 83–91. Springer, 2017.
31. Thomas Lampert, Odyssée Merveille, Jessica Schmitz, Germain Forestier, Friedrich Feuerhake, and Cédric Wemmert. Strategies for training stain invariant cnns. *arXiv preprint arXiv:1810.10338*, 2018.
32. Yann LeCun, Yoshua Bengio, and Geoffrey E. Hinton. Deep learning. *Nature*, 521(7553):436–444, 2015.
33. Yann LeCun, Léon Bottou, Yoshua Bengio, Patrick Haffner, et al. Gradient-based learning applied to document recognition. *Proceedings of the IEEE*, 86(11):2278–2324, 1998.
34. Tsung-Yi Lin, Michael Maire, Serge Belongie, James Hays, Pietro Perona, Deva Ramanan, Piotr Dollár, and C Lawrence Zitnick. Microsoft coco: Common objects in context. In *European conference on computer vision*, pages 740–755. Springer, 2014.
35. Jonathan Long, Evan Shelhamer, and Trevor Darrell. Fully convolutional networks for semantic segmentation. In *Proceedings of the IEEE conference on computer vision and pattern recognition*, pages 3431–3440, 2015.
36. Anant Madabhushi and George Lee. Image analysis and machine learning in digital pathology: Challenges and opportunities, 2016.
37. Raphaël Marée, Loïc Rollus, Benjamin Stévens, Renaud Hoyoux, Gilles Louppe, Rémy Vandaele, Jean-Michel Begon, Philipp Kainz, Pierre Geurts, and Louis Wehenkel. Collaborative analysis of multi-gigapixel imaging data using cytomine. *Bioinformatics*, 32(9):1395–1401, 2016.
38. James L McClelland, David E Rumelhart, PDP Research Group, et al. Parallel distributed processing. *Explorations in the Microstructure of Cognition*, 2:216–271, 1986.

39. Peter Naylor, Marick Laé, Fabien Reyal, and Thomas Walter. Nuclei segmentation in histopathology images using deep neural networks. In *2017 IEEE 14th International Symposium on Biomedical Imaging (ISBI 2017)*, pages 933–936. IEEE, 2017.
40. Peter Naylor, Marick Laé, Fabien Reyal, and Thomas Walter. Segmentation of nuclei in histopathology images by deep regression of the distance map. *IEEE transactions on medical imaging*, 38(2):448–459, 2018.
41. Hyeonwoo Noh, Seunghoon Hong, and Bohyung Han. Learning deconvolution network for semantic segmentation. In *Proceedings of the IEEE international conference on computer vision*, pages 1520–1528, 2015.
42. Augustus Odena, Christopher Olah, and Jonathon Shlens. Conditional image synthesis with auxiliary classifier gans. In *Proceedings of the 34th International Conference on Machine Learning-Volume 70*, pages 2642–2651. JMLR. org, 2017.
43. Baochuan Pang, Yi Zhang, Qianqing Chen, Zhifan Gao, Qinmu Peng, and Xinge You. Cell nucleus segmentation in color histopathological imagery using convolutional networks. In *2010 Chinese Conference on Pattern Recognition (CCPR)*, pages 1–5. IEEE, 2010.
44. Luis Perez and Jason Wang. The effectiveness of data augmentation in image classification using deep learning. *arXiv preprint arXiv:1712.04621*, 2017.
45. G. Quellec, G. Cazuguel, B. Cochener, and M. Lamard. Multiple-instance learning for medical image and video analysis. *IEEE Reviews in Biomedical Engineering*, 10:213–234, 2017.
46. Shaoqing Ren, Kaiming He, Ross Girshick, and Jian Sun. Faster r-cnn: Towards real-time object detection with region proposal networks. In *Advances in neural information processing systems*, pages 91–99, 2015.
47. Olaf Ronneberger, Philipp Fischer, and Thomas Brox. U-net: Convolutional networks for biomedical image segmentation. In *Medical Image Computing and Computer-Assisted Intervention - MICCAI 2015 – 18th International Conference Munich, Germany, October 5 – 9, 2015, Proceedings, Part III*, pages 234–241, 2015.
48. Frank Rosenblatt. The perceptron: a probabilistic model for information storage and organization in the brain. *Psychological review*, 65(6):386, 1958.
49. Marie-Christine Rousselet, Sophie Michalak, Florence Dupré, Anne Croué, Pierre Bedossa, Jean-Paul Saint-André, and Paul Calès. Sources of variability in histological scoring of chronic viral hepatitis. *Hepatology*, 41(2):257–264, 2005.
50. David E Rumelhart, Geoffrey E Hinton, and Ronald J Williams. Learning internal representations by error propagation. Technical report, California Univ San Diego La Jolla Inst for Cognitive Science, 1985.
51. Monjoy Saha and Chandan Chakraborty. Her2net: A deep framework for semantic segmentation and classification of cell membranes and nuclei in breast cancer evaluation. *IEEE Transactions on Image Processing*, 27(5):2189–2200, 2018.
52. Monjoy Saha, Chandan Chakraborty, and Daniel Racoceanu. Efficient deep learning model for mitosis detection using breast histopathology images. *Computerized Medical Imaging and Graphics*, 64:29–40, 2018.
53. Joel Saltz, Rajarsi Gupta, Le Hou, Tahsin Kurc, Pankaj Singh, Vu Nguyen, Dimitris Samaras, Kenneth R Shroyer, Tianhao Zhao, Rebecca Batiste, et al. Spatial organization and molecular correlation of tumor-infiltrating lymphocytes using deep learning on pathology images. *Cell reports*, 23(1):181–193, 2018.
54. Hoo-Chang Shin, Holger R Roth, Mingchen Gao, Le Lu, Ziyue Xu, Isabella Nogues, Jianhua Yao, Daniel Mollura, and Ronald M Summers. Deep convolutional neural networks for computer-aided detection: Cnn architectures, dataset characteristics and transfer learning. *IEEE transactions on medical imaging*, 35(5):1285–1298, 2016.
55. Karen Simonyan and Andrew Zisserman. Very deep convolutional networks for large-scale image recognition. *arXiv preprint arXiv:1409.1556*, 2014.
56. Korsuk Sirinukunwattana, Josien PW Pluim, Hao Chen, Xiaojuan Qi, Pheng-Ann Heng, Yun Bo Guo, Li Yang Wang, Bogdan J Matuszewski, Elia Bruni, Urko Sanchez, et al. Gland segmentation in colon histology images: The glas challenge contest. *Medical image analysis*, 35:489–502, 2017.

57. Jie Song, Liang Xiao, Mohsen Molaei, and Zhichao Lian. Multi-layer boosting sparse convolutional model for generalized nuclear segmentation from histopathology images. *Knowledge-Based Systems*, 176:40–53, 2019.
58. Yang Song, Ju Jia Zou, Hang Chang, and Weidong Cai. Adapting fisher vectors for histopathology image classification. In *2017 IEEE 14th International Symposium on Biomedical Imaging (ISBI 2017)*, pages 600–603. IEEE, 2017.
59. Nikolas Stathonikos, Mitko Veta, André Huisman, and Paul J van Diest. Going fully digital: Perspective of a dutch academic pathology lab. *Journal of pathology informatics*, 4, 2013.
60. Hai Su, Fujun Liu, Yuanpu Xie, Fuyong Xing, Sreenivasan Meyyappan, and Lin Yang. Region segmentation in histopathological breast cancer images using deep convolutional neural network. In *2015 IEEE 12th International Symposium on Biomedical Imaging (ISBI)*, pages 55–58. IEEE, 2015.
61. Z Swiderska-Chadaj, T Markiewicz, J Gallego, G Bueno, B Grala, and M Lorent. Deep learning for damaged tissue detection and segmentation in ki-67 brain tumor specimens based on the u-net model. *Bulletin of the Polish Academy of Sciences. Technical Sciences*, 66(6), 2018.
62. C. Szegedy, V. Vanhoucke, S. Ioffe, J. Shlens, and Z. Wojna. Rethinking the inception architecture for computer vision. In *2016 IEEE Conference on Computer Vision and Pattern Recognition (CVPR)*, pages 2818–2826, June 2016.
63. Christian Szegedy, Wei Liu, Yangqing Jia, Pierre Sermanet, Scott Reed, Dragomir Anguelov, Dumitru Erhan, Vincent Vanhoucke, and Andrew Rabinovich. Going deeper with convolutions. In *Proceedings of the IEEE conference on computer vision and pattern recognition*, pages 1–9, 2015.
64. Jing Tang, Jun Li, and Xiangping Xu. Segnet-based gland segmentation from colon cancer histology images. In *2018 33rd Youth Academic Annual Conference of Chinese Association of Automation (YAC)*, pages 1078–1082. IEEE, 2018.
65. Maja Temerinac-Ott, Germain Forestier, Jessica Schmitz, Meyke Hermsen, JH Bräsen, Friedrich Feuerhake, and Cédric Wemmert. Detection of glomeruli in renal pathology by mutual comparison of multiple staining modalities. In *Proceedings of the 10th International Symposium on Image and Signal Processing and Analysis*, pages 19–24. IEEE, 2017.
66. Jakub M Tomczak, Maximilian Ilse, Max Welling, Marnix Jansen, Helen G Coleman, Marit Lucas, Kikki de Laat, Martijn de Bruin, Henk Marquering, Myrtle J van der Wel, et al. Histopathological classification of precursor lesions of esophageal adenocarcinoma: A deep multiple instance learning approach. In *MIDL 2018*, 2018.
67. Margarita Udall, Maria Rizzo, Juliet Kenny, Jim Doherty, SueAnn Dahm, Paul Robbins, and Eric Faulkner. Pd-l1 diagnostic tests: a systematic literature review of scoring algorithms and test-validation metrics. *Diagnostic pathology*, 13(1):12, 2018.
68. Yves-Rémi Van Eycke, Cédric Balsat, Laurine Verset, Olivier Debeir, Isabelle Salmon, and Christine Decaestecker. Segmentation of glandular epithelium in colorectal tumours to automatically compartmentalise ihc biomarker quantification: A deep learning approach. *Medical image analysis*, 49:35–45, 2018.
69. John N Weinstein, Eric A Collisson, Gordon B Mills, Kenna R Mills Shaw, Brad A Ozenberger, Kyle Ellrott, Ilya Shmulevich, Chris Sander, Joshua M Stuart, Cancer Genome Atlas Research Network, et al. The cancer genome atlas pan-cancer analysis project. *Nature genetics*, 45(10):1113, 2013.
70. Tian Xia, Ashnil Kumar, Dagan Feng, and Jinman Kim. Patch-level tumor classification in digital histopathology images with domain adapted deep learning. In *2018 40th Annual International Conference of the IEEE Engineering in Medicine and Biology Society (EMBC)*, pages 644–647. IEEE, 2018.
71. Xinpeng Xie, Yuexiang Li, Menglu Zhang, and Linlin Shen. Robust segmentation of nucleus in histopathology images via mask r-cnn. In *International MICCAI Brainlesion Workshop*, pages 428–436. Springer, 2018.

72. Jun Xu, Xiaofei Luo, Guanhao Wang, Hannah Gilmore, and Anant Madabhushi. A deep convolutional neural network for segmenting and classifying epithelial and stromal regions in histopathological images. *Neurocomputing*, 191:214–223, 2016.
73. Yan Xu, Zhipeng Jia, Yuqing Ai, Fang Zhang, Maode Lai, I Eric, and Chao Chang. Deep convolutional activation features for large scale brain tumor histopathology image classification and segmentation. In *2015 IEEE international conference on acoustics, speech and signal processing (ICASSP)*, pages 947–951. IEEE, 2015.
74. Farhad Ghazvinian Zanjani, Svitlana Zinger, Babak Ehteshami Bejnordi, Jeroen AWM van der Laak, and Peter HN de With. Stain normalization of histopathology images using generative adversarial networks. In *2018 IEEE 15th International Symposium on Biomedical Imaging (ISBI 2018)*, pages 573–577. IEEE, 2018.

Innovative Deep Learning Approach for Biomedical Data Instantiation and Visualization

Ryad Zemouri and Daniel Racoceanu

Abstract One of the main problems that most of biomedical applications face, is represented by the massive amount of unlabeled data. Manually analyzing and classifying massive database by human expert is mostly unfeasible, being—in certain limited conditions (still, extremely time-consuming)—partially been done, only for simple signatures, easily recognizable by an expert. Concerning this aspect, medical experts face two challenging problems: how to select the most significant data for labeling, and what is the minimum size of the data set—but sufficient to define each pathology—to perform the training of the classifier. In this chapter, we propose a new method, based on a visual data analysis, to build an efficient classifier with a minimum of labeled data. An encoder, part of a *Convolutional Variational Autoencoder* (CVAE), is used as a data projection for a 2D-visualization. The input vectors are encoded into a 2D-latent space, which helps the expert to visually analyze the spatial distribution of the training data set.

Keywords Data visualisation · Deep learning · Variational autoencoders · Breast cancer diagnosis

1 Introduction

Artificial Neural Networks (ANNs) and *Deep Learning* (DL) are actually the leading *Machine Learning* (ML) tools in biomedical fields, as reported by recent survey publications [2, 7, 22, 29, 30, 34, 46, 48]. With the technological and scientific advances, the biomedical data used by the medical practitioners are very heteroge-

R. Zemouri (✉)
CEDRIC Laboratory of the Conservatoire National des Arts et Metiers (CNAM), HESAM University, Paris, France
e-mail: ryad.zemouri@cnam.fr

D. Racoceanu
Sorbonne University and Paris Brain Institute, Paris, France
e-mail: daniel.racoceanu@sorbonne-universite.fr

M. Elloumi (ed.), *Deep Learning for Biomedical Data Analysis*,
https://doi.org/10.1007/978-3-030-71676-9_8

neous, such as a wide range of clinical analyses, biological parameters and medical imaging modalities. By the multitude of these data as well as the completeness of certain atypical diseases, biomedical data are usually imbalanced [20, 42] and nonstationary [11], being characterized by a high complexity [20]. In this context, ML represents a tremendous opportunity: (1) to support physicians, biologists and medical authorities to exploit and significantly improve big medical data analysis; (2) to reduce the risk of medical errors; and (3) to generate a better harmonization of the diagnosis and prognosis protocols.

The applications of the DL in the biomedical fields cover all the medical levels, i.e. from the genomic applications to the public medical health management, and are structured according to three main orientations [48]:

- *Computer-Aided Diagnosis*: to help the physicians for an efficient and early diagnosis, with a better harmonization and less contradictory diagnosis;
- *Patients Medical care*: to enhance the medical care of patients with better personalized therapies; and
- *Human wellbeing*: to improve the human wellbeing, for example by analyzing the spread of disease and social behaviors in relation with environmental factors, or to implement a brain–machine interface for controlling a wheelchair [1].

One of the main problems that most of biomedical applications face, is represented by the massive amount of unlabeled data. Manually analyzing and classifying massive database by a human expert is mostly unfeasible, being—in certain limited condition (still, extremely time-consuming)—partially been done, only for simple signatures, easily recognizable by the expert. Concerning this aspect, medical experts face two challenging problems: how to select the most significant data for labeling, and what is the minimum size of the data set—but sufficient to define each pathology—to perform the training of the classifier. In this chapter, we propose a new method, based on a visual data analysis, to build an efficient classifier with a minimum of labeled data. An encoder, part of a *Convolutional Variational Autoencoder* (CVAE) [25, 49, 50], is used as a data projection for a 2D-visualization. The input vectors are encoded into a 2D-latent space, which helps the expert to visually analyze the spatial distribution of the training data set.

The rest of the chapter is organized as follows: In Sect. 2, we give a brief introduction to the ANNs with a particular focus to some of the weaknesses usually encountered. Then, Sect. 3 presents some of the emerging architectures that have recently find a great success in the biomedical applications. Section 4 is dedicated to the *Variational Autoencoder* (VAE) [23, 24, 49] applied to data visualization and data analysis. An innovative DL approach for biomedical data instantiation and visualization is then presented. In Sect. 5, we give some results around a practical case study which is the *Breast Cancer Wisconsin dataset*, available by anonymous ftp from ice.uci.edu [12]. Finally, critical discussion and open challenges are given in Sect. 6.

2 Deep Neural Networks: A Brief Introduction

In this section, we first introduce a brief history of ANNs and present the main concepts of *Deep Neural Networks* (DNNs). Then, we develop the two major weaknesses of ANNs. The first one is the difficulty to find the best neural structure while the second is the lack of interpretability of the obtained results.

2.1 From Shallow to Deep Neural Networks

ANNs were inspired—in the 1960s—by biological neural networks in the brain. The feed forward ANNs are composed by layers of interconnected units (*neurons*). The mathematical point of view of ANNs consists of a non-linear transformation $y = F(x)$ of the input x (Fig. 1a). Compared to shallow architectures, ANNs with more hidden layers, called DNNs [37], offer much higher capacity to learn fitting and feature extracting from high complexity input data (Fig. 1b). The starting point of DL was in 2006, with the greedy layer-wise unsupervised learning algorithm used for *Deep Belief Networks* (DBNs) [5, 21].

The interconnection between two units or neurons, has an associated connection weight w_{ji}, which is fitted during the learning phase. The input data are propagated from the input layer, neuron after neuron, until the output layer. This propagation will transform these data from a given space to another one, by the neurons of the layers, in a nonlinear way. Each neuron computes a weighted sum of its inputs and applies a nonlinear activation function to calculate its output $f(x)$ (Fig. 1c). The most used activation functions are the *sigmoid function* [37] and its variant the *hyperbolic tangent function* for the *shallow architectures*, the *Rectified Linear Unit function* (ReLU) and its variant the *softplus function* for the deep architectures, and the *softmax function* commonly used for the final layer in classification tasks.

The two main applications of the ANNs are $classification$ and $regression$. The objective of the classification is to organize the input data space into several classes by supervised or unsupervised learning techniques. In the regression applications or function approximation, the objective is to predict an unknown output parameter, usually by supervised learning.

In supervised learning, the predicted label is compared with the true label, for the current set of model weights θ_w, to compute the output error (also called *loss function* $L(\theta_w)$) (Fig. 2a). The loss function is high-dimensional and non-convex, with many local optimums. The learning phase consists of tuning the connection weights at each learning step, in order to minimize $L(\theta_w)$, by backward propagating the gradient of the loss function, through the network. This back propagation gradient was the renew of the ANNs in the mid of the 80s, when the *Back Propagation* (BP) algorithm was used for classification [36]. During the learning procedure, two sets of data are usually used: training test and test set

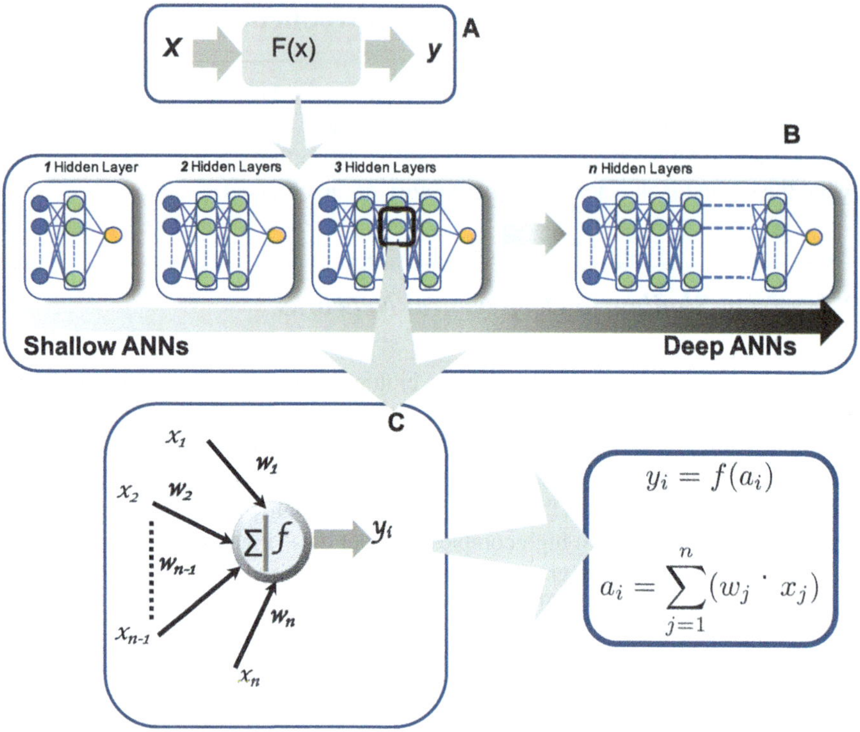

Fig. 1 ANNs [48]

(sometime a third set is used for validation). The training set is used for the learning while the test set is used for the ANNs performances evaluation. An efficient learning algorithm is to converge towards a global optimum while avoiding all the local optimums of the loss function, which looks like a landscape, with many hills and valleys. A learning rate η is used to jump over valleys at the beginning of the training and fine-tune the weights in later stages of the learning process. If the learning rate is too low (little jump), it may take forever to converge with a high risk of jamming in a local optimum. Conversely, a too high value (big jump) can cause a non-convergence of the learning algorithm (Fig. 2b). Varying and adapting the learning rate during the training process, produces better template update.

When deep architectures are used, the magnitude of the back propagated error derivative decreases rapidly along the layers, resulting in slight update of the weights in the first layers (Fig. 2c).

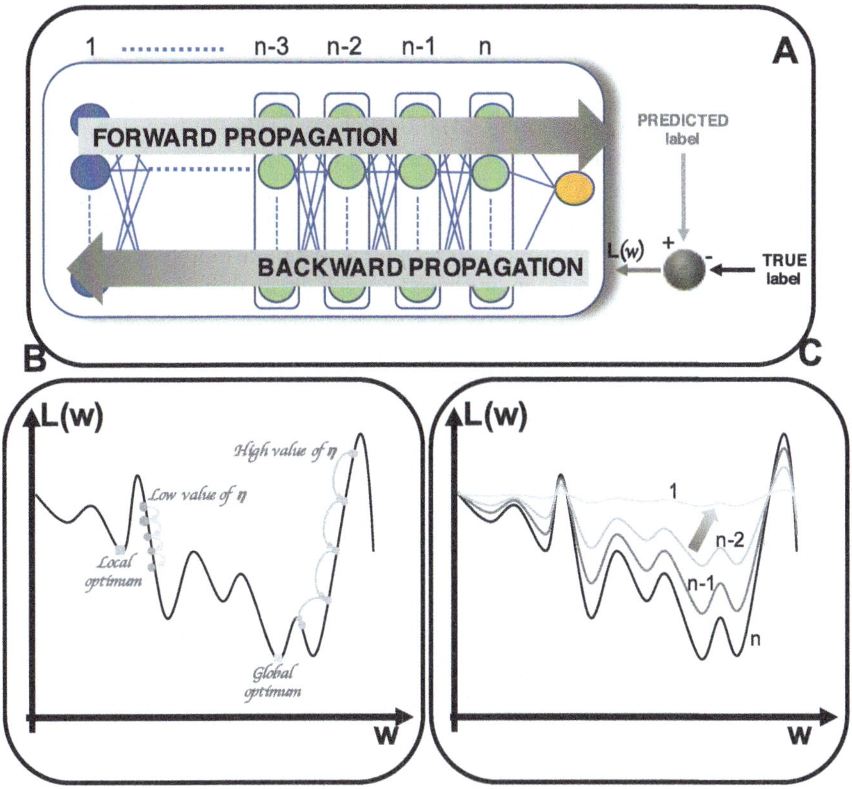

Fig. 2 The Loss function and the backward propagation of its gradient. (**a**) The predicted label is compared with the true label for the current set of model weights θ_w to compute the output error (also called *loss function*) $L(\theta_w)$. (**b**) The learning rate η is used to jump over valleys. If the learning rate is too low (little jump), it may take forever to converge with a high risk of jamming in a local optimum. Conversely, a too high value (big jump) can cause a non-convergence of the learning algorithm. (**c**) For deep architectures, the magnitude of the back propagated error derivative decreases rapidly along the layers, resulting in slight update of the weights in the first layers [48]

This drawback was partially solved by using the ReLU and the *softplus* activation function which allow faster learning and superior performances compared to the conventional activation function (e.g. *sigmoid* or *hyperbolic tangent*). Other solution is to consider the learning rate as a hyper parameter where different learning rates are used for different layers. However, few works in the literature use this concept (see [8] for a review). The most popular method used to create deep architectures and solve the problem of the random initialization of the weight parameters, consists in an unsupervised pre-training phase used before the supervised fined-tuned learning phase. *Auto-Encoders* (AEs) and *Restricted Boltzmann Machine* (RBM) are stacking in a layer-wise as the basic building blocks [37].

2.2 *Some Weaknesses*

The different steps to follow when using an ANN are: (1) model choosing, (2) model building, (3) model learning, (4) model checking. In the first step, we must choose one neural architecture (CNN, AE, DBN, . . .). In the second step, we have to define the size of the ANN: how many layers, how many units per layer, how many convolution filters and what is their size. In the third step, the ANN will be trained by unsupervised or supervised techniques while avoiding over-fitting and under-fitting. During the last step, we have to check the quality of the ANN.

2.2.1 Finding the Best Neural Structure

The main difficulty with the ANNs is the model building. What are the criteria that define the number of hidden layers and neurons per layer? The user often proceeds by trying several ANN topologies to find the best structure and try to avoid the oversized and undersized structure.

> Finding the best ANN architecture for a given problem is still challenging, especially for the DNN learning, which remains an active research area.

This is a real and computationally expensive drawback, especially when deep architectures are used. There is no guarantee that the selected number of hidden layers/units is optimal. To prevent the network from over-training (usually caused by the oversized design), some regularization techniques are used, as the *dropout* [39], *Maxout* [17] or the *weight decay* [32], which is a penalty added to the error function.

Evolutionary learning procedures give also interesting solutions where the ANN evolved gradually during the training procedure until an optimum structure that satisfy some evolutionary criteria. These adaptive ANNs are divided into three categories: *constructive* or *growing* algorithms [47], *pruning* algorithms [14] and *hybrid* methods [18]. A good alternative to this drawback is to consider the neural architecture as a hyper-parameter evolving during the learning process. The ANN is built step by step during the learning process, until convergence. To avoid an oversized architecture, some of the parameters as non-significant units or connections between neurons can be removed. Recently, several promising studies about the constructive and pruning algorithms were published (see [33, 47] for a complete survey).

2.2.2 The Interpretability of the Obtained Results

The other difficulty is the interpretability of the obtained results. It is very hard to understand what happen in the hidden layers and why a trained ANN gives a positive diagnosis for a certain pathology.

> ANNs learn to associate an output according to a given input, but they do not learn to give any reason or interpretation associated to this response.

This black-box aspect is very restrictive, specially in medicine, where a decision interpretability is very important and can have serious legal consequences [27]. When the convolution networks are used for image processing, several methods have been developed to visualize what happen in the intermediate layers. Some of these algorithms are, for example, visual explanations from CNN networks via gradient-based localization [38]; a visualization technique of the input stimuli that excite individual feature maps at any layer in a CNN model [44] or; a *deep Taylor decomposition method* [31] for interpreting generic multilayer ANNs by decomposing the network classification decision into contributions of its input elements [31]. When the input data are not images, the interpretability of the hidden layers activities is less obvious. Some visualization techniques, as the *t-Distributed stochastic Neighbor Embedding projection* (t-SNE) [28], converts a high-dimensional data set into a 2D-matrix of pairwise similarities. Feature maps of the model are then obtained; all the difficulty is to explain the classification decisions, according to these maps.

3 Emergent Architectures: Generative Networks

One of the most emerging architectures used in the biomedical applications are the *Generative Networks* (GNs). The GNs provide a way of data augmentation to enlarge the deep representations without extensively annotated training data [9]. Two kinds of GNs exist:

- *Generative Adversarial Networks* (GANs) [16, 45]
- *Variational Auto-Encoders* (VAEs) [23, 24, 49].

3.1 The Generative Adversarial Networks

Proposed in 2014 by Goodfellow [16], a GAN includes two DNNs: a generator and a discriminator. The first network is seen in a common analogy, as an art forger that creates forgeries with the aim of making realistic images. The discriminator represents the "art expert" which should distinguish between the synthetic and the authentic images. The training of a GAN requires both finding the parameters of a discriminator and a generator, the discriminator having to maximize its classification accuracy and the generator having to confuse the discriminator as much as possible [9, 45]. During the training process, only one of the two networks is concerned by the parameters updating. The second one keeps its own parameters, frozen.

The GANs were recently applied in all the biomedical fields such as in Omics for a protein modeling [26], where the loop modeling is seen as the image in-painting problem and the generative network has to capture the context of the loop region with a prediction of the missing area.

In the *Brain and Body Machine Interface* (BBMI) applications [48] such as cardiac *ElectroCardioGram* (ECG) [41], a novel concept that embeds a generative VAE into the objective function of Bayesian optimization is applied to estimating tissue excitability in a cardiac electrophysiological model by [10]. In [15], a deep generative model is trained to generate the spatiotemporal dynamics of *TransMembrane Potential* (TMP).

In bioImaging histology applications, in order to classify the newly given prostate datasets into low and high *Gleason grade*, an adversarial training is used to minimize the distribution discrepancy at the feature space, with the loss function adopted from the GAN [35]. In [19], a cascaded of refinement GANs for phase contrast microscopy image super-resolution is proposed.

Most of the publications using the GANs, concern the medical imaging application for image quality enhancement, image reconstruction, crafted images generation or image registration and segmentation (see [48] for a extensive survey).

3.2 Variational Auto-Encoders

An AE is an unsupervised ANN trained to recreate or reproduce the input vector $\mathbf{x}$ [4, 13, 43]. The AE is composed by two main structures: an encoder and a decoder (Fig. 3) which are multilayered ANNs parameterized by ϕ and θ, respectively. The first one encodes the input data $\mathbf{x}$ into a latent representation $\mathbf{z}$ by the encoder function $\mathbf{z} = f_\phi(\mathbf{x})$, whereas the second one decodes this latent representation onto $\hat{\mathbf{x}} = h_\theta(\mathbf{z})$ which is an approximation or reconstruction of the original data. In an AE, an equal number of units are used in the input/output layers while less units are used in the latent layer (Fig. 3). The AEs are usually used for data compression (i.e., feature extraction/reduction), noise removal and pre-trained parameters for a complex network.

A VAE has the same functions as the AE; in the sense that it is composed by an encoder and a decoder (Fig. 3). VAE becomes a popular generative model by combining *Bayesian inference* [23, 24] and the efficiency of the ANNs to obtain a nonlinear low-dimensional latent space. The Bayesian inference is obtained by an additional layer used for sampling the latent vector $\mathbf{z}$ with a prior specified

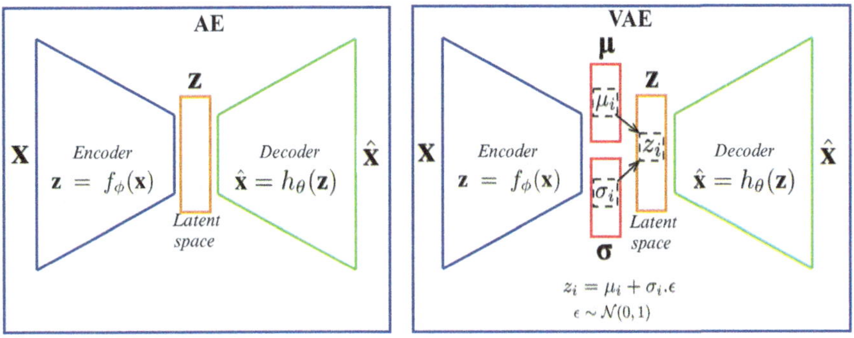

Fig. 3 Schematic architecture of a standard deep autoencoder and a variational deep autoencoder. Both architectures have two parts: an encoder and a decoder

distribution $p(\mathbf{z})$, usually assumed to be a *standard Gaussian* $\mathcal{N}(0, \mathbf{I})$ [23, 24], where $\mathbf{I}$ is the identity matrix. Each element z_i of the latent layer is obtained as follow:

$$z_i = \mu_i + \sigma_i.\epsilon \tag{1}$$

where μ_i and σ_i are the ith components of the mean and standard deviation vectors, ϵ is a random variable following a *standard normal distribution* [23, 24] ($\epsilon \sim \mathcal{N}(0, 1)$). Unlike the AE, which generates the latent vector $\mathbf{z}$, the VAE generates vector of means μ_i and standard deviations σ_i. This allows to have more continuity in the latent space than the original AE. The VAE loss function given by the Eq. 2 has two terms. The first term $\mathcal{L}_{rec}$ is the reconstruction loss function (Eq. 3). Usually, the negative expected log-likelihood (e.g., the cross-entropy function) is used but also the mean squared error. The second term $\mathcal{L}_{KL}$ (Eq. 4) corresponds to the *Kullback-Liebler* (KL) [23, 24] divergence loss term that forces the generation of a latent vector with the specified *normal distribution* [23, 24]. The KL divergence is an information theoretic measure of proximity between two densities $q(x)$ and $p(x)$. It is asymmetric ($KL(q \parallel p) \neq KL(p \parallel q)$) and non-negative. It is minimized when $q(x) = p(x)$ [6]. Thus, the KL divergence term measures how closely the conditional distribution density $q_\phi(\mathbf{z} \mid \mathbf{x})$ of the encoded latent vectors is from the desired Normal distribution $p(\mathbf{z})$. The value of KL is zero when the two probability distributions are the same, which forces the encoder of VAE $q_\phi(\mathbf{z} \mid \mathbf{x})$ to learn the latent variables that follow a multivariate normal distribution over a k-dimensional latent space.

$$\mathcal{L} = \mathcal{L}_{rec} + \mathcal{L}_{KL} \tag{2}$$

$$\mathcal{L}_{rec} = -E_{q_\phi(\mathbf{z}|\mathbf{x})}(log(p_\theta(\mathbf{x} \mid \mathbf{z}))) \tag{3}$$

$$\mathcal{L}_{KL} = KL(q_\phi(\mathbf{z} \mid \mathbf{x}) \parallel p(\mathbf{z})) \tag{4}$$

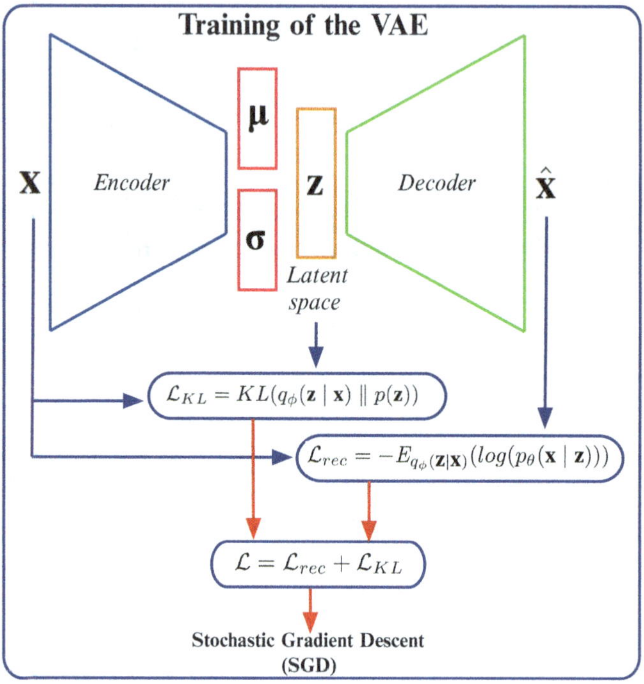

Fig. 4 The VAE loss function. The first term $\mathscr{L}_{rec}$ is the reconstruction loss function. The second term $\mathscr{L}_{KL}$ corresponds to the Kullback-Liebler divergence loss term that forces the generation of a latent vector with the specified Normal distribution. When the VAE is trained, the two functions encoder/decoder can be used separately even to reduce the space dimension by encoding the input data or to generate synthetic samples by decoding new variables from the latent space

When the VAE is trained, the two functions (encoder and decoder) can be used separately, even to reduce the space dimension by encoding the input data, or to generate synthetic samples by decoding new variables from the latent space (Fig. 4).

4 Variational Autoencoders for Data Visualization and Analysis

In this section, we present an innovative DL approach for biomedical data instantiation and visualization. This approach is based on the use of the VAE to solve the problem of the massive high dimensionality data that is usually encountered in biomedical applications. This section is an extension of our previous study [49]. We begin by a brief introduction on the importance of the spatial distribution of the training data. Next, we present the framework of the visual classification methodology as a support for data labeling.

4.1 Massive Amount of Unlabeled Data

One of the main issue faced by most of the biomedical applications, concerns the massive amount of unlabeled data. Manually analyzing and classifying huge database by a human expert is mostly unfeasible, being partially been done, only for simple signatures, easily recognizable by the expert. Concerning this matter, the medical experts face two challenging problems: how to select the most significant data for labeling, and what is the minimum size of the data set—sufficient to define each pathology—in order to perform a proper training of the classifier.

Usually, one consider that classification improves with the number of labeled data, but this is usually, not enough. In fact, the quality of the classification depends on the spatial distribution of the training data set, which is a very important parameter to be considered before processing the training.

Figure 5 gives a basic illustration of two different training data sets. The first set has more samples than the second one, but the spatial distribution of the second

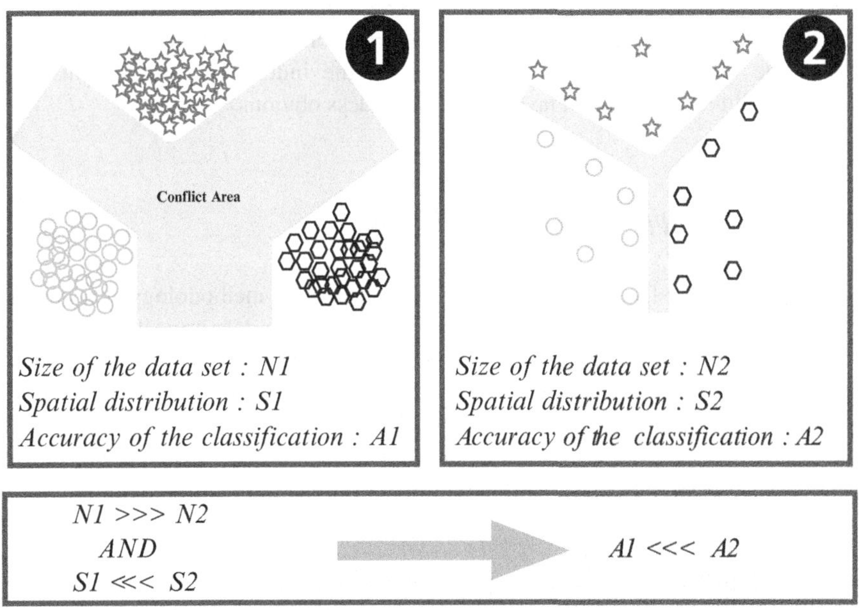

Fig. 5 Basic illustration of an arbitrary 2D-representation of two different training data sets with two different spatial distributions. The data set #1 has more samples than the data set #2 but the accuracy of the classification is better for the data set #2. The spatial distribution of the data is more important than the size of the dataset [49]

data set is better than the first one. The grey zone represents the conflict area, where most of the false positive predictions are produced by the classifier. Better the spatial distribution of the training data set, better the accuracy of the prediction. To reduce this "dead zone", the expert must choose some new points belonging or being near to this conflict area, for labeling.

Many research have been recently focusing on improving learning from imbalanced data, but none of these methods rely on a visual analysis of the learning data—one of the most intuitive approach. We consider that the visualization of the learning data to understanding its nature and to extracting more information from the 2D-space distribution, is an essential step during the design of the diagnosis model.

One of the main limitations of ANNs, is the lack of interpretability of the obtained results. It is very hard to understand what happens in the hidden layers and why a trained ANN gives a positive diagnosis for a given input sample. This "black-box" aspect is very restrictive in many medical application fields, where a decision interpretability can lead to serious legal consequences [27].

When the CNNs are used for medical image processing, several methods have been developed to visualize what happen in the intermediate layers. Some of these algorithms are for example visual explanations from CNNs via gradient-based localization [38]; a visualization technique of the input stimuli that excites individual feature maps at any layer in a CNN model [44] or; a deep Taylor decomposition method for interpreting generic multilayer ANNs by decomposing the network classification decision into contributions of its input elements [31]. When the input data are not images, as for the industrial measurements, the interpretability of the hidden layers activities is less obvious.

4.2 *Visual Support for Data Labeling*

Figure 6 shows the framework of the visual classification methodology as a support for data labeling. An encoder, part of a VAE, is used as a data projection for a 2D-visualization. The input features vectors are encoded into a 2D-latent space, which helps the expert to visually analyze the space distribution of the training data set. At the beginning, few data points from the big unlabeled database are selected from the 2D-latent space. These points are labeled by the expert and used to train a neural classifier. The obtained classifier is then tested over all the unlabeled data set.

To identify the conflict area, i.e., the gray zones illustrated in the Fig. 5, several classifiers are trained on the same initial labeled data set, and tested over all the unlabeled data set. The conflict zones are identified on the 2D-space by analyzing the conflicts results of all the trained classifiers.

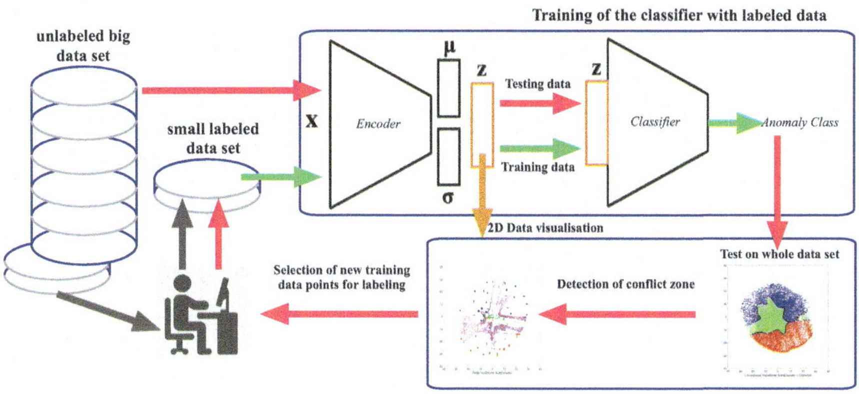

Fig. 6 The framework of the visual classification methodology using an encoder function of a VAE as a data projection for a 2D-visualization. The input features vectors are encoded into a 2D-latent space. If the conflict zone, identified on the 2D-space, is too large, new points near these conflict zones are selected and labeled by the expert. These new labeled points are then added to the initial training set for new training/testing iterations [49]

As we can see on the Fig. 5, if the conflict zone is too large, the boundary between classes, fixed by the classifier, is too uncertain. Usually, most of the false predictions occur near this conflict zone. To reduce the proportion of these false predictions, new points—near or belonging to these conflict zones—are then selected and labeled by the expert. These new labeled samples are added to the initial training set for a new training iterations. Therefore, several interactions between the diagnostic expert and the DL expert are necessary, in order to refine and reduce the boundary zone.

4.3 *Identification and Reduction of the Conflict Area*

As described in the previous section, the conflict zone is the area in the data space definition where most of the false predictions occur. This is due to the poor coverage of the learning data space caused by the selection of only clear examples when designing the classifier. Because the human brain can only perceive two or three-dimensional space, it is impossible for humans to process the n-features dimensional space used to define the input vector of a biomedical data. It is even worse when trying to compare data between them. Therefore, projecting data in the 2D-latent space will help the expert to easily identify clusters of similar data and locate boundary or conflict zone on which he needs to work. When the conflict zone is too large, the boundary between classes, fixed by the classifier, leaves a large number of data with uncertain classification. This means that two different classifiers C_i and C_j with $i \neq j$ will definitely have two opposite responses for the same input data k:

$$\Psi_i(k) \neq \Psi_j(k) \tag{5}$$

where $\Psi_i(k)$ and $\Psi_j(k)$ are respectively the output class obtained by the classifiers C_i and C_j for the input sample k. To identify these conflict zones, several classifiers C_i were trained with the same 2D training data set Ω^{2D}_{train}. Subsequently, all the trained classifiers have been tested over the entire data set. For each input sample k of the data set, if two classifiers have two opposite responses, the input sample k is then considered as a conflict sample.

To reduce the proportion of false predictions, the size of the conflict zone must be reduced.

Therefore, the expert will choose new learning points in these poorly covered areas, in order to adjust the learning of the classifier. These new samples are then labeled by the expert and added to the previous training set. The entire procedure is then repeated until the conflict zone is considered as acceptable by the medical expert, without having to consider all the data but just a few additional points located in the conflict zone.

5 Case Study: The Breast Cancer Wisconsin Dataset

In this section, we present a step-by-step practical case study which concerns the diagnosis of the breast cancer [3]. We show how the VAE can be practically used to solve the problem of labeling the minimum of data to quickly improve the learning process. The proposed DL approach has been developed using Keras[1] DL framework on a PC machine with 4.0-GHz Intel Xeon CPU and 32-GB memory.

5.1 Description of the Dataset

The *Breast Cancer Wisconsin dataset* used as a case study is available by anonymous ftp from ice.uci.edu [12]. This dataset consists of 569 breast cancer patterns from the university of Wisconsin. Each pattern has 30 attributes and the dataset is divided into two classes, 212 are malignant and 357 are benign. The original dataset was proposed by Nick Street, Wolberg and Mangasarian [40].

[1]https://keras.io/.

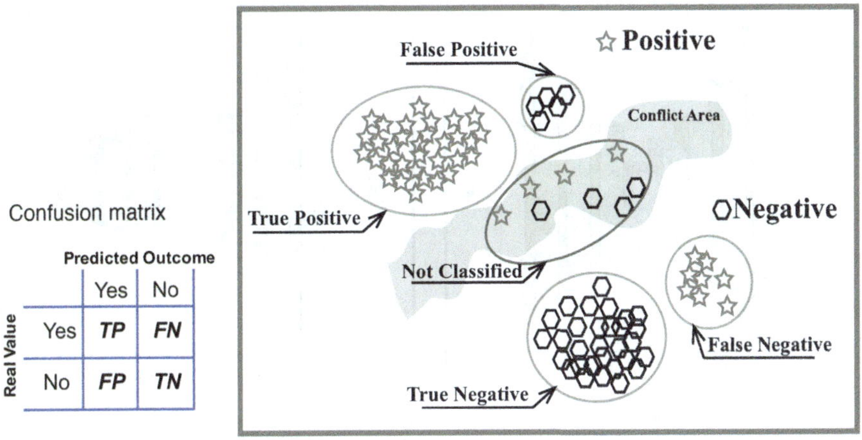

Fig. 7 Confusion matrix

5.2 *Evaluation Metrics*

To be able to evaluate the performance of various classification methods, there is a need to introduce quantitative criteria. A confusion matrix is commonly used to calculate these performance parameters (Fig. 7). This matrix contains information about the actual and the predicted classifications:

- *True Positive* (TP) values are the number of Positive classification correctly classified as Positive,
- *True Negative* (TN) values represent the number of Negative classifications correctly classified as Negative,
- *False Positive* (FP) values are the number of Negative classifications incorrectly classified as Positive,
- *False Negative* (FN) values represent the number of Positive classification incorrectly classified as Negative,
- *Not Classified* (NC) values represent the number of samples belonging to the conflict area.

Based on these values, the following metrics are thus calculated:

- *Accuracy* (Acc) = $\frac{TP + TN}{TP + TN + FP + FN + NC}$
- *Negative Predictive Value* (NPV) = $\frac{TN}{FN + TN}$
- *Positive Predictive Value* (PPV) = $\frac{TP}{FP + TP}$
- *True Negative Rate* (TNR) = $\frac{TN}{FP + TN}$
- *True Positive Rate* (TPR) = $\frac{TP}{TP + FN}$

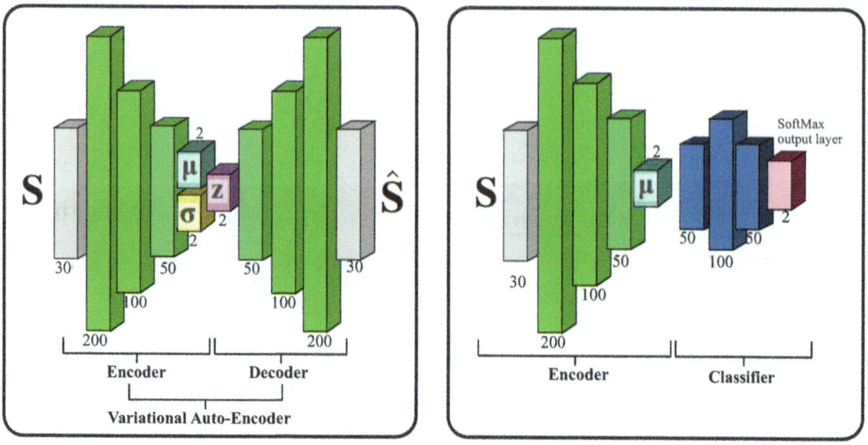

Fig. 8 The VAE architecture used. The encoder used jointly with a neural classifier. The mean layer of the convolutional encoder is considered as the input 2D-vector of the classifier

5.3 The Variational Auto-Encoder Used as Visual Support

The VAE architecture used is illustrated by the Fig. 8. This architecture includes two parts, an encoder and a decoder, which are two symmetrical and reversed structures. Each one is composed by three fully connected layers. The latent two-dimensional space is represented by two layers for the encoder: the mean and the standard deviation layers (i.e., μ and σ), and one sampling layer ($\mathbf{Z}$) for the decoder.

Training the VAE does not need the label information of the input data. However, for an efficient data encoding, we used an indirect labeling, since all training samples belong to one of the classes described above.

The first step was to train the whole VAE architecture for the reconstruction of the feature vector ($\hat{\mathbf{S}} = \mathscr{F}(\mathbf{S})$). When the training process of the VAE is done, the encoder part is then used jointly with a neural classifier, as presented in Fig. 8. The mean layer μ of the variational encoder is considered as the input 2D-vector of the classifier. The second step is then to train the classifier for the diagnosis. The encoder parameters obtained by the previous step are then frozen during the classifier training step.

All the details of DNNs architectures used, i.e. the VAE and the classifier, are presented in the Table 1.

Table 1 Proposed DNNs architectures

#	Layer type	Neurons	#	Layer type	Neurons
Encoder			Classifier		
0	Input feature vector	30	0	Input (mean layer)	2
1	Fully connected	200	1	Fully connected	50
2	Fully connected	100	2	Fully connected	100
3	Fully connected	50	3	Fully connected	50
4	Mean layer	2	4	SoftMax output layer	2
4	Standard deviation layer	2			
Decoder					
0	Sampling layer	2			
1	Fully connected	50			
2	Fully connected	100			
3	Fully connected	200			
4	Output reconstructed vector	30			

5.4 *Visualization of the Latent Space*

Considering that the latent vectors are the encoding representation of the input feature vectors, it is interesting to visualize the 2-dimensional representation of the original feature and to evaluate the similarities within each class. Figure 9 shows the 2D-latent representation at different iterations of the training process of the variational auto-encoder.

The 2D-latent representation is the output of the sampling layer Z of the encoder illustrated in the Fig. 8.

At the beginning of the training process (i.e. iteration 1), the 2-D representation looks like a compact cluster. As the learning progresses, the cluster extends into the 2-D latent space until covering the entire space (iteration 1000), with almost a uniform Normal distribution—forced by the Kullback-Liebler divergence loss term (see Sect. 3.2). Two clusters are then formed, the green for the benign class and the red for the malignant class.

5.5 *Selection of the Training Samples from the 2D-Latent Space*

Considering that all the used dataset is unlabeled, we will see how the 2D-latent space used as a visualization support can help the expert to select the most

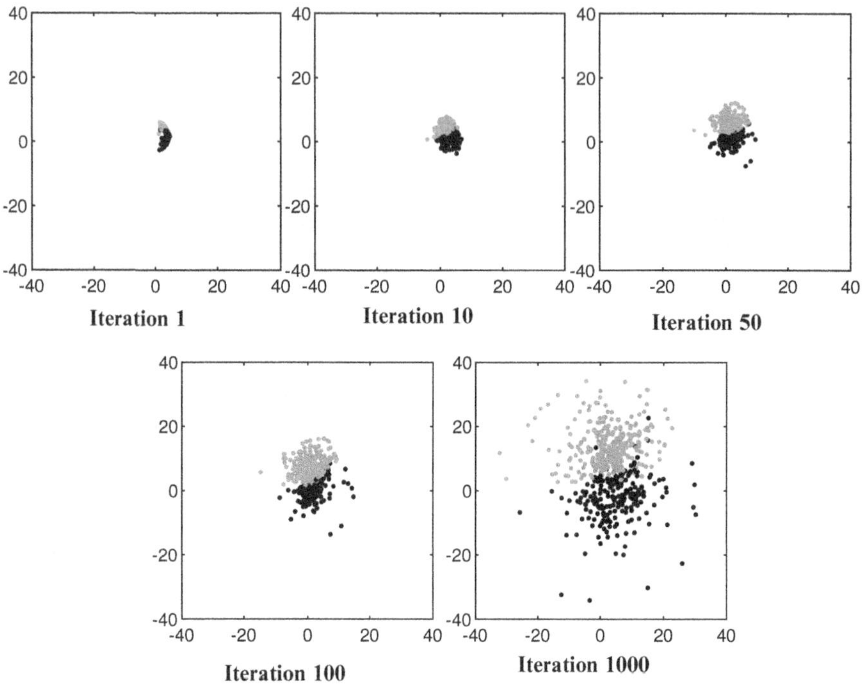

Fig. 9 The progression of the latent space throughout the learning process of the VAE

significant data for labeling. Figure 10a shows the 2D-projection of all the dataset obtained by the variational encoder. Since all the dataset is unlabeled, all the samples belong to the same cluster.

5.5.1 Step1: Selection of the First Training Set

At the beginning, the expert must choose some samples to label for the training process. Instead of choosing randomly the samples, we can use the 2D-projection as a visualization support to choose the samples with respect to a certain spatial distribution. For example, as shown by the Fig. 10a, a first set of six samples (Set #1) is selected from the 2D-latent space: 3 benign samples and 3 malignant samples. These samples are used for the training process of the classifiers.

5.5.2 Step2: Identification of the Conflict Area

As described by the Sect. 4.3, to identify the conflict area between the two classes (i.e. benign and malignant), we train different classifiers with the same training

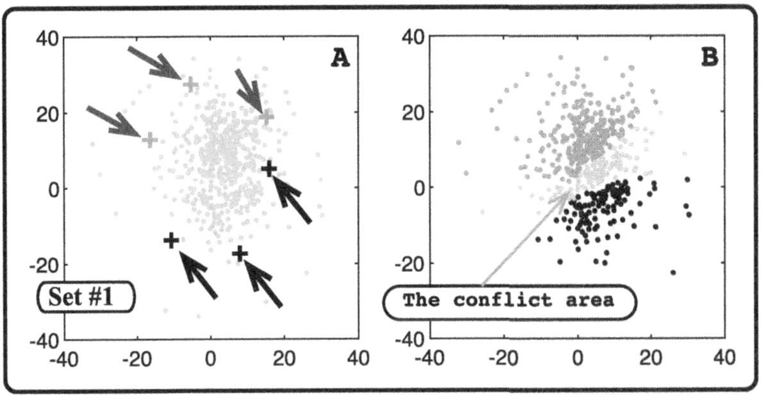

Fig. 10 (**a**) First training set selection from 2D-projection, (**b**) Identification of the conflict area

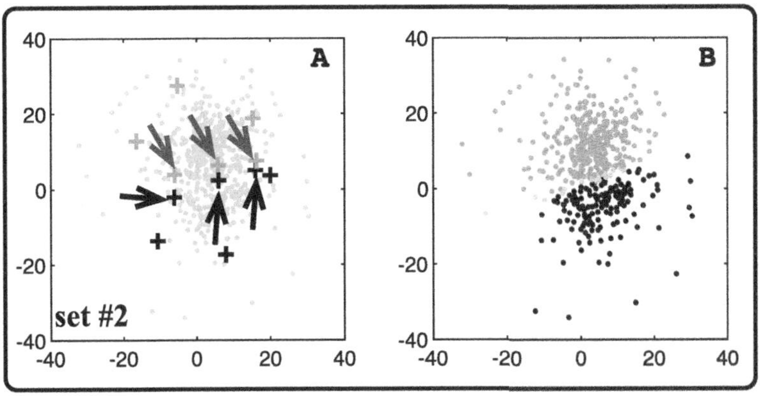

Fig. 11 (**a**) Selection of new training samples nearby or belonging to the conflict area from the 2D-Projection, (**b**) Reduction of the new conflict area

samples. The conflict area, in the Fig. 10b, represents the cases where the classifiers are in disagreement.

5.5.3 Step3: Reduction of the Conflict Area

To reduce the conflict area, new samples are selected nearby or belonging to the conflict area. As shown by the Fig. 11a, a second set of six new samples (Set #2) is selected from the 2D-latent space and added to the initial training set. After a new training process of the classifiers, the new obtained conflict area is then reduced, as we can see on the Fig. 11b.

Table 2 Results of the classification performances obtained on the whole dataset for each one of the training set #1 and #2. The malignant set is the Positive class and the benign is the Negative one

Training set	TP	TN	FP	FN	Not classified
Set #1	116 (20.39%)	325 (57.12%)	0 (0%)	16 (2.81%)	112 (19.68%)
Set #2	148 (26.01%)	354 (62.21%)	1 (0.18%)	45 (7.91%)	21 (3.69%)
	Acc	NPV	PPV	TNR	TPR
Set #1	77.50%	95.31%	100%	100%	87.88%
Set #2	88.22%	88.72%	99.33%	99.72%	76.68%

5.5.4 Results Analysis

Table 2 shows the entire classification performances obtained on the whole dataset for each of the training set #1 and #2. The metrics used to analyze the obtained results are those presented by the Sect. 5.2. We suppose that the malignant set is the Positive class and the benign is the Negative class. We can observe that the conflict area (i.e. the not classified patterns) has been scaled down between the training set #1 and #2. Indeed, this disagreement zone decreases from 112 samples to 21, which represents respectively 19.68 and 3.69% compared to the whole dataset.

By reducing the proportion of the uncertainty area, the size of the true and false predictions increases. In fact, the number of the True Positive and the True Negative samples has been raised, which gives a better accuracy for the training set #2 (88.22% instead of 77.50% for the set #1). On the other hand, the proportion of the false predictions have been also raised by reducing the conflict area, which proves that most of the false predictions are near or within the conflict area. For this reason, the NPV, PPV, TNR and the TPR values are better for the training set #1.

5.6 Random Selection of Training Samples

This section provides comparison results between the proposed 2D-visual selection method and a random selection of the training samples. Figure 12 gives the 2D-representation of four training sets (set #3, 4, 5 and 6) randomly selected from the entire dataset. These sets are respectively shown in the figures a, c, e and g. For each example, the conflict areas obtained by the confrontation tests of the classifiers are illustrated in the figures b, d, f and h.

Table 3 summarizes the results of the classification performances obtained for each training group on the whole dataset using the performances metrics listed in the previous section. The analysis of these comparative results reveals several strong points. First, it is evident that the performances are very disparate for the random selection, as we can see for the TP values (47 instances for the sets #3 against 120 for the set #6). For each random selection, thanks to the 2D-representation, we can visually evaluate the spatial distribution of the chosen training instances (Fig. 12).

Table 3 Comparison results between the proposed 2D-visual selection method (Sets #1 and 2) and a random selection of the training samples (Sets #3, 4, 5 and 6). For each set, we give between brackets the number of instances per class (# of benign, # of malignant)

Training set	TP	TN	FP	FN	Not classified
Set #1 (3, 3)	116 (20.39%)	325 (57.12%)	0 (0%)	16 (2.81%)	112 (19.68%)
Set #2 (6, 6)	148 (26.01%)	354 (62.21%)	1 (0.18%)	45 (7.91%)	21 (3.69%)
Set #3 (5, 1)	47 (8.26%)	344 (60.46%)	0 (0%)	37 (6.50%)	141 (24.78%)
Set #4 (3, 3)	63 (11.07%)	317 (55.71%)	0 (0%)	11 (1.93%)	178 (31.28%)
Set #5 (2, 4)	114 (20.04%)	253 (44.46%)	50 (8.79%)	16 (2.81%)	136 (23.90%)
Set #6 (3, 3)	120 (21.09%)	274 (48.15%)	1 (0.18%)	9 (1.58%)	165 (29%)
	Acc	NPV	PPV	TNR	TPR
Set #1 (3, 3)	77.50%	95.31%	100%	100%	87.88%
Set #2 (6, 6)	88.22%	88.72%	99.33%	99.72%	76.68%
Set #3 (5, 1)	68.71%	90.29%	100%	100%	55.95%
Set #4 (3, 3)	66.78%	96.64%	100%	100%	85.13%
Set #5 (2, 4)	64.49%	94.05%	69.51%	83.49%	87.69%
Set #6 (3, 3)	69.24%	96.81%	99.17%	99.63%	93.02%

Qualitatively, the distributions of the sets #1 and #2 (shown by the Figs. 10 and 11) are better than those obtained by the sets #3, 4, 5 and 6. Indeed, the data space is better covered by the learning sets #1 and #2. This better coverage of the data space necessarily leads to better classification results. It is clearly observed that the performances obtained by these two groups outperform the others by a large margin with the highest accuracy. The comparison results between our 2D-visual selection method and the conventional random selection method demonstrate that our framework benefits of a better data visibility and a better results interpretation.

6 Conclusion

In this chapter, we investigated the use of VAE as a support for data visualization and diagnosis interpretation. However, after training of the VAE, the encoder part is used as a data projection from an input features mapping to a latent 2D-mapping. The VAE is trained without using class labels to learn properties in the data. The power of VAE falls in capturing the complicated data features from the multi-dimensional input space and compressing them into a smaller 2D-latent space, more easy for a visualization by a human expert. The VAE reveals a promising tool to produce a 2-dimensional embedding of high dimensional data with the goal of simplifying the identification of clusters when used jointly with a classifier.

As articulated in the results section, it is quite easy to understand the diagnosis given by the neural classifier by visually analyzing the spatial distribution of the classified samples. Therefore, the knowledge area of the ANN and the boundaries between the classes are perceived by the expert. The conflict regions which are

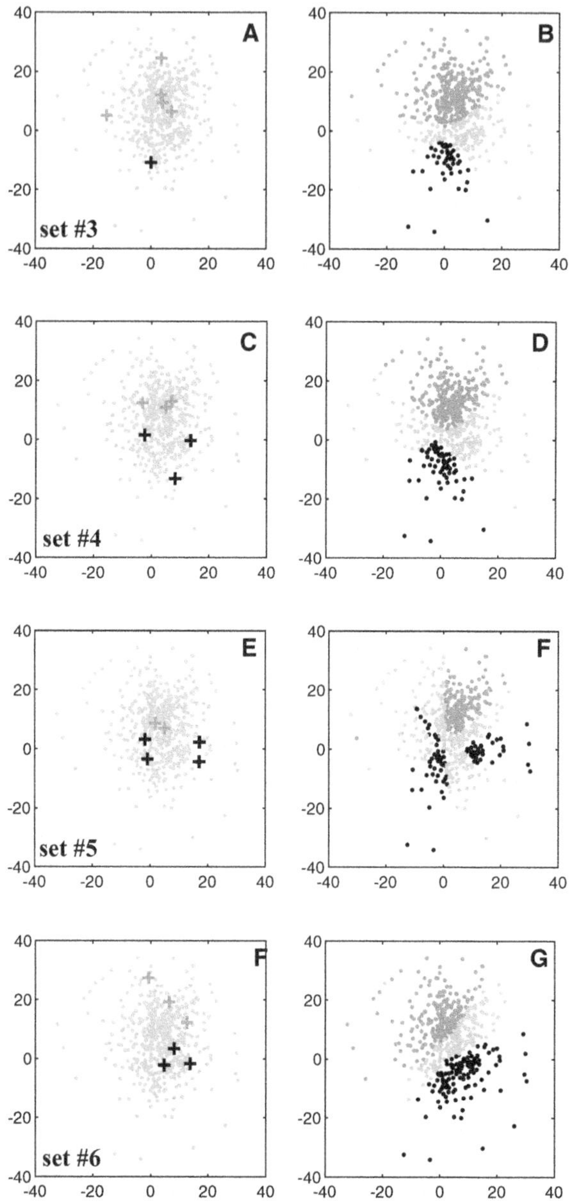

Fig. 12 Four examples of a random selection. The figures (**a**), (**c**), (**e**) and (**g**) gives four examples of four training sets (set #3, 4, 5 and 6) randomly selected from the entire dataset. For each training set, the obtained classifiers are tested on the whole dataset where different conflict areas are obtained, as shown by the figures (**b**), (**d**), (**f**) and (**h**)

poorly covered by the training samples are then easily identified. The ANN is then less perceived as a "black-box" in the sense that its knowledge area is visible, the interpretation of the false predictions and their understanding become realizable.

The ending condition of the proposed algorithm is the diagnosis model evaluation through the testing of a performance criterion. The criterion used is the reduction of the conflict area, which is visually perceived by the expert, at each round of the algorithm. To improve the performances of the proposed method, other metrics than the reduction of the conflict area must be defined and used to evaluate the performances of a such model during the building process. These metrics should give some answering elements to these non-exhaustive questions:

- How to evaluate the quality of the spatial distribution of such an input training dataset?
- How to evaluate the features used which are closely related to the quality of the input vector?
- How to evaluate the performances of the VAE, especially the encoder function?
- A more fundamental question concerns the minimum size of the conflict area tolerated by the medical expert—a real dilemma for biomedical applications. For some borderline cases, an ambiguity response from the ANN can be more acceptable than a false prediction. For these critical cases, It is then preferable that the ANN lets the medical expert decide, rather than providing a catastrophic false prediction.

References

1. Alvarado-Díaz W, Lima P, Meneses-Claudio B, Roman-Gonzalez A (2017) Implementation of a brain-machine interface for controlling a wheelchair. In: 2017 CHILEAN Conference on Electrical, Electronics Engineering, Information and Communication Technologies (CHILECON), pp 1–6, DOI 10.1109/CHILECON.2017.8229668
2. Angermueller C, Pärnamaa T, Parts L, Stegle O (2016) Deep learning for computational biology. Molecular Systems Biology 12(7), DOI 10.15252/msb.20156651, URL https://msb.embopress.org/content/12/7/878, https://msb.embopress.org/content/12/7/878.full.pdf
3. Baltres A, Zeina AM, et al RZ (2020) Prediction of oncotype dx recurrence score using deep multi layer perceptrons in estrogen receptor-positive, her2 negative breast cancer. Breast Cancer 27(5):1007–1016, DOI https://doi.org/10.1007/s12282-020-01100-4
4. Bengio Y (2014) How auto-encoders could provide credit assignment in deep networks via target propagation. CoRR URL https://arxiv.org/abs/1407.7906
5. Bengio Y, Lamblin P, Popovici D, Larochelle H (2007) Greedy layer-wise training of deep networks. In: Schölkopf B, Platt J, Hoffman T (eds) Advances in Neural Information Processing Systems (NIPS 06), MIT Press, pp 153–160, DOI https://www.iro.umontreal.ca/~lisa/pointeurs/BengioNips2006All.pdf
6. Blei DM, Kucukelbir A, McAuliffe JD (2016) Variational inference: A review for statisticians. arXiv e-prints arXiv:1601.00670, https://arxiv.org/abs/1601.00670
7. Cao C, Liu F, Tan H, Song D, Shu W, Li W, Zhou Y, Bo X, Xie Z (2018) Deep learning and its applications in biomedicine. Genomics, Proteomics & Bioinformatics 16(1):17–32, DOI https://doi.org/10.1016/j.gpb.2017.07.003, URL https://www.sciencedirect.com/science/article/pii/S1672022918300020
8. Chandra B, Sharma RK (2016) Deep learning with adaptive learning rate using laplacian score. Expert Systems with Applications 63:1–7, DOI https://doi.org/10.1016/j.eswa.2016.05.022
9. Creswell A, White T, Dumoulin V, Arulkumaran K, Sengupta B, Bharath AA (2018) Generative adversarial networks: An overview. IEEE Signal Processing Magazine 35(1):53–65, DOI 10.1109/MSP.2017.2765202

10. Dhamala J, Ghimire S, Sapp JL, Horáček BM, Wang L (2018) High-dimensional bayesian optimization of personalized cardiac model parameters via an embedded generative model. In: Frangi AF, Schnabel JA, Davatzikos C, Alberola-López C, Fichtinger G (eds) Medical Image Computing and Computer Assisted Intervention – MICCAI 2018, Springer International Publishing, Cham, pp 499–507
11. Ditzler G, Roveri M, Alippi C, Polikar R (2015) Learning in nonstationary environments: A survey. IEEE Computational Intelligence Magazine 10(4):12–25, DOI 10.1109/MCI.2015.2471196
12. Dua D, Graff C (2017) UCI machine learning repository. URL https://archive.ics.uci.edu/ml
13. Fan YJ (2019) Autoencoder node saliency: Selecting relevant latent representations. Pattern Recognition 88:643–653, DOI https://doi.org/10.1016/j.patcog.2018.12.015, URL https://www.sciencedirect.com/science/article/pii/S0031320318304369
14. Fnaiech N, Fnaiech F, Jervis BW (2011) Feedforward NeuralNetworks Pruning Algorithms, Industrial Electronics Handbook, vol 5, j.d. irwin, 2nd edn, chap 15, pp 15–1 to 15–15
15. Ghimire S, Dhamala J, Gyawali PK, Sapp JL, Horacek M, Wang L (2018) Generative modeling and inverse imaging of cardiac transmembrane potential. In: Frangi AF, Schnabel JA, Davatzikos C, Alberola-López C, Fichtinger G (eds) Medical Image Computing and Computer Assisted Intervention – MICCAI 2018, Springer International Publishing, Cham, pp 508–516
16. Goodfellow I, Pouget-Abadie J, Mirza M, Xu B, Warde-Farley D, Ozair S, Courville A, Bengio Y (2014) Generative adversarial nets. In: Ghahramani Z, Welling M, Cortes C, Lawrence ND, Weinberger KQ (eds) Advances in Neural Information Processing Systems 27, Curran Associates, Inc., pp 2672–2680, URL https://papers.nips.cc/paper/5423-generative-adversarial-nets.pdf
17. Goodfellow IJ, Warde-farley D, Mirza M, Courville A, Bengio Y (2013) Maxout networks. In: In ICML
18. Han HG, Qiao JF (2013) A structure optimisation algorithm for feedforward neural network construction. Neurocomputing 99:347–357, DOI https://dx.doi.org/10.1016/j.neucom.2012.07.023
19. Han L, Yin Z (2018) A cascaded refinement gan for phase contrast microscopy image super resolution. In: Frangi AF, Schnabel JA, Davatzikos C, Alberola-López C, Fichtinger G (eds) Medical Image Computing and Computer Assisted Intervention – MICCAI 2018, Springer International Publishing, Cham, pp 347–355
20. He H, Garcia EA (2009) Learning from imbalanced data. IEEE Transactions on Knowledge and Data Engineering 21(9):1263–1284, DOI 10.1109/TKDE.2008.239
21. Hinton GE, Osindero S, Teh YW (2006) A fast learning algorithm for deep belief nets. Neural Comput 18(7):1527–1554, DOI 10.1162/neco.2006.18.7.1527
22. Jones W, Alasoo K, Fishman D, Parts L (2017) Computational biology: deep learning. Emerging Topics in Life Sciences 1(3):257–274, DOI 10.1042/ETLS20160025, URL https://www.emergtoplifesci.org/content/1/3/257, https://www.emergtoplifesci.org/content/1/3/257.full.pdf
23. Kingma D (2017) Variational inference & deep learning: A new synthesis. PhD thesis, Faculty of Science (FNWI), Informatics Institute (IVI), University of Amsterdam, URL https://hdl.handle.net/11245.1/8e55e07f-e4be-458f-a929-2f9bc2d169e8
24. Kingma DP, Welling M (2013) Auto-Encoding Variational Bayes. arXiv e-prints arXiv:1312.6114, https://arxiv.org/abs/1312.6114
25. Lee S, Kwak M, Tsui KL, Kim SB (2019) Process monitoring using variational autoencoder for high-dimensional nonlinear processes. Engineering Applications of Artificial Intelligence 83:13–27, DOI 'https://doi.org/10.1016/j.engappai.2019.04.013, URL https://www.sciencedirect.com/science/article/pii/S0952197619300983
26. Li Z, Nguyen SP, Xu D, Shang Y (2017) Protein loop modeling using deep generative adversarial network. In: 2017 IEEE 29th International Conference on Tools with Artificial Intelligence (ICTAI), pp 1085–1091, DOI 10.1109/ICTAI.2017.00166

27. Litjens G, Kooi T, Bejnordi BE, Setio AAA, Ciompi F, Ghafoorian M, van der Laak JA, van Ginneken B, Sánchez CI (2017) A survey on deep learning in medical image analysis. Medical Image Analysis 42:60–88, DOI https://doi.org/10.1016/j.media.2017.07.005, URL https://www.sciencedirect.com/science/article/pii/S1361841517301135
28. van der Maaten L, Hinton G (2008) Visualizing data using t-sne. Journal of Machine Learning Research 9(11):2579–2605
29. Mahmud M, Kaiser M, Hussain A, Vassanelli S (2018) Applications of deep learning and reinforcement learning to biological data. IEEE Trans Neural Netw Learn Syst DOI 10.1109/TNNLS.2018.2790388.
30. Min S, Lee B, Yoon S (2016) Deep learning in bioinformatics. CoRR abs/1603.06430, URL https://arxiv.org/abs/1603.06430
31. Montavon G, Lapuschkin S, Binder A, Samek W, Müller KR (2017) Explaining nonlinear classification decisions with deep taylor decomposition. Pattern Recognition 65:211–222, DOI https://doi.org/10.1016/j.patcog.2016.11.008, URL https://www.sciencedirect.com/science/article/pii/S0031320316303582
32. Nakamura K, Hong B (2019) Adaptive weight decay for deep neural networks. CoRR abs/1907.08931, URL https://arxiv.org/abs/1907.08931
33. Pérez-Sánchez B, Fontenla-Romero O, Guijarro-Berdiñas B (2016) A review of adaptive online learning for artificial neural networks. Artificial Intelligence Review DOI 10.1007/s10462-016-9526-2
34. Raví D, Wong C, Deligianni F, Berthelot M, Andreu-Perez J, Lo B, Yang GZ (2017) Deep learning for health informatics. IEEE Journal of Biomedical and Health Informatics 21(1):4–21, DOI 10.1109/JBHI.2016.2636665
35. Ren J, Hacihaliloglu I, Singer EA, Foran DJ, Qi X (2018) Adversarial domain adaptation for classification of prostate histopathology whole-slide images. In: Frangi AF, Schnabel JA, Davatzikos C, Alberola-López C, Fichtinger G (eds) Medical Image Computing and Computer Assisted Intervention – MICCAI 2018, Springer International Publishing, Cham, pp 201–209
36. Rumelhart DE, Hinton GE, Williams RJ (1986) Learning representations by back-propagating errors. Nature 323:533 EP –, URL https://dx.doi.org/10.1038/323533a0
37. Schmidhuber J (2015) Deep learning in neural networks: An overview. Neural Networks 61:85–117, DOI https://doi.org/10.1016/j.neunet.2014.09.003
38. Selvaraju RR, Das A, Vedantam R, Cogswell M, Parikh D, Batra D (2016) Grad-cam: Why did you say that? visual explanations from deep networks via gradient-based localization. CoRR abs/1610.02391, URL https://arxiv.org/abs/1610.02391
39. Srivastava N, Hinton G, Krizhevsky A, Sutskever I, Salakhutdinov R (2014) Dropout: A simple way to prevent neural networks from overfitting. Journal of Machine Learning Research 15:1929–1958, DOI https://jmlr.org/papers/v15/srivastava14a.html
40. Street WN, Wolberg WH, Mangasarian OL (1993) Nuclear feature extraction for breast tumor diagnosis. International Symposium on Electronic Imaging Science and Technology 1905, DOI 10.1117/12.148698, URL https://doi.org/10.1117/12.148698
41. Tan JH, Hagiwara Y, Pang W, Lim I, Oh SL, Adam M, Tan RS, Chen M, Acharya UR (2018) Application of stacked convolutional and long short-term memory network for accurate identification of cad ecg signals. Computers in Biology and Medicine 94:19–26, DOI https://doi.org/10.1016/j.compbiomed.2017.12.023, URL https://www.sciencedirect.com/science/article/pii/S0010482517304201
42. Yu H, Yang X, Zheng S, Sun C (2018) Active learning from imbalanced data: A solution of online weighted extreme learning machine. IEEE Transactions on Neural Networks and Learning Systems pp 1–16, DOI 10.1109/TNNLS.2018.2855446
43. Yu S, Príncipe JC (2019) Understanding autoencoders with information theoretic concepts. Neural Networks 117:104–123, DOI https://doi.org/10.1016/j.neunet.2019.05.003, URL https://www.sciencedirect.com/science/article/pii/S0893608019301352
44. Zeiler MD, Fergus R (2014) Visualizing and understanding convolutional networks. In: Fleet D, Pajdla T, Schiele B, Tuytelaars T (eds) Computer Vision – ECCV 2014, Springer International Publishing, Cham, pp 818–833

45. Zemouri R (2020) Semi-supervised adversarial variational autoencoder. Machine Learning and Knowledge Extraction (MAKE) 2(3):361–378, DOI https://doi.org/10.3390/make2030020
46. Zemouri R, Devalland C, Valmary-Degano S, Zerhouni N (2019) Intelligence artificielle : quel avenir en anatomie pathologique ? Annales de Pathologie 39(2):119–129, DOI https://doi.org/10.1016/j.annpat.2019.01.004, URL https://www.sciencedirect.com/science/article/pii/S0242649819300203, l'anatomopathologie augmentée
47. Zemouri R, Omri N, Fnaiech F, Zerhouni N, Fnaiech N (2019) A new growing pruning deep learning neural network algorithm (gp-dlnn). Neural Computing and Applications DOI 10.1007/s00521-019-04196-8, URL https://doi.org/10.1007/s00521-019-04196-8
48. Zemouri R, Zerhouni N, Racoceanu D (2019) Deep learning in the biomedical applications: Recent and future status. Applied Sciences 9(8), DOI 10.3390/app9081526, URL https://www.mdpi.com/2076-3417/9/8/1526
49. Zemouri R, Lévesque M, Amyot N, Hudon C, Kokoko O, Tahan SA (2020) Deep convolutional variational autoencoder as a 2d-visualization tool for partial discharge source classification in hydrogenerators. IEEE Access 8:5438–5454, DOI 10.1109/ACCESS.2019.2962775
50. Zhang Z, Jiang T, Zhan C, Yang Y (2019) Gaussian feature learning based on variational autoencoder for improving nonlinear process monitoring. Journal of Process Control 75:136–155, DOI https://doi.org/10.1016/j.jprocont.2019.01.008, URL https://www.sciencedirect.com/science/article/pii/S095915241930037X

Convolutional Neural Networks in Advanced Biomedical Imaging Applications

Daniel A. Greenfield, Germán González, and Conor L. Evans

Abstract *Deep learning* (DL), in particular *Convolutional Neural Networks* (CNNs), can be used to build powerful quantitative image analysis tools. To take advantage of these capabilities and establish tractable goals and problem solving approaches, it is crucial to understand how both imaging and computational tools have been developed and used together. Different advanced optical imaging methods, whether invasive or non-invasive, are applicable to a wide variety of biomedical research, and CNN algorithms can be tailored to assist with extracting meaningful results from imaging data.

Keywords Biomedical optics · Convolutional neural networks · Model interpretability · Optical coherence tomography · Multiphoton fluorescence microscopy · Second harmonic generation · Coherent Raman imaging · Edge computing

1 Introduction

This chapter aims to provide knowledge on *Convolutional Neural Networks* (CNNs) and demonstrate their use in solving complex problems in biomedical imaging. It provides the reader with a foundational knowledge of *Artificial Intelligence* (AI) methods, optical microscopy, and examples of research at the interface of the two fields. The work presented in this chapter is by no means an exhaustive list of the

D. A. Greenfield
Harvard University Biophysics Program, Boston, MA, USA
e-mail: dgreenfield@g.harvard.edu

G. González
Sierra Research S.L.U., Alicante, Spain
e-mail: ggonzale@sierra-research.com

C. L. Evans (✉)
Wellman Center for Photomedicine, Boston, MA, USA
e-mail: Evans.Conor@mgh.harvard.edu

M. Elloumi (ed.), *Deep Learning for Biomedical Data Analysis*,
https://doi.org/10.1007/978-3-030-71676-9_9

complex problems in biomedical optics that can be explored by *Machine Learning* (ML), yet provides insight into how research problems can be framed such that computational solutions can be valuable. As technological advances increase the capability of both computation and imaging, novel problems can be tackled by more sophisticated algorithms to turn complex data into useful information.

The rest of this chapter is organized as follows: in Sect. 2, we discuss the timeline of published research using both advanced optics and CNNs, along with interspersed computational and microscopy developments. In Sect. 3, we analyze the use of *Deep Learning* (DL) in advanced imaging. Section 4 describes data management strategies for the use of CNNs in imaging applications. Finally, in the last section, we present the conclusions in addition to potential future research directions.

2 Scientific Foundations

Optical microscopy tools have the ability to offer many insights into problems ranging across the fields of biology, medicine, chemistry, and physics. Numerous imaging tools have been developed that create contrast from physical phenomena such as light scattering and absorption, fluorescence and phosphorescence, and interactions with vibrational resonances. Advanced optical imaging platforms can draw on multiple sources of contrast simultaneously, providing large, multiprocess, hyperspectral, time-lapse, volumetric datasets. The ability to gather these multitudes of data holds the promise of offering immense insight into biomedical problems, in particular. A major challenge, however, arises from how to parse and process these reams of data into interpretable and actionable information.

ML methods, and DL approaches in particular, hold significant promise in solving this central challenge. The ability to train and leverage *Artificial Neural Networks* (ANNs) to perform analysis on complex datasets has already been demonstrated to capture new insights into problems ranging from cellular processes to cancer diagnosis. The application of DL methods to advanced optical imaging tools is still at its early stage, making this an exciting time of exploration and discovery. To understand how these two fields can work together, it is first necessary to provide background on both. The fields of ML and microscopy have rich backgrounds based on decades of development that would be impossible to cover in even one book, let alone a chapter. Instead, the goal here is to provide a focused review tailored towards the application of DL methods in optical imaging analysis. To accomplish this, we will first provide a brief review of important background information regarding ANNs followed by a review of common advanced optical imaging tools.

2.1 *Artificial Neural Networks*

This subsection covers a general foundation of ANNs and their architectures. At a high level, the power of ANNs can be understood from the *Universal Approximation Theorem* [57, 58].

> . . .multilayer feedforward networks are, under very general conditions on the hidden unit activation function, universal approximators provided that sufficiently many hidden units are available. However, it should be emphasized that our results do not mean that all activation functions ψ will perform equally well in specific learning problems. [58]

The capabilities of ANNs for non-linear problem solving have been understood for nearly three decades. However, optimizing ANNs through learning rates, activation functions, and other hyperparameters is still an active area of research. Furthermore, defining proper model architectures for solving specific problems is arguably as important. The *Universal Approximation Theorem* above puts no bound on the time needed to universally approximate solutions, yet the trade-off between network accuracy and training requirements is crucial.

2.1.1 Basics of Artificial Neural Networks

ANNs are constructed from individual units, known as *neurons*, that are connected by operations. As starting examples, two basic ANNs are presented: *perceptron* and *multi-layer perceptron*. These ANNs are composed of distinct layers consisting of individual neurons (see Fig 1a. for basic neuron structure). These neurons, also called *nodes*, are the main operational unit of any ANN. Nodes can have arbitrarily many inputs which are combined through an input function. The resulting value of the input function is put through an activation function to determine node output (Fig. 1a). The first layer of a perceptron, referred to as the input layer, reads in data, performs a linear combination of its input and applies a non-linear activation function (see Fig. 1b for common activation functions) [98]. In multi-layer perceptrons, the information housed in input neurons is sent to the next layer, the *hidden layer*, where the process is repeated and the output is stored. Finally, in both ANN structures, the output layer reports either a value or probability depending on whether the ANN is applied to a classification or regression problem. Traditional perceptrons are better suited towards binary classification problems, while multi-layer perceptrons, which are really multiple basic perceptrons stacked on top of one another, are typically used for non-linear problems that benefit from stochastic solutions to solve complex optimization problems.

ANNs are capable of learning through updating the weights between nodes (neurons) across the different layers. For both ANN structures, the goal is to minimize the error between the label associated with the input data and the label estimated at the output layers. Such error is quantified through the use of a loss function. The selection of the loss function depends on the given problem and the structure of the

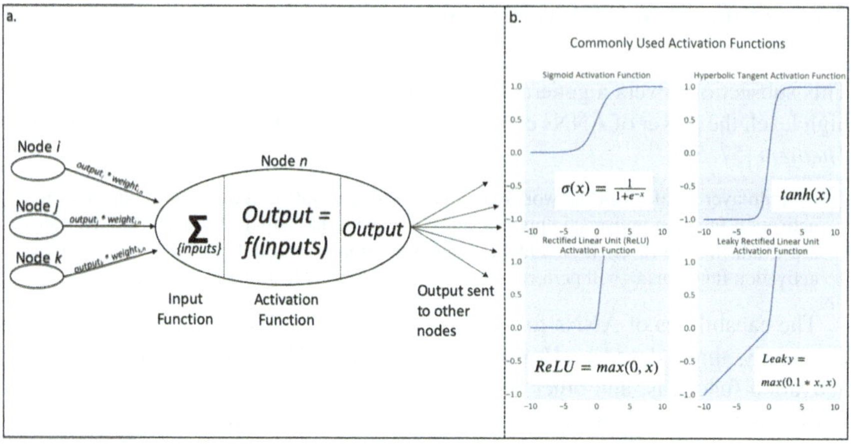

Fig. 1 (**a**) Example of the input, calculations, and output of a node. (**b**) Common activation functions used to calculate node output. Different functions operate on distinct domains and are optimal for various problem types

ANN. As such, ANNs used for classification often use the categorical *cross-entropy* [78], segmentation ANNs may use the *Dice coefficient*[122] and regression ANNs may use either the $L1$ [86] or $L2$ [86] distance between the sample and estimated labels. The simplest ANN architectures, perceptrons with no hidden layers, update their weights by calculating the error between input and output layers through the sum of squared error. Assuming an expected output y and ANN estimate $h_w(x)$, error is defined by:

$$E = \frac{1}{2}Err^2 = \frac{1}{2}(y - h_w(x))^2 \tag{1}$$

Once the loss function between the ANN output and the input labels is defined, a gradient descent approach can be applied to reduce E by taking the partial derivative of E at each weight. This requires a differentiable activation function (such as sigmoid), resulting in a weight update formula:

$$W_i^{t+1} = W_i^t + \alpha \times E \times f'(in) \times x_i \tag{2}$$

with learning rate α, activation function f, node value x_i, and iteration counter t.

Multi-layer Perceptrons (MLPs), the foundation of modern neural nets, use backpropagation for weight updates and learning. The backpropagation algorithm was created in order to account for error being added to the output of the ANN coming from hidden layers. For example, hidden node h contributes a fraction of the total error δ_t to each output node to which it is connected. In order to properly attribute error backwards to the hidden layer from the output layer, more highly weighted connections assign a proportionally higher amount of error δ_h to the hidden node. Mathematically, the error is similar to the traditional perceptron case:

$$\delta_h = f'(in_h) \times \sum_t W_{h,t}\delta_t \tag{3}$$

And the weight update rule from the input layer with value a_i to the hidden layer becomes:

$$W_{i,h} = W_{i,h} + \alpha + o_i + \delta_h \tag{4}$$

with output unit o_i.

2.1.2 From Feed Forward Artificial Neural Networks to Convolutional Neural Networks

The first CNN was published by Yann LeCun in 1989, as a result of trying to optimize the balance between the amount of time needed to train an ANN and that ANN's generalizability across a diverse set of inputs [71]. To do this, LeCun set out to build an ANN with fewer free (trainable) parameters in order to reduce the sample space of functions that could be generated by the ANN and tested as solutions. This was implemented through weight space transformations which allowed for multiple links between nodes in an ANN to be controlled by one weight. Furthermore, that weight is updated only in the parameter space needed to solve the problem, rather than across every possible set of weights, through the convolution operation. LeCun compared the performance of a perceptron, multi-layer perceptron, basic image feature extractor ANN (with downsampling), image feature extractor with weight (or parameter) sharing, and finally, a hierarchical feature extractor, the basis of today's convolutional neural nets. With the increase in downsampling (information compaction), weight sharing, and feature extraction, ANN accuracy and speed increased.

Multi-Layer Perceptrons (MLPs) [49] laid the foundation for modern neural networks. Traditionally, MLPs operate through matrix multiplication to calculate node-wise output and propagate error forward and backward through the ANN. CNNs, however, operate through the **convolution** operation. Mathematically, this operation is defined as the integral of the product of an input and kernel function after one of them is reversed and shifted. This is shown below, where $e(t)$ is the estimate of our image region, $i(p)$ is the input measured at point p, and $w(p)$ is the weight function (kernel) [49].

$$e(t) = \int i(p)w(t-p)dp \tag{5}$$

Another common notation of the convolution operation uses $*$:

$$e(t) = (i * w)(t) \tag{6}$$

In the context of image analysis, a convolution takes the product of an input image and a kernel (an array of parameters with smaller dimensionality than the input image) and outputs a feature map describing how the shape of the input is modified by the kernel. The reversing and shifting of the input function allows for CNNs to be translation invariant with respect to the image. With more rounds of training or increased training data, translation invariance becomes more robust. This means that the same image patch will result in the same CNN output irrespective of the image location where the patch appears. In CNNs, the parameters of the kernel are the weights of the neuron, and are modified through the training process.

In successful CNNs, the feature map is an accurate encoding of the important aspects of the input data needed for the CNN to learn. After applying a kernel function, the resulting CNN only stores node connections that correspond to specific regions of an image, allowing for sparser CNNs with fewer highly weighted unnecessary connections. This means that the CNNs do not contain spurious important features and are more likely to focus on truly important aspects of the data. CNNs with sparse connectivity require fewer operations to compute output and backpropagate error. Thus, learning in a CNN is dependent on optimizing the accuracy of encoded information.

In Fig. 2, the first 2D convolutional layer (far left) takes the image data in as input and passes that data to each of the 1000 different filters (nodes) in that layer.

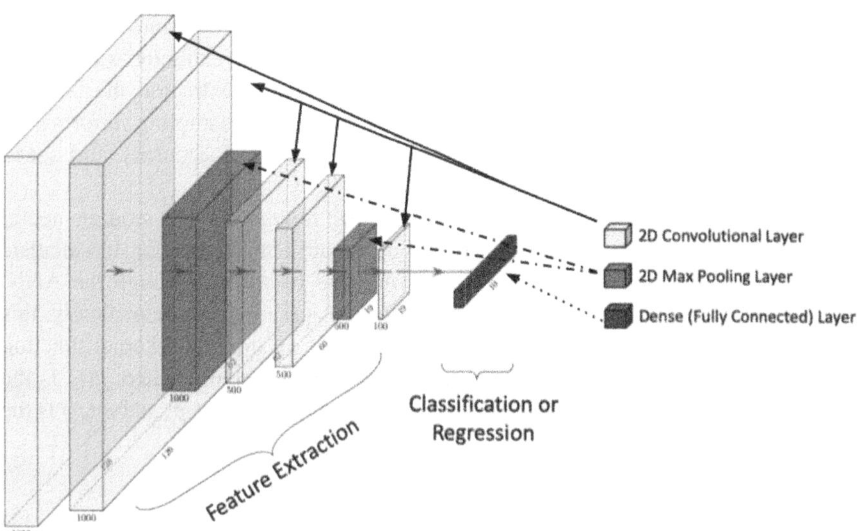

Fig. 2 A sample representation of a CNN

The convolution operation then operates on each of the 1000 filters, downsizing the image input by 2 pixels in each dimension—from a 128×128 input image to 126×126 and applying a nonlinear activation function to each of the 1000 filters to generate values corresponding to image-area importance. The next convolutional layer operates similarly, downsizing the original image to 124×124, again with a nonlinear activation function. This new image, or feature map, is put through a 'Max Pooling' layer, grouping 2×2 grids of pixels throughout the image and extracting the pixel with maximum information to propagate further into the CNN. This effectively downsizes the image by a factor of 2, to 62×62. After another series of convolutional layers and max pooling, the final feature space ends up being 100 different feature maps of 19×19 pixels. This information is flattened into a 1-dimensional vector and put through a series of dense layers with nonlinear activations. Finally, the features are put through the last dense layer with a SoftMax activation function to extract image classes or a Linear activation to extract a regression value. The ideal output of this CNN could be a binary classification for each pixel, corresponding to whether that pixel belongs to the puppy or to the background. The CNN diagram was generated by modifying the *PlotNeuralNet* implementation [61].

While convolution itself is one linear operation, other operations are necessary to propagate information throughout the CNN efficiently. The outputs of convolutional layers are passed through non-linear activation functions that allow for more accurate feature maps and better generalizability across different input and problem types. Typically, following the activation function is a pooling layer. This allows for local value averaging over regions of feature maps in order to decrease the amount of information passed to the next convolution layer. Outside of decreasing the amount of information passed between layers, pooling strengthens translation invariance by allowing the CNN to learn, for example, what should generally be in a region of pixels for a certain image type and also by augmenting the receptive field of the neuron. This method of summarizing what goes on in different neighborhoods of the input to each layer allows for fewer inputs that need to be processed deeper in the CNN, lowering memory requirements and increasing statistical efficiency of the CNN. Following a series of convolution and pooling layers is at least one densely (fully) connected layer of nodes, where all features extracted from the input are included and where the CNN learns to decide the output of the classification/regression problem. Indeed, the convolutional part of the model is often regarded as a feature extraction method, while the fully connected part combines such features to match the label associated with the sample. The exclusion of the dense layer would result in just features being extracted from data, without classification/regression being done on it. An example of a CNN is shown in Fig. 2.

2.1.3 Transfer Learning and Algorithm Interpretability

Coding competitions on large databases of natural images have brought forth numerous high-performing algorithms such as *GoogLeNet*, *Inception*, *ResNet*, and others [55, 109, 110]. A common feature of these networks is that they tend to be wide, meaning that multiple different channels for passing information exist alongside operations that combine data from different regions to optimally learn from inputs. These CNNs are structured to have high resolution feature extractors at higher layers and sufficient low level encodings of those features to pass information efficiently to the classification or regression arm of the CNNs.

Transfer Learning (TL) [87] refers to the adaptation of an ANN whose weights have been trained in a problem to another different problem. The flexible design of these CNNs have enabled their implementation in countless TL approaches that apply the earlier stages of high-performing algorithms as feature extractors to a variety of different datasets. In TL, the learned weights at the higher layers are held constant as they are already trained over a wide sampling of inputs to learn how to extract features. To specify the performance of a model to the exact problem at hand, new nodes and layers are added in and connected to the previously existing ones. This allows for the completion of high accuracy classification tasks, and relatively little training overhead due to weight updates only occurring at the few new convolutional and/or dense layers. A further advantage of TL is that, by having the feature extraction part of the CNN trained, a lower amount of training data is required to achieve high performance.

A growing concern with CNNs, however, is the interpretability of the model. CNNs and more generally ANNs have been referred to as *black boxes*, and unpacking that box is a nontrivial task. A common approach to 'unpacking the black box' is through **post-hoc analysis**. Some these approaches use saliency maps to highlight image regions containing information relevant to correct performance on the trained task, while others look at high dimensional embeddings of output data to clarify CNN operations [105]. Other algorithms, such as *Local Intepretable Model-Agnostic Explanations* (LIME), aim to explain the decision making of a classifier for each prediction as a function of nearby predictions on similar inputs through a linear model approximation. LIME can quantify the important features the model used to make its decision by looking at each prediction in the context of the predicted data point's local neighborhood of similarly predicted data [93]. Shapley values, a game theory based approach, quantifies the contribution of each possible feature to the overall prediction of an input. Compared to LIME, Shapley values give a more complete interpretation of a model by considering how all features play into a model's decision making, whereas LIME focuses on the important ones as defined by the linear model approximation. However, calculating Shapely scores explicitly is computationally demanding and sometimes impossible in complex models, especially compared to the computations needed to compute LIME explanations. *SHapely Additive exPlanations* (SHAP) solves this problem through approximating Shapley values, and still computes the contribution of individual features to a model's prediction. SHAP is related to LIME as it uses a linear model approximation to calculate Shapley values once features and groups of features (coalitions) are sampled. Each feature is weighted by the SHAP kernel, which provides a weight to coalitions based on the number of features that group together. This results in small coalitions (i.e. size = 1) and large coalitions being weighted the most, as they represent either important individual features or important large correlations between features that help the model make decisions[75].

The general consensus across interpretability theorists is that the trade-off between more transparent ML algorithms like *logistic regression* [56] or *Support Vector Machines* (SVMs) [56] and *black-box* algorithms [22] is equivalent to balancing high interpretability versus high accuracy [97]. On some tasks, such as those that do not involve clinical decision making, it might be more acceptable to not know how an algorithm is working as long as it is correct. However, when humans make decisions based on these algorithms, it is crucial to be able to understand CNN functionality in order to enforce regulations or performance thresholds so that the results can be interpreted safely.

CNNs provide cutting edge performance on regression and classification tasks and typically excel at image analysis. With the biologically and clinically relevant tools provided by imaging and the wealth of existing data in both realms, it is only a matter of time before CNNs become highly integrated into research and clinical decision making. Importantly, the biomedical optics field is rapidly developing, with more powerful and efficient imaging modalities being created at a fast rate. The union of these two fields will inevitably provide novel opportunities for extracting meaning and making decisions based on imaging data.

2.1.4 Relevant Terms

Here we outline some of the important terms related to ANNs, CNNs, and DL as mentioned above.

Generalizability: In the context of models, the ability to successfully operate across similar problem types with different input data [65].

Trainable Parameter: The ability of nodes, learning rates, or decay rates to be updated during a model's training [50].

Downsampling: In CNNs, decreasing the size of features in order to force the network to focus on specific aspects of input data that are important for learning [120].

Convolution: A mathematical operation on two functions that produces a third function expressing how the shape of one is modified by the other [108].

Transfer Learning (TL): Applying a model trained on an input dataset to new input data by keeping trained feature extraction layers frozen (untrainable by new input data) and adding new nodes to account for specific features of new input [111].

(continued)

Post-hoc analysis: Approximating a black box through a more easily interpretable algorithm [83].
Translation Invariance: The characteristic of an ANN allowing it to predict slightly perturbed but generally similar data in the same way [50].

2.2 Advanced Optical Imaging Tools

Optical imaging methods provide a powerful toolkit for imaging and quantifying complex systems, ranging from subcellular-level dynamics in biological systems to whole-body imaging of potentially cancerous lesions. Optical methods in biomedicine leverage the myriad of interactions between light and tissue to generate imaging contrast. The majority of optical imaging methods make use of three major light-matter interactions: absorption, scattering, and emission [27]. For example, a simple light microscope primarily provides absorption contrast, with features visualized by their colors and relative brightness (or darkness) in suitably thin samples. Importantly, these light-tissue interactions can be both linear and non-linear, with non-linear interactions capable of yielding new and unique insights. While there are numerous optical imaging modalities, this section will focus on a selection of imaging methods where DL tools are finding special use.

2.2.1 Optical Coherence Tomography

Optical Coherence Tomography (OCT), developed in the late early 1990s [101], is an interferometric optical technique based on light scattering contrast. OCT is the optical analogue of ultrasound imaging, where the round-trip travel time of light to different layers of tissue is effectively measured to construct cross-sectional and tomographic images. As light is substantially faster that sound, OCT makes use of optical interference and signal processing to extract volumetric imaging data. OCT has been used widely in biology and medicine, with clinical deployments in the eye [41], skin [116], cardiovascular system [9], and the gastrointestinal tract [38].

OCT systems all make use of interferometric configurations. In the most common scheme, a broadband light source is split into two paths: a "reference path" and the "sample" path, which are set approximately equal in length. Light returning from these paths is interfered, and the interference is measured on a detector [5, 101]. Modern OCT system operate in the so-called *spectral domain*, using either fixed broadband light, such as a diode, or a swept-source light source [33]. The interferogram is typically processed by a Fourier transform to recover depth information; sweeping the light source over the sample allows for the construction of cross-sectional and 3D tomographic images. OCT datasets can be rather large, in particular OCT image data gathered from *Optical Frequency Domain Imaging* (OFDI) systems, where the interferogram is over-sampled to improve image quality.

The contrast in OCT images is derived from changes in tissue index of refraction, which occurs at the boundary between tissue layers, fibers, and cells. This enables OCT images to capture structural information, such as the individual layers of the retina [85], the structure of the esophagus [39], or details of the placement of arterial stents [13]. As OCT is an optical ranging technique with signal derived via interferometry, OCT has the capability of peering deep within tissues at depths exceeding a few millimeters. Additional information, such as the speed of blood flow and nerve bundle parameters, can be assessed using more specialized OCT toolkits. Coherent noise, also known as speckle, is common in OCT images, generating a “grainy” appearance [5]. This graininess can be removed by averaging pixel values over local neighborhoods, but potentially removes important tissue features. Accounting for scatter and denoising OCT images has been and is still an active area of research.

Computational analysis of OCT is crucial, from the initial processing of the interferograms to image segmentation and feature recognition. OCT data is routinely processed to generate quantitative structural maps of tissue in diagnostic settings, making accurate image processing essential. This is of particular importance in functional OCT imaging, where tissue dynamics are derived from 4D datasets (x,y,z, and time), such as in OCT elastography and Doppler OCT methods.

2.2.2 Multiphoton Fluorescence Microscopy

Fluorescence contrast is used widely throughout biomedical research and medicine, as fluorophores can be leveraged to target specific molecules and structures. For example, antibody-fluorophore conjugates can be synthesized to specifically image where a given protein resides in a cell. Fluorescence imaging is generally carried out using conventional wide-field and confocal fluorescence microscopes, which can visualize multiple fluorescence species in the same sample for multiplexed image acquisition. A challenge with standard fluorescent tools is that the wavelengths used for fluorescence excitation lie largely in the UV to green range of the visible spectrum of light. These wavelengths have limited penetration depth into tissues, allowing for only shallow imaging of specimens.

Multiphoton fluorescence microscopy approaches enable deeper imaging in the turbid tissue environment by making use of longer-wavelength near-infrared light. In multiphoton fluorescence, ultrashort pulsed near-infrared laser light is focused down to a small focal volume. Within this small volume, the photon flux is instantaneously high—in some cases brighter than the surface of the sun. At these photon densities, chromophores can simultaneously absorb multiple photons non-linearly, exciting the molecule into a higher state from which it can fluoresce. The most common form of multiphoton microscopy is *Two-Photon Excitation Fluorescence* (TPEF) microscopy. For example, the molecule fluorescein is typically excited with single-photon excitation at 488 nm, but can be two-photon excited with 750–800 nm near-infrared light [21, 30]. TPEF has been the focus of rapid technological development, with commercial systems widely available today.

TPEF, and to a lesser extent three-photon excitation fluorescence, have found applications for *in vivo* imaging applications, such as those focusing on the skin [79] and the brain [59]. While the majority of multiphoton fluorescence imaging applications are found in research, these tools have some clinical applications. The use of dyes limits real-time applicability in the clinic, but, nonetheless, both modalities have been used to image biopsied tissues, especially tumor samples [89]. For example, multiphoton excitation of endogenous fluorophores can be used to visualize structural elements and metabolic rates [92].

Multiphoton images can be used to generate rich, volumetric datasets of multiple simultaneous fluorescent labels over time. Multiphoton imaging can have resolutions on the hundreds of nanometers, and can be scaled through mosaicing methods up to centimeter spatial scales, making these 5D datasets challenging to process quantitatively. Image segmentation, image feature recognition, enumeration of specific-labeled elements, and tracking objects in time-lapse data are some of the most common computational problems.

2.2.3 Second Harmonic Generation

Second Harmonic Generation (SHG) [43] is a nonlinear process where two photons of the same wavelength interact with a material to be converted into a single photon at half of the original wavelength. The second harmonic process occurs only in special materials that are so-called non-centrosymmetric. Many collagen proteins, as well as myosins, have this unique property, making them attractive targets for imaging based on SHG [20].

Like multiphoton fluorescence imaging, SHG uses laser wavelengths in the near infrared range (700–1300 nm), allowing for imaging hundreds of microns deep with high resolution [21]. SHG and TPEF are commonly carried out together, with SHG providing structure details of tissue alongside the specific target labeling offered by TPEF. SHG and TPEF find extensive use in the imaging of skin, where the distribution and pattern of collagen can be used to understand skin structure and disease state [23]. Like TPEF, SHG datasets can be challenging to quantitatively analyze and often require extensive manual or semi-automated routines.

2.2.4 Coherent Raman Imaging

The *Raman phenomenon* [8] is an inelastic scattering mechanism where light interacts with a molecule's vibrational modes. In spontaneous Raman scattering, light scattering off a molecule can take or leave a quantum of vibrational energy. These interactions can be measured via acquisition of the spectrum of the scattered light to measure a vibrational spectroscopic "fingerprint" of a given molecule. As they have the capability to detect specific molecular signatures without the need for labels, Raman tools are extremely promising for biomedical research and diagnostic medicine. One challenge, however, is that the Raman process is rather weak, such that spontaneous Raman images are acquired in hours or even days [8].

Coherent Raman imaging comprises a set of nonlinear Raman processes that provide the same degree of chemical specificity, but with significantly greater signal strength. The two most commonly used coherent Raman imaging methods are *Coherent Anti-Stokes Raman Scattering* (CARS) [36] and *Stimulated Raman Scattering* (SRS) [44], both of which can be carried out at video-rate speeds [35, 100]. Both imaging methods generate images with vibrational contrast, with CARS being most appropriate for chemical structural imaging and SRS being ideal for chemical quantification. The CARS and SRS processes themselves were discovered decades ago, and were recently implemented as efficient imaging techniques starting in 1999. Both methods are nonlinear imaging approaches that make use of two colors of pulsed laser light tuned such that their difference frequency matches that of the vibrational mode of interest. In this way, coherent Raman imaging tools can specifically tune into and visualize the distribution of vibrational modes, and thus chemical species, within a sample.

Coherent Raman imaging has found many applications in the research space, with significant studies focused on the skin [36, 44, 100, 115] and brain [37, 45, 46, 63]. These tools are gradually being translated into clinical use: a portable CARS imaging system was recently commercialized [16], as was an SRS imaging tool special-built for the detection of challenging tumor margins in brain cancer [45]. Based on much of the same imaging hardware as TPEF, coherent Raman imaging can generate massive 5D datasets. Where in TPEF the dimensions were X,Y, Z, time, and fluorescence channel, in coherent Raman the last dimension now becomes vibrational energy, substantially expanding the size of experimental datasets. This is now a particular challenge with the advent of new, fast-tunable laser systems for coherent Raman imaging [17].

2.2.5 Advanced Imaging Analysis Challenges and Opportunities

These imaging tools, while having different sources of contrast and differing underlying mechanisms, all share similar image analysis challenges.

1. These tools have the capacity to generate extremely large datasets. For example, a coherent Raman imaging system capturing drug uptake data in skin can easily generate 30 GB in a single session. A cardiac OCT system during acquisition can generate tens of GB of interferometric data in seconds. Immediate analysis of that data is required to make diagnostic information regarding issues as complex as stent placement.
2. Advanced image tools generate highly complex and heterogeneous data, especially when images are acquired from tissues and living systems. Spatial information can scale from 100s of nanometers to centimeters, and the data cannot be easily simplified or broken down into reduced forms.

3. Data can be acquired over time, meaning that analysis tools not only need to capture or quantify structural and/or chemical elements, but also temporal changes. This can pose substantial challenges in recognizing and tracking objects of importance, or quantifying key components throughout dynamic image alterations.

Image analysis of advanced microscopy and image system data has traditionally been carried out using computer vision libraries, such as *Open CV* [14] and MATLAB. These approaches often fail on edge cases, as deterministic algorithms suffer from the heterogeneity contained in many images. Many analysis routines end up incorporating manual or semi-automated components, which take time, reduce throughput, include an operator bias, and do not lend themselves towards clinical translation. New approaches via advances in ML address these challenges well and fill in a much needed gap in processing and quantifying advanced imaging data.

3 Interdisciplinary Research Applications Involving Advanced Imaging and Deep Learning

DL approaches have the ability to provide breakthroughs in image interpretation and image analysis. Here we discuss selected examples where DL, and CNNs in particular, have been applied to advanced imaging tools. As these imaging methods are primarily deployed in biomedical and clinical environments, the focus will be on biomedical research applications.

3.1 Applied Optical Coherence Tomography Paired with Deep Learning Analysis

An common goal in OCT image analysis is the extraction of parameters that correspond to the scattering and absorption properties of the sample, referred to as the scattering coefficient and the absorption coefficient. This is typically posed as an inverse problem, where researchers must recover the coefficients from the data acquired by the imaging system. Solving this inverse problem allows for image reconstruction at high resolution, but is a computationally intensive operation.

3.1.1 Analysis of Retinal Degeneration in Disease States

Though optical coherence tomography has been applied to medical imaging since the 1990s, the extent of quantification in the early research was largely through approximations of the scattering and absorption coefficients through complex mathematical models and experimental measurements of tissue composition. Interdisciplinary approaches involving OCT and CNNs were reported in mid-2018 by Camino et al. [3], with the goal of quantifying photoreceptor loss in two types inherited retinal degeneration in humans, choroideremia and retinitis pigmentosa, through a series of depth-resolved retinal tissue images. Retinal images were collected from a 70-kHz, 840-nm-wavelength spectral-domain OCT system. Inherently, the output of this system is a 1-dimensional depth profile (an A-scan), with 300 A-scans sampled along a line to generate a 2-dimension xz profile (a B-scan). Two B-scans were acquired at each lateral location of the retina, with motion correction technology used to merge the data and remove noise from the collected B-scan images.

The team focused their efforts on determining *Ellipsoid Zone* (EZ) differences between photoreceptors of healthy and diseased retinas. The EZ is an eye region clinically used to estimate photoreceptor damage; retinal degeneration diseases can cause the formation of lesions and other abnormalities within the EZ. Each retinal image collected and analyzed depicted a thickness map between the EZ and another neighboring region of the retina, known as Bruch's membrane. The thickness map was then given to a human expert to determine whether there was EZ loss and to grade the extent of EZ loss in diseased patients. The human analysis was saved and used as a benchmark for CNN training, which aimed to segment each B-scan image to establish the EZ boundary and consequently determine EZ thickness.

The CNN implemented by Camino et al. was fed 33×33 pixel patches of B-scan images as input, with the expected output being a set of 2 probabilities corresponding to whether that patch suffered EZ loss or EZ preservation. The CNN architecture was comprised of 3 alternating convolutional and max pooling layers (6 layers total), used to downsample the images and force the CNN to discern important aspects of each retinal image corresponding to the EZ boundary. Following the 6 feature extracting layers, the features were flattened and put through a fully connected (dense) layer with 64 nodes and then to a final dense layer with 2 nodes. These final two nodes correspond to an output vector containing the binary classification of patches, specifically the probabilities of EZ loss or preservation. This 2D probability vector was used to generated *en face* probability maps of EZ loss and preservation. To quantitatively analyze the CNN accuracy, the *Jaccard similarity index* [53] was used to compare the overlap between post-processed (segmented) classified images and their ground truth counterpart. The post-processing took the probability maps generated by the classification algorithm and turned it into a segmentation of the image region corresponding to EZ defects in disease states. Compared to the human grader, the accuracy of the CNN's segmentation of the EZ region was $0.894 \pm .0102$ for retinitis pigmentosa and 0.912 ± 0.055 for choroideremia. Qualitatively, Camino et al. reported fairly high correspondence between the segmentation maps of diseased retinas generated by the human grader and the images classified as diseased by the algorithm (see Fig. 3). In Fig 3. regions in green are areas of the manual segmentation that were not encompassed by the CNN's segmentation (false negative) and the regions in pink are false positive regions. Input image size begins at 33×33 pixels and the final features extracted are an 8×8 pixel region. The network diagram was generated

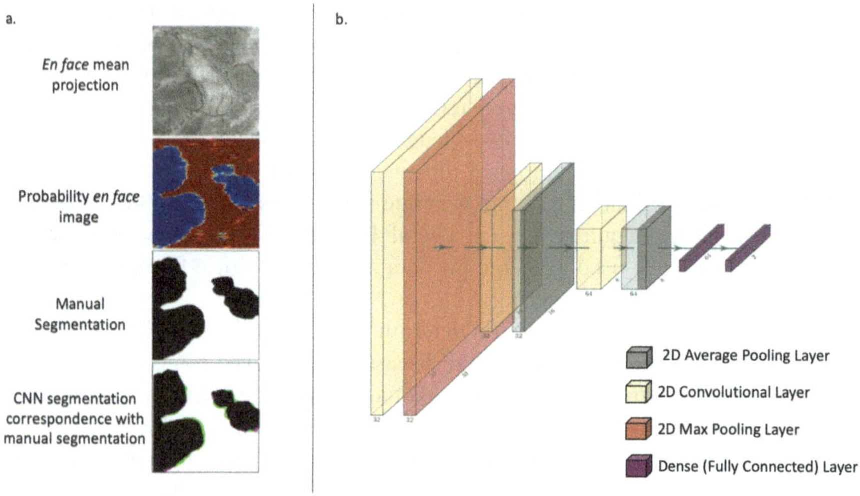

Fig. 3 (**a**) Comparison of OCT image, manual segmentation, and correspondence between manual segmentation and CNN's classified image segmentation. (**b**) Visualization of CNN structure, parameters, and dimensions

by modifying the PlotNeuralNet implementation [61]. All DL computations were carried out a desktop computer with an *Intel* CPU, 16 GB of RAM and a NVIDIA *GeForce Quadro* K420 GPU with 1GB VRAM.

Overall, this exploration of DL with OCT images provided a significant foundation for future studies. While generally successful as a supervised learning algorithm for segmentation, the CNN implemented could be more robust with a larger training set of healthy and diseased images. Using only 10 scans of diseased eyes did not provide enough variation in EZ thickness to give clinically useful results. Specifically, the CNN lacked power to quantitatively discern healthy and diseased retina images. The CNN, furthermore, was unable to detect tubulations in the choroideremia images that, in the future, might be useful for diagnosing or quantifying the extent of disease. However, with advances in computational power, a CNN taking larger images as input might be more adept in extracting high concentrations of tubulations and other small details of images that could correspond to clinically relevant information.

3.1.2 Segmenting Clinically Relevant Retinal Layers in OCT Images

s Retinal imaging was one of the first major clinical applications of OCT, with decades of research and development. The volumetric imaging data offered by OCT can generate 3D renderings of retinal structure completely non-invasively, yielding great clinical value in the diagnosis and monitoring of numerous diseases. While OCT data can be used to visualize the individual layers of the retina, automatic recognition and segmentation of these structures has been a long-standing challenge [42]. Retinal layer segmentation remains a challenge due to image noise, detectable morphological features, image contrast, and retinal layer(s) captured [81].

Shah et al. [102] aimed to analyze retinal B scans from both healthy eyes and eyes with *Age-related Macular Degeneration* (AMD), similar to Camino et al. [3], Shah et al. [102] used 380 OCT volumes with accompanying manual tracings from a publicly available dataset repository [40]. Shah et al. focused on segmenting three surfaces: the *Internal Limiting Membrane* (ILM), the *Inner aspect of Retinal Pigment Epithelium* (IRPE) drusen complex, and the *Outer aspect of Bruch's Membrane* (OBM). In contrast to the approach of Camino et al., Shah et al. included the IRPE in their training and analysis, which was a challenge due to its high reflectivity. The team took an algorithmic approach to determining the best ANN for their segmenting needs. Three different CNN-based structures were tested on the curated dataset: CNN-S, a CNN derived from *Alex-Net* by Shah et al., and two *U-Net* structures that only varied in the final output convolution layer [69, 95]. Benefits of using the *U-Net* included strong segmentation power, fast runtime, and robustness (in terms of not needing many training images for accurate segmentation). CNN-S had upsampling and downsampling layers through varying node density per layer starting at 16 nodes and increasing to 32, then 64, before decreasing back to 16 nodes. Two fully connected layers with 2500 and 1200 nodes were used to carry out the segmentation post feature extraction. The segmentation approach attempted to assign different values to different retina layer volumes shown in the OCT image corresponding to ILM, IRPE, and OBM. To account for error and allow for accurate weight updates, a canonical pixel-wise mean square error function was used for the *U-Net*, while a quadratic-penalty loss function was applied to surface position errors in CNN-S during training. The CNN was implemented via CAFFE [64], and was trained and tested using an NVIDIA *Titan* X GPU. A *Linux* desktop (3.4 GHz, 16 GB memory) was used to generate the manual segmentations.

Prior to training, Shah et al. used data augmentation to increase the variation in training data images while still providing biologically feasible images to the tested CNNs. Their augmentation technique involved multiple translations and rotations to each image, and through saving intermediate images allowed for a 4-fold expansion in training set size with over 75,000 images. To assess segmentation error, Unsigned Mean Surface Positioning error [72] and *Unsigned Average Symmetric Surface Distance* (UASSD) error [102] were employed, but only UASSD was calculable for the *U-Nets*. Generally, these metrics quantify differences between predicted and ground truth segmentation of images.

Quantitatively, CNN-S outperformed both *U-Nets* and an additional non-DL method tested. However, it should be noted that each image tested was compared against manual tracings for a 5 mm diameter region, serving as the ground truth or ideal segmentation that the CNN aimed to achieve. Consequently, larger regions were not tested, and given *U-Net* performance on both cell and tissue level segmentation tasks, it is quite possible that performance between CNN-S and the *U-Nets* with larger images would be comparable.

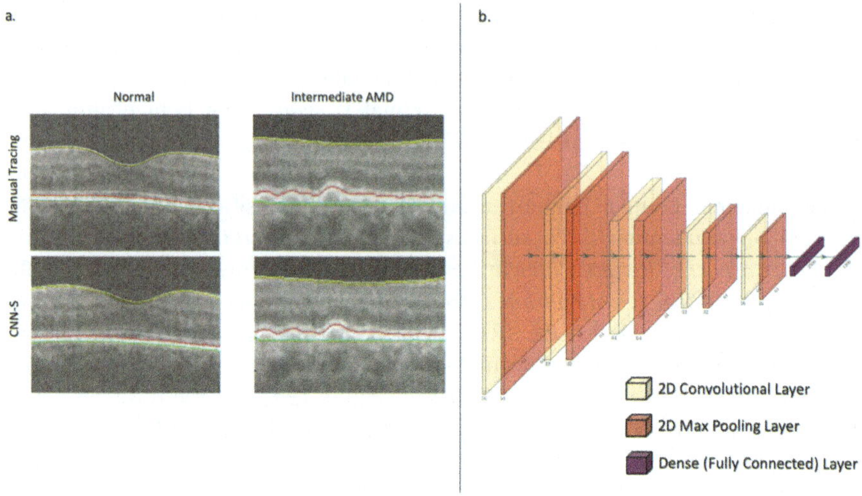

Fig. 4 (**a**) Comparison of segmentation in healthy and diseased retinal images done by both humans and CNN-S. (**b**) Visualization of CNN structure, parameters and dimensions

While concern about small image size is valid, the work in Shah et al. [102] demonstrated a robust comparison of various DL approaches for retinal OCT image analysis. Due to the focus on different retinal layers, no 1-to-1 comparison between the work published in Camino et al. [3] to Shah et al. 's work is possible; however, it seems as though the CNN-S model is less human-dependent than the method implemented by Camino et al. The approach employed by Camino et al. is dependent on pre- and post-processing steps for image input and CNN output respectively. However, the CNN-S approach is entirely automated after data augmentation. The similarity between the top performing CNNs used (Figs. 3 and 4) speaks to the dynamic and somewhat unpredictable changes between CNN performance across even mildly different tasks.

3.1.3 CNN Analysis of Non-Retinal OCT Volumes

Clinically relevant research and medical imaging extend far beyond ophthalmology applications, including tissues where biopsy is risky, such as cardiovascular [47]. Mojahed et al. [82] reported a CNN-based analysis of clinically relevant OCT data not related to opthalmology. To date, this appears to be the first application of OCT outside of the eye that involved DL to both quantify and analyze the imaging. This effort focused on the analysis of breast tissue OCT images from both cancerous and non-cancerous regions, and employed a CNN trained to discern between healthy and diseased samples to extract clinically useful results. Furthermore, the CNN model was tested on its ability to classify individual tissue classes from one another.

Mojahed et al. leveraged the optical penetration depth of OCT imaging to visualize 4 different breast tissue types—Stroma, Adipose, *Ductal Carcinoma In Situ* (DCIS), and *Invasive Ductal Carcinoma* (IDC) from human samples with an average size of 1.2 cm^2. They implemented a custom-built OCT setup with an axial resolution of 2.7 μm, lateral resolution of 5.5 μm, and low-noise super-continuum laser centered at 840 nm. The group chose to work with this custom set up as it generates better quality images (in terms of human interpretability and classification ability) compared to commercial OCT systems.

Many aspects of the architecture and computational hardware used were not disclosed, but it is apparent that the CNN operates through varying stride sizes and normalizing information passed between layers. The structure of their CNN includes 5 iterations of 3 directly connected convolutional layers with non-linear activation functions for feature extraction. Following the 15 convolutional layers is a dropout layer to avoid overfitting and finally the output layer of 4 nodes containing classification probability for the 4 tissue types. The output of the model was tested by measuring the *Dice coefficient* [122], which is used to measure similarity/overlap between sets of data. This score was applied to both the binary split of healthy/diseased images as well as across the split of four tissue types tested for classification ability.

The model showed the highest *Dice coefficient*[122] for classifying IDC in the tissue classification task and for classifying images with cancerous samples for the binary classification. By demonstrating the OCT with automated analysis outside of the eye, it serves as an example that could be used to build improved analysis methods to process data for clinical applications.

3.2 Non-linear Microscopy with CNN Analysis

Histological analysis of biopsy specimens is the gold-standard in clinical diagnosis. Trained pathologists manually inspect thin-sliced, labeled portions of tissue for cellular and morphological characteristics including cellular and nuclear appearance, abnormal cell counts and density, alterations to tissue structure, and the abnormal presence of cells, amongst others.

There has been a considerable push to automate and quantify histological analysis, which is a potentially apt application of ML approaches. Currently these algorithms still operate with pathologist input as 'ground truth' for diagnosis, but often operate orders of magnitude faster than pathologists with sometimes minimal differences in accuracy [32, 55, 69, 106, 110, 112, 114].

Optical imaging approaches, which can acquire multiplexed, multi-label images of prepared tissue specimens, seem to be ripe for applications to augment histological analysis. Chemically-specific imaging, whether derived from fluorescence or vibrational contrast, can be carried out at high speeds even *in situ* potentially obviating the need for biopsy. The following series of studies detail advanced microscopy approaches paired with DL analysis that compete with the diagnostic accuracy of pathologists.

3.2.1 CARS and CNNs for Automated Lung Cancer Diagnosis

One of the earliest publications detailing a combined advanced microscopy-CNN approach was in lung cancer diagnosis. Weng et al. [117] made use of CARS microscopy to identify pathologically relevant lung tissue features for real-time diagnosis. CARS, a coherent Raman imaging method, enables chemically-specific imaging by tuning into specific vibrational modes of molecules.

At the time of this work, a joint biopsy and histopathology analysis from *Computed Tomography* (CT) images served as the diagnostic standard for lung cancer. However, the drawback of this biopsy approach is that lung lesion biopsy and imaging at a potential cancerous site in lung tissue is difficult due to limited image quality as a result of the breathing motion of patients. Weng et al. [117] used 2-D and 3-D CARS images of lung tissue along with a CNN classifier system that can discern between healthy and diseased lung tissue, as well as separately classify three lung cancer subtypes from normal tissue: small cell carcinoma, squamous cell carcinoma, and adenocarcinoma. In order to operate with high accuracy, the algorithm needed to associate differences in fibrous elastin and collagen structures between the images of different cancers. This is a crucial observation for pathologists to make during diagnosis, as healthy lungs are composed of up to 50% elastin to allow for the lung to stretch during respiration. Collagenous lung structures allow for the maintenance of tensile strength in the lung as it operates. Cancerous lungs have less fibrous structures, corresponding to higher cell density seen in CARS images [19, 34].

Segmenting densely packed from loosely packed cells is a fairly easy ask of CNNs, as demonstrated by many groups [6, 62, 113]. However, to replicate pathologist training and performance, the algorithm needed to perceive higher complexity cellular features associated with the different cancer subtypes. For example, in adenocarcinoma, cells typically appear round and contain vesicular nuclei, while the surrounding elastin and collagen fibers are not degraded [18]. Squamous cell carcinoma cells have cross-cellular bridges and dense layers of cell and cytoplasm. Finally, small-cell carcinoma appears as round cells with densely packed nuclei and relatively less visible cytoplasm. Distinguishing between these differences is critical in accurate lung cancer diagnosis.

Adequate CNN performance is often a function of the number of images available for training, as well as the CNN's feature extraction ability. Weng et al. repurposed *GoogleNet Inception v3* [110], a tournament winning algorithm in image recognition, through TL in order to analyze the CARS images. Similar TL approaches to medical data showed promising results, and simple to implement. In order to apply *GoogleNet* to this task, Weng et al. [117] made minor modifications such that at least one feature extraction layer was trained on the CARS images and the classification layer had 4 outputs—corresponding to the 4 classes of possible images (healthy or the three cancers). After augmenting their image dataset by rotating and mirroring many of their images, which provides more biologically feasible images to the CNN, Weng et al. used nine-fold cross validation to train their CNN to minimize categorical cross-entropy. The TL was implemented using *Tensorflow* [4], with training executed on the *Amazon Web Services* (AWS) cloud, using an instance which contains 4 vCPU with 61 GiB memory and 1 GPU with 12 GiB memory.

The classification accuracy reported by the CNN included perfect classification of healthy lung images from diseased images, a reasonable milestone given the stark differences between healthy and diseased tissue CARS images. However, of note is the ability of the CNN to discern not only between the types of cancers, but also within the two most prevalent *Non-Small Cell Lung Carcinoma* (NSCLC), squamous cell and adenocarcinoma. While the algorithm's average accuracy in classifying squamous cell carcinoma and adenocarcinoma is roughly 88%, squamous cell is more typically misclassified as adenocarcinoma, which is expected given their similar morphologies. Given that the algorithm's adenocarcinoma detection accuracy is 92%, it is possible that the algorithm detected minor unique characteristics in images from that disease state. However, these characteristics may not be sufficiently robust to be a main deciding factor between the NSCLC types. The algorithm was able to determine small-cell carcinoma images from the other 3 images types with 95% accuracy, speaking to the distinct characteristics separating normal tissue, small cell carcinoma, and NSCLC. In addition to their 2-D image analysis, Weng et al. [117] used 3-D CARS images to label tissue types which worked at an overall higher accuracy than the four-image classifier system on 2-D images (92.3% accurate), but this accuracy in the 3-D approach could be inflated by only testing their method on 39 of images, compared to over 700 used for testing the 2-D CNN approach.

This landmark paper provided a strong foundation for future automated diagnosis work. By not only having an algorithm that performs well, but also visualizing hidden layers of the CNN, Weng et al. [117] demonstrated a more interpretable black-box (CNN-based) approach to image classification. This is quite significant, as setting interpretability thresholds, such a representation of the hidden layers of a CNN, or quantifying differences between hidden layers in different image type is crucial to deploying algorithms in the clinic trusted by patients, doctors, and governing bodies. Developing a portable CARS system, a goal mentioned by Weng et al. [117], paired with DL analysis and pathologist proof reading of disease outcomes from images could drastically change clinical practice. Overall, the results and interpretation provided in this work pave the way for CARS and other non-toxic microscopy applications in the clinic.

3.2.2 CNN and Multiphoton Microscopy for Automated Ovarian Tissue Analysis

In a similar style to the work done by Weng et al. [117] on lung tissue CARS images, Huttunen et al. [60] applied TL approaches to analyze ovarian tissue in infected mice. Huttunen et al. [60] used two types of two photon microscopy, *Second-Harmonic Generation* (SHG) and *Two-Photon Excitation Fluorescence* (TPEF) to image collagen fibers and elastin, the same hallmarks of cancerous tissue visualized by Weng et al. [117]. In the ovarian tissue case, collagen and elastin are indicators of disease through their presence or absence in the extracellular matrix [68, 84]. SHG is specifically focused on these two components, while TPEF visualized tissue fluorescence and thus cellular morphology and structure. These two imaging modalities have been used in tandem to study ECM restructuring various cancer progressions, but no robust or automated approach to quantifying these images and structural changes existed until this approach by Huttunen et al. [12, 25, 68, 84, 91, 118].

Huttunen et al. [60] used four high-performance pre-trained CNNs in their analysis, *AlexNet*, VGG-16, VGG-19, and *GoogLeNet* [69, 107, 110]. These four CNNs all consecutively won the *ImageNet Large Scale Visual Recognition Challenges* [99]. Similar to the approach by Weng et al. [117], many of the layers in the pre-trained CNN were left 'frozen', and were not updated by training on the set of ovarian tissue images. The only layers accessible to training corresponded to the classification portion of the CNN—in which Huttunen et al. [60] fed the features extracted by the convolutional layers to either a support vector machine or fully connected dense ANNx layers. These classification layers learn weights associated with the ovarian tissue images to determine whether an image corresponds to a healthy or diseased tissue sample.

To create samples to image using SHG and TPEF, 5 healthy mice and 5 ovarian cancer model mice were imaged. The generated tissue images were originally 800×800 pixels, which would have been extremely large to pass through a CNN. Instead, the four CNNs were trained with images that were 227×227 or 224×224 cropped image patches taken from the main image, which functioned as a data augmentation mechanism as well. Although patch extraction is beneficial to the CNN by giving more diverse images to train and test on, it is possible that gaps between patches resulted in missing information that could impact CNN learning speed or overall accuracy. Training and testing of the models was done on individual image types and on a combination dataset with both multiphoton microscopy types. Overall CNN classification accuracy was high across all four CNNs both with a densely connected layer based classification and SVM classification. Consistently, the dense layer approach to classification outperformed SVMs [60], a common feat, but SVMs show more transparency in terms of where the algorithm can and cannot perform well. SVMs allow for visualization of model decision making by using the parameters of the model's fit to generate a decision boundary that maximizes the distance between the possible classifications. The trade-off between increased accuracy through dense layer-based classification approaches versus increased comprehension of algorithm decision making is one that has yet to be optimized. Typically, it is up to the researcher to decide whether the performance boost is worth the decrease in transparency. However, the 3% average accuracy increase in the CNN approach paired with that 5% accuracy increase through using both TPEF and SHG data does appear to be worth prioritizing the dense layer method to achieve consistently better performance.

It is interesting to consider why, in this case, using both TPEF and SHG images resulted in the best algorithm performance. It appears that the CNN is dependent on seeing both the extracellular matrix and matrix morphology to accurately diagnose ovarian tissue. Typically, pathologists base their diagnosis decisions on the collagen and elastin presence in the ECM, raising an interesting point that additional tissue morphology image data could further increase clinician diagnostic accuracy. The CNNs tested here all classify data with high sensitivity and specificity (low false positive and false negative rates), and with overall accuracies around 95%. The work by Huttunen et al. [60] demonstrates another valuable label-free imaging approach that is capable of extremely high resolution images, and potentially applicable to clinical tasks. Though CNNs as large as the ones used by both Weng et al. [117] and Huttunen et al. [60] are memory intensive, developing smaller CNNs tailored to specific tasks rather than generally good image recognition would allow for better real time imaging analysis in the clinic to pair with standard pathology analysis.

3.2.3 Summary

The discussed research is by no means an exhaustive list of unique and productive results obtained by advanced microscopy paired with CNNs. However, they do show two high quality TL approaches applied to clinically and biologically relevant data created by label-free nonlinear imaging approaches that can be directly applicable in the clinic. Label-free approaches result in no toxic elements being required for acceptable image generation, making organizational approval for these tasks applied to humans less cumbersome, and non-invasive approach safer and less complicated for patients. These examples of TL demonstrate it to be a powerful and generalizable data analysis tool.

3.3 U-Nets and Advanced Microscopy Applications

This section introduces the application of a specific type of CNN, known as a *U-Net*, to a number of advanced imaging modalities [95]. The *U-Net*, first introduced in 2015, is a robust and efficient CNN used for segmentation among other biologically relevant tasks. It has powerful feature extraction abilities, allowing it to be deployed in across different scenarios. As originally published, *U-Net* training requires original images as well as 'image masks' depicting what an accurate segmentation should look like. Then, the *U-Net* learns the relationship between

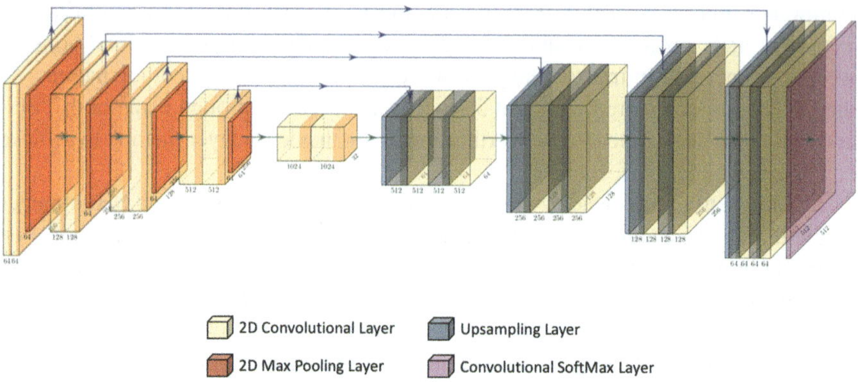

Fig. 5 Visualization of *U-Net* architecture

those masks and the original image to perform segmentation. While it is a supervised learning method, after the initial energy to create a rigorously annotated dataset, the implementation of the CNN is straightforward and the quality of the output is typically high. The works covered in this section are not by any means an exhaustive list of the applications or potential uses of the *U-Net*, but are intended to showcase the versatility of a broadly applicable CNN that can perform at high levels on many different tasks with limited training data.

Figure 5 shows how uniquely, *U-Nets* combine features generated from traditional convolutional layers that downsample and convolutional layers that 'upsample'. Preserving these features on either side of the bottleneck (the consecutive 1024×32 blocks) may be the reason *U-Nets* are so adept at image segmentation and denoising. By iteratively downsampling, only important image aspects remain in the bottleneck, then these features are made larger in the upsampling region to the right of the bottleneck. The CNN diagram was generated by modifying the *PlotNeuralNet* implementation [61].

3.3.1 U-Net-Inspired Algorithms for Image Segmentation

The *U-Net* was originally created for cellular segmentation tasks. A strong suit of the *U-Net* is its ability to operate with augmented images, namely mirror images and rotations applied to original images, in order to learn translation invariance and accurate segmentation. This allows for less time to be spent on potentially laborious imaging and more time to be devoted to letting the CNN learn on both real and augmented data. Originally, the *U-Net* was applied to 3 different cell image datasets, one cell type imaged through *Electron Microscopy* (EM), and two through light microscopy. In all cases, and by all reported metrics, the *U-Net* outperformed the previous state of the art algorithm used for image segmentation.

In 2019 alone, hundreds of research articles have been published that use the *U-Net* in some capacity. Here, we cover a few unique publications [31, 66, 76] detailing *U-Net*-inspired algorithms for different segmentation tasks. These tasks aimed to detect small discolorations in retina images that correspond to *microaneurysms*, to segment histopathology images, and segment mouse skin layers from OCT images [31, 66, 76]. All three of these approaches use custom built *U-Net*s with the same general encoding/decoding approach that allowed the original *U-Net* to learn high level and low level image features, but the incorporated changes to the algorithm increase the performance relative to the original *U-Net*.

In what is the most unique application of the *U-Net* out of the three covered in this section, Mahmood et al. [76] created an generative adversarial *U-Net* to both synthesize histopathology images and segment them. This implementation was based on *cycle Generative Adversarial Networks* (GANs) and *conditional GAN*s [80], the details of which are beyond the scope of this chapter. Briefly, a GAN is composed of a generator, *G*, which creates synthesized data, and a discriminator *D* that determines whether data passed through the ANN is real or fake. Their CNN used histopathology images sampled from different organs to create synthetic *Hematoxylin & Eosin* (H & E) stained images [76]. Additionally, the adversarial *U-Net* learns an optimal loss function from training, and is capable of segmenting a test set of slides, both true slides and synthesized ones, with over 94% accuracy.

This GAN can truly be decomposed into sub-CNNs, one that creates the synthetic images, and one that does the segmenting. In short, to maintain synthetic data fidelity from the cycle GAN portion used for data generation, multiple loss terms were created that penalize the GAN for the creation of unrealistic histopathology images, L_{GAN} and L_{CYC}. These terms were maximized over the discriminator portion of the GAN and minimized over the generator part of the GAN, which in essence maximizes the ability of the GAN to detect synthetic images and minimizes the likelihood that a generated histopathology image is biologically unfeasible. The conditional GAN portion is used to segment images that pass the quality thresholds imposed by the cycle GAN generator and discriminator. Here, the GAN aims to minimize L_{GAN} and a per-pixel loss term L_{L1} over the minimum quality segmenter, *S*, and maximum discriminator *D* to effectively penalize poor pixel configurations allowed by the discriminator and poor segmentation output of the GAN. This model was qualitatively analyzed across the different organ nuclei tested, colorectal, prostate, bladder, stomach, kidney, and breast, and overall showed remarkably good overlap between ground truth segmentation and predicted segmentation by the GAN. Quantitatively, this adversarial *U-Net* approach outperformed the previous state of the art reported by Kumar et al. [70] across four metrics, most notably the *Dice coefficient*[122] measuring overlap between ground truth and predicted segmented output.

Two interesting publications came out two weeks apart, both applying dense recurrent residual blocks within the *U-Net* to perform segmentation: *Dilated-Residual U-Net* (DRU-Net) [31] and *Densely Connected U-Net* (DCU-Net) [66]. The DCU-Net was created by Kepp et al. [66] in order to segment OCT images of mouse skin. Kepp et al. [66] imaged 20 OCT volumes of mouse skin at a central wavelength of 1310 nm, an axial resolution of 3.4 μm and a transversal resolution of 6.5 μm. The total field of view was 3.5 × 8.5 × 8.5 mm, with an image volume size of 1024 × 1307 × 1307 voxels [66]. In order to increase image quality by stabilizing the mouse, a spacer was attached to the mouse prior to image acquisition to dampen the breathing motion of the mouse.

The *U-Net*-inspired algorithm implemented by Kepp et al. [66] uses the same general principle of upsampling and downsampling data while incorporating connections that both follow the downsampling and upsampling through the network, but also connect across the 'U' to retain high level features as well as low level ones (see Fig. 5). However, rather than use plain convolutions as originally in the *U-Net*, the group replaced the convolutional and max pooling layers within a *Densely Connected* (DC) block. Each DC block saves the original data fed into it, but also performs a downsampling before concatenating the original input data with the newly extracted features from the downsampling. Each DC block effectively preserves and reuses features which allows for accurate learning with less training.The CNN was implemented in *PyTorch* [88] and trained using a *NVidia GeForce 1080 TI* using CUDA v8.0 with cuDNN v5.

To measure the quality of predicted segmentation relative to ground truth, *Average Symmetric Surface Distance* (ASSD) [102] and the *Dice coefficient* [122] were used, along with the *Hausdorff Distance* [54] of outlier detection. Overall, the DCU-Net generally outperformed benchmark algorithms such as *Random Forests* [15] and the original *U-Net* in all metrics across the 4 regions imaged. There are a few cases, however, where previous algorithms are better. First, the original *U-Net* resulted in the best *Dice coefficient* [122] for segmenting the deepest region imaged, additionally the Random Forest had the best *Hausdorff Distance* for the 2 most shallow regions of tissue images. The only metric where the DCU-Net won across the board was ASSD.

The second modified *U-Net* published around mid 2019 [31] used recurrent blocks to segment small regions of retina images in a publicly available database to find small retinal regions corresponding to microaneurysms, a common complication of diabetic retinopathy that can cause blindness if not diagnosed early. The CNN used here, a DRU-Net, replaces the typical convolutional layers of a *U-Net* with a *Recurrent Residual CNN* (RRCNN) block, which uses recurrent convolutional layers to not only propagate information forward but also stack that information within the same node. Each RRCNN block has 3 recurrent convolutions that operate sequentially and concatenate with the original data input to the node. Similar to above, this allows for the preservation of information passed to the CNN, increasing the number of times the same feature is used in learning which speeds up the overall training process. The CNN was created using *Keras Tensorflow* [24] and trained using an *NVIDIA GeForce GTX 1080Ti*. The performance of the CNN was measured through calculating the *sensitivity*, *specificity*, *overall accuracy*, and *Area Under Receiver Operator Characteristic* (AUROC). This CNN was tested on two publicly available datasets, *E-Opatha* [28] and IDRiD [90]. The DRU-Net performed the best compared to a fully connected CNN, the original *U-Net*, and a modified *U-Net* using residual learning called *ResU-Net*, except for having the second best specificity on the *E-Opatha* dataset of .963 compared to *ResU-Net* at .967. As originally created, the *U-Net* is a powerful segmenter. However, slight modifications to allow for the reuse of features seems to bolster the segmentation ability of the CNN. Whether through recurrent nodes or dense blocks, concatenating information passed to a layer to new features created by downsampling/upsampling that data passed results in more robust performance according to a variety of metrics. Using a *U-Net* in a GAN did not impact the ability of the CNN to segment, but more quantitative analysis on that work would vastly strengthen confidence in such an approach, especially given that it was tested on images from six different organs. Overall, the *U-Net*'s ability to segment images makes it useful for biological and clinical goals, but as demonstrated in the next section, it can perform other tasks with high accuracy.

3.3.2 U-Net for Image Denoising

Lin et al. [74] and Manifold et al. [77] independently explored some of the earliest applications of DL to denoising *Stimulated Raman Scattering* (SRS) images. SRS is capable of visualizing individual cells at ideal conditions, but under challenging settings such as high-speed data acquisition or deep tissue imaging, the signal-to-noise ratio can rapidly decline. This makes denoising SRS images extremely valuable for cleaning up poor image quality for downstream quantitative analysis. This approach has the capacity to be especially powerful for safety in the clinic by reducing laser exposure time. Fast denoising algorithms would allow for SRS to be used for near real-time medical imaging analysis along with visualization of biologically relevant samples.

The first reported research using the *U-Net* for image denoising came from Lin et al. [74]. Their group aimed to first create a fast scanning SRS imaging system, capable of imaging speeds of 20 μs per spectrum. This imaging speedup came at the expense of reducing signal integration time, which lowers image quality. To overcome this, Lin et al. [74] used DL to remove image noise from rapidly acquired imaged data to generate improved SRS images of PaCa-2 cancer cells with quality comparable to that acquired at slower imaging speeds.

In this case, the goal of the *U-Net* was to take high speed low-SNR images and map them to low speed high-SNR images. The training dataset consisted of high speed and low speed images generated at the same *Field Of View* (FOV). This allows for the CNN to know what the low-quality, high speed image should look like upon refinement. By feeding in both image quality types, the model could learn spatial and spectral correlations that exist between the two images by keeping the FOV constant. Low speed images were created by averaging 100 images taken of the same field of view to ensure consistency in image quality. High speed images were extracted by using just 1 out of these 100 images taken of the cells. Lin et al. [74] rationalized the number of images averaged (N) by claiming that SNR increased by $\sqrt{N}$ times. Nonetheless, the *U-Net* implemented in *Keras Tensorflow* [24] and trained with an *Nvidia GTX 1070* GPU was capable of refining low-fidelity (high speed) images to appear almost identical to ground truth (averaged low speed images). No quantitative metrics such as a *Dice coefficient* [122] or *Pearson correlation coefficient*[67] was calculated, thus no error was reported in this work, yet it is clear through the images and qualitative analysis that the *U-Net* is well suited to denoising SRS images.

Roughly six months after the Lin et al. [74] publication, Manifold et al. [77] claimed to report the first application of DL to SRS images, to their knowledge. Manifold et al. [77] applied a much more rigorous approach to analyzing *U-Net* performance on denoising SRS images. Compared to Lin et al. [74], Manifold et al. [77] publication samples the two possible SRS image geometries—epi and transmission. Epi SRS images typically are lower in quality due to the backwards collection geometry.

Manifold et al. [77] applied SRS to two different samples, *HeLa* cells and murine brain cells. The *HeLa* cells were imaged by using a pump pulse of 800 nm with a Stokes pulse centered at 1040 nm, with images collected at 2913 cm^{-1} and a spectral resolution of 15 cm^{-1}. Murine brain images were collected in two colors, at 2913 cm^{-1} and 2994 cm^{-1} with a spectral resolution of 45 cm^{-1}. These two image types appear vastly different, and serve as good test-case data for *U-Net* denoising. Similar to the approach of Lin et al. [74], images were collected at high and low quality with FOV held constant. This provides the *U-Net* with sufficient information to understand how to refine the low quality images. The image quality is controlled by the Stokes power used in SRS imaging. For example, images collected at 20 mW have 20-fold higher SNR than those collected at 1mW, but the 1mW images are created much faster [121]. Images collected at high and low stokes mW values were denoised by a Poisson approach and by PURE-LET [73] as benchmarks for the *U-Net* approach.

To determine denoising quality, Manifold et al. [77] constructed plots measuring pixel values across a line drawn in the FOV. In addition to these plots, three metrics were computed—*Peak Signal to Noise Ratio* (PSNR), *Root Mean Square Error* (RMSE), and Pearson's *Correlation Coefficient* (CC). PSNR is a logarithmic measure of image quality compared to a ground truth image (in this case, compared to the high Stokes power image). RMSE quantifies denoising algorithm accuracy compared to the noise of the ground truth image. Finally, CC measures colocalization of salient image features in the low quality and high quality images. In the *HeLa* imaging case, the DL approach to image denoising resulted in a pixel value plot showing as smooth, if not smoother, changes in pixel values across the image as the high Stokes power image showed. Furthermore, across all three metrics, the *U-Net* outperformed the other denoising methods. In the two color murine brain imaging case, the *U-Net* outperformed the other algorithms, but not by as much as in the cell imaging case. This could be due to the two colors in the image providing better references for denoising algorithms than the 1-color cell imaging case. Nonetheless, this DL approach to SRS image denoising qualitatively and quantitatively outperform other denoising methods.

These two examples indicate the power of *U-Net* denoising. Whether applied to cells or tissues, the *U-Net* can adequately improve image quality. Compared to the custom *U-Net*-inspired algorithms used for segmentation approaches, these canonical *U-Net* applications scratch the surface of possible performance ability. More customized ANNs to specific problems that integrate recurrent layers or dense layers, as described for segmentation examples, could very likely result in much better denoising ability.

4 Data Management for CNNs in Imaging Applications

The choice of the model architecture, the use of TL, a proper loss function and an optimization algorithm are key for the performance of a CNN. However, the availability of a large and correctly annotated training dataset may have more influence on the CNN performance than the above-mentioned parameters.

What happens when an algorithm trained to learn how to best solve a specific problem using one dataset is evaluated for accuracy in a different dataset? What if the datasets were acquired differently, such as images acquired in distinct imaging sites with different protocols or from patients with the same disease but different symptoms? In these tests of ANN generalizability, the performance often decreases. Yet, for a DL algorithm to be useful in practice it should be able to perform under different image acquisition conditions, on different instrumentation and, more importantly, be able to account for the variability of the subjects in which the system is going to be tested without suffering dropping its performance.

Ideally, a large annotated dataset encompassing all possible image variability could be used to test the limits of algorithm performance. Unfortunately, such is rarely the case in most clinical and research applications. In natural image processing, strong efforts have been devoted to the generation of such annotated datasets through crowd-sourcing mechanisms. Such is possible in natural images, since there is no need of expert training to be able to label the images. Unfortunately such a paradigm is not applicable to biomedical images, as years of expertise are often required to properly analyze the image. Systems using standard optical imaging techniques are widespread, and electronic health records and picture archiving and communication systems are common in health institutions. The combination of both systems have led to the possibility of generating large multi-institution datasets in which to train and test the ML systems and has set the standard for data management in CNN applications.

4.1 Case Studies

To explore these applications, this section will focus on four DL systems that have been trained on clinical datasets: diabetic retinopathy diagnosis from retinal fundus images, melanoma diagnosis from skin pictures, macular disease diagnosis from OCT images and cervical cancer diagnosis from CARS.

4.1.1 Detection of Diabetic Retinopathy in Retinal Fundus Photographs

In 2016 Gulshan et al. [51] published an outstanding work on the use of DL for the automated diagnosis of diabetic retinopathy and diabetic macular edema in retinal fundus photographs. Perhaps the most impressive part of this work is how they used and managed the training and validation data. The authors used a very large training database of 128,175 retinal images acquired at several sites in the U.S. and at three different hospitals in India. The evaluation of the algorithm was performed on a dataset acquired in the same U.S. hospitals (EyePACS-1 [26]) and on a second one in three hospitals in France (Messidor [29]). The acquisition cameras for the training dataset and the *EyePACS-1* dataset were the same; images for the *Messidor* dataset were taken with a camera that had been used to acquire other images in the training dataset. There were small variabilities on the performance of the system in the *EyePACS-1* or the *Messidor* dataset. The performance augmented with respect to the number of samples used to train the system, plateauing at 60,000 cases.

Several important aspects of this training dataset are important to point out: it is multisite, it is multi-instrument and it includes subjects from Caucasian, Asian, African and Hispanic origin. With such design the authors shielded themselves with respect to imaging protocol, imaging device and subject variability.

One could criticize that the high performance on the EyePACS-1 database is related to the fact that they are acquired under the same conditions. Noteworthy, however, was that a replication database unrelated to the development database, showed this to not be the case. The inclusion of a replication database here was therefore critical, and highlights a need for developing CNNs in optical applications. It is important to note that a variation of the system was subsequently evaluated in a prospective manner on two further datasets [52], achieving similar performance metrics as in the two other databases.

4.1.2 Melanoma Diagnosis from Skin Photographs

Melanoma detection from pictures of nevi is another problem suitable for DL techniques. A commercial system, *Moleanalyzer-Pro* [104] was trained with more than 120,000 dermoscopic images to detect melanoma. The work of Winkler et al. [119] posed a test to this system: does the fact of having surgical skin markings affect the probabilities of the lesion being a melanoma? To answer such question, they created a database of 130 lesions (103 benign and 23 melanomas) that were imaged with and without skin markings and then cropped. With respect to unmarked nevi, marked lesions had an increase in probability score, which resulted in higher sensitivity, and a drop both in specificity and in area under the curve. Cropped images had both higher sensitivity and specificity. This work shows that unrelated image artifacts or modifications could have a strong effect on AI systems interpretation.

4.1.3 Macular Disease Detection from OCT Scans

Pegasus-OCT [11] is a clinical decision support system to detect anomalous features in the macula. The system was developed by *Visulytic Ltd.* [94] and uses CNNs based on the VGG16 convolutional neural model. The training database consists of OCT data acquired from different manufacturers. The work of Bhatia et al. [10], has focused on the evaluation of such commercial systems on five different datasets acquired in different countries (USA, China, Iran, Greece and U.K.), with four different devices, achieving device-specific areas under the curve greater than 0.98. This external validation study shows high performance under different imaging conditions.

4.1.4 Cervical Cancer Diagnosis by Coherence Anti-Stokes Raman Scatttering

In Aljakouch et al. [7] the authors study the use of CARS and DL for the diagnosis of cervical cancer. Pap smears were stained and analyzed using standard methods, and then imaged using CARS. A DL network was trained to distinguish between normal and cancerous cells. Such CNN could classify cancerous pap smears with 100% accuracy. However, the dataset used for the experiments consisted of 30 subjects, 10 healthy women, 10 patients with low grade and 10 patients with high grade intraepithelial lesions. The authors did not clarify how the data was handled for training, validation and testing. Furthermore, the authors did not use a replication cohort, another dataset unrelated to algorithm development where the performance is tested. For such reasons, the results presented in this paper is tantalizing, but requires a followup larger clinical trial.

4.2 Building Systems with Small Training Datasets

The cost of developing a training dataset for DL algorithms for a system that is not used in clinical practice is daunting. As we have seen in the case studies, often one needs thousands of subjects to achieve a good performance of the methods.

Biomedical image analysis experts have resorted to TL and data augmentation strategies to deal with this situation. As discussed in previous sections, the initial layers of a CNN act as feature extractors. Low-level features of natural images resemble those of biomedical images, typically consisting of edges, ridges and textures, and can therefore be translated to biomedical image analysis tasks. Higher level features, as well as the fully connected parts of the CNNs can be built upon such low-level features.

Data augmentation relates to the generation of artificial, but biologically and clinically plausible samples with corresponding labels created by modifying the original labeled training set. An example of a trivial data augmentation strategy is the application of linear transformations to both the image and the segmentation mask when training a CNN to perform segmentations. Non-linear image deformations have also been applied to microscopy images as a strategy for data augmentation, improving the performance of the CNNs.

It is almost as equally important to have a large training dataset as to have a large validation and testing dataset. The validation dataset is used to train the meta-parameters of the ANN and to estimate performance on unseen images, serving to estimate if the ANN is overfitting the training data. It should be noted that if the validation data has been used for training, the performance on validation may be higher than the performance on truly unseen data.

The test dataset is ideally large, and representative of all image and patient variability, and should be used as little as possible: never during the algorithm development phase and only once to report the performance of the ANN. This database is critical for judging the robustness and reproducibility of the algorithm with respect to imaging properties, and most importantly, to patient variability and pathology.

4.3 Data Management Summary

The imaging data collected via optical imaging tools are largely compatible with DL approaches, enabling automatic image analysis and inference in most applications. While the ultimate goal of DL-based image analysis is to achieve or surpass the performance of a trained human operator, this requires extensive clinical validation of image databases. Systems deployed in clinical practice have access to such databases thanks to electronic health records and picture archival and communication systems. However, many of the methods highlighted in this chapter are not clinically widespread, and thus can only afford to collect smaller databases. These smaller datasets can make validation testing more straightforward, but at the same time, requires tools such as domain transfer and data augmentation techniques that must be carefully implemented. As the use of these optical tools grow and expand, it is important to set up coordinated efforts to create multi-site testing databases that encompass combinations of imaging devices and race, gender and pathologies. The development of such cross-site, shared strong databases will facilitate the development of improved, validated CNNs that will support future clinical translation.

5 Conclusions

Advanced imaging tools have immense utility in biomedical research and growing potential in the areas of clinical care and clinical diagnostics. Their central challenge continues to be the immense stream of collected data and the need for routine, automated extraction of actionable information. As we have discussed, CNNs are well suited to tackle these demanding applications. From image segmentation to feature recognition, CNNs are capable of recognizing image elements across a variety of contexts in ways that would be challenging or otherwise impossible using traditional image analysis tools. The ability to couple CNN feature extraction layers with dense ANN layers enable their use in classification settings, making them important future tools in interpreting research and clinical images.

A somewhat understated feature of these ANNs is their accessibility. CNNs and ANNs make use of common, open source and free libraries. Numerous guides are available in setting up and using these resources, whether it be a TL approach or training a *U-Net* from scratch. Online repositories such as on *GitHub* make sharing these tools easy across companies and academic institutions. While efficient training does require GPU cards, the cost of these items has dropped immensely: high capability boards with more than 4000 cores and 24GB of memory are under $3000, with smaller sized cards available for hundreds. The availability of inexpensive online GPU computing is also expanding: *Amazon*, *Google*, and *Microsoft* all offer relatively inexpensive pay-as-you-go instances, and companies like *Paperspace* [2] cater to academic research. What was an immensely costly and involved undertaking just a few years ago is now more open and accessible than ever.

CNNs and their ilk can be used in stand-alone fashion, or incorporated into data analysis pipelines for robust, reproducible research. ANNs can be easily incorporated as modules within the data analysis pathway [48], and can be applied alongside tools such as virtual environments and containers. This combination of ANNs with traditional coding tools and containerization offers considerable capabilities, with the ability to package and ship containerized analysis routines that work everywhere for everyone, regardless of computer make or operating system.

Still, while the tools themselves might be more accessible, there is a growing concern over the nature of the ANNs. Unlike traditional deterministic code, which can be reviewed to understand its behavior, edge cases, and potential failures, ANN training largely creates *black boxes* whose layers and nodes cannot be readily inspected or understood. While this a somewhat natural outcome in the evolution of larger and larger ANNs, it creates a potentially problematic scenario when these tools are applied to mission-critical biomedical and clinical situations. Some of these challenges can be seen in IBM's Watson initiative [1], where the developed toolkit was found to suggest therapies or courses of action that could have been fatal for patients [96]. The ANNs discussed here not as large or encompassing as IBM's, but still prone to many other challenges. Initiatives to peek under the hood and map trained ANNs are growing and will be crucial when these tools are applied to clinical care.

Another area where considerable work is needed is in *Edge Computing* [103]. The examples in this chapter are almost all in post-hoc data analysis, where CNNs are used in the hours and days following data acquisition. To unlock many of the exciting capabilities of these tools, computational platforms are required for real-time analysis of captured images. *Edge Computing* modules, such as those being developed by *Google* and others, offer such an approach, where CNNs can be applied on day at the bedside.

References

1. IBM Watson's Initiative. https://www.ibm.com/academic/home. Accessed: 2020-07-07.
2. Paperspace. GPU cloud tools built for developers. Powering next-generation workflows and the future of intelligent applications. https://www.paperspace.com/. Accessed: 2020-07-07.
3. C. A, W. Z, W. J, and et al. Deep learning for the segmentation of preserved photoreceptors on en face optical coherence tomography in two inherited retinal diseases. *Biomed Opt Express*, 9(7):3092–3105, 2018.
4. M. Abadi, A. Agarwal, P. Barham, E. Brevdo, Z. Chen, C. Citro, G. S. Corrado, A. Davis, J. Dean, M. Devin, S. Ghemawat, I. Goodfellow, A. Harp, G. Irving, M. Isard, Y. Jia, R. Jozefowicz, L. Kaiser, M. Kudlur, J. Levenberg, D. Mané, R. Monga, S. Moore, D. Murray, C. Olah, M. Schuster, J. Shlens, B. Steiner, I. Sutskever, K. Talwar, P. Tucker, V. Vanhoucke, V. Vasudevan, F. Viégas, O. Vinyals, P. Warden, M. Wattenberg, M. Wicke, Y. Yu, and X. Zheng. TensorFlow: Large-scale machine learning on heterogeneous systems, 2015. Software available from tensorflow.org.
5. M. Adhi and J. S. Duker. Optical coherence tomography–current and future applications. *Current opinion in ophthalmology*, 24(3):213, 2013.
6. S. U. Akram, J. Kannala, L. Eklund, and J. Heikkilä. Cell segmentation proposal network for microscopy image analysis. In *Deep Learning and Data Labeling for Medical Applications*, pages 21–29. Springer, 2016.
7. K. Aljakouch, Z. Hilal, I. Daho, M. Schuler, S. D. Krauß, H. K. Yosef, J. Dierks, A. Mosig, K. Gerwert, and S. F. El-Mashtoly. Fast and noninvasive diagnosis of cervical cancer by coherent anti-stokes raman scattering. *Analytical Chemistry*, 2019.
8. L. A. Austin, S. Osseiran, and C. L. Evans. Raman technologies in cancer diagnostics. *Analyst*, 141(2):476–503, 2016.
9. H. G. Bezerra, M. A. Costa, G. Guagliumi, A. M. Rollins, and D. I. Simon. Intracoronary optical coherence tomography: a comprehensive review: clinical and research applications. *JACC: Cardiovascular Interventions*, 2(11):1035–1046, 2009.
10. K. Bhatia, M. Graham, L. Terry, A. Wood, P. Tranos, S. Trikha, and N. Jaccard. Disease Classification of Macular Optical Coherence Tomography Scans Using Deep Learning Software: Validation on Independent, Multicenter Data. *Retina*, page 1, 2019.
11. K. K. Bhatia, M. S. Graham, L. Terry, A. Wood, P. Tranos, S. Trikha, and N. Jaccard. Disease classification of macular optical coherence tomography scans using deep learning software: validation on independent, multi-centre data. *arXiv preprint arXiv:1907.05164*, 2019.
12. C. Bonnans, J. Chou, and Z. Werb. Remodelling the extracellular matrix in development and disease. *Nature reviews Molecular cell biology*, 15(12):786, 2014.
13. B. Bouma, G. Tearney, H. Yabushita, M. Shishkov, C. Kauffman, D. D. Gauthier, B. MacNeill, S. Houser, H. Aretz, E. F. Halpern, et al. Evaluation of intracoronary stenting by intravascular optical coherence tomography. *Heart*, 89(3):317–320, 2003.
14. G. Bradski. The OpenCV Library. *Dr. Dobb's Journal of Software Tools*, 2000.
15. L. Breiman. Random forests. *Machine Learning*, 45(1):5–32, Oct 2001.
16. H. G. Breunig, M. Weinigel, R. Bückle, M. Kellner-Höfer, J. Lademann, M. E. Darvin, W. Sterry, and K. König. Clinical coherent anti-stokes raman scattering and multiphoton tomography of human skin with a femtosecond laser and photonic crystal fiber. *Laser Physics Letters*, 10(2):025604, 2013.
17. M. Brinkmann, A. Fast, T. Hellwig, I. Pence, C. L. Evans, and C. Fallnich. Portable all-fiber dual-output widely tunable light source for coherent raman imaging. *Biomedical optics express*, 10(9):4437–4449, 2019.
18. P. T. Cagle, R. Barrios, and T. C. Allen. *Color atlas and text of pulmonary pathology*. Lippincott Williams & Wilkins, 2008.
19. R. Cairns, R. Khokha, and R. Hill. Molecular mechanisms of tumor invasion and metastasis: an integrated view. *Current molecular medicine*, 3(7):659–671, 2003.

20. P. Campagnola. Second harmonic generation imaging microscopy: applications to diseases diagnostics, 2011.
21. R. Carriles, D. N. Schafer, K. E. Sheetz, J. J. Field, R. Cisek, V. Barzda, A. W. Sylvester, and J. A. Squier. Invited review article: Imaging techniques for harmonic and multiphoton absorption fluorescence microscopy. *Review of scientific instruments*, 80(8):081101, 2009.
22. D. Castelvecchi. Can we open the black box of ai? *Nature News*, 538(7623):20, 2016.
23. J. Chen, S. Zhuo, X. Jiang, X. Zhu, L. Zheng, S. Xie, B. Lin, and H. Zeng. Multiphoton microscopy study of the morphological and quantity changes of collagen and elastic fiber components in keloid disease. *Journal of biomedical optics*, 16(5):051305, 2011.
24. F. Chollet. keras. https://github.com/fchollet/keras, 2015.
25. T. R. Cox and J. T. Erler. Remodeling and homeostasis of the extracellular matrix: implications for fibrotic diseases and cancer. *Disease models & mechanisms*, 4(2):165–178, 2011.
26. J. Cuadros and G. Bresnick. Eyepacs: an adaptable telemedicine system for diabetic retinopathy screening. *Journal of diabetes science and technology*, 3(3):509–516, 2009.
27. M. W. Davidson and M. Abramowitz. Optical microscopy. *Encyclopedia of imaging science and technology*, 2002.
28. E. Decencière, G. Cazuguel, X. Zhang, G. Thibault, J.-C. Klein, F. Meyer, B. Marcotegui, G. Quellec, M. Lamard, R. Danno, et al. Teleophta: Machine learning and image processing methods for teleophthalmology. *Irbm*, 34(2):196–203, 2013.
29. E. Decencière, X. Zhang, G. Cazuguel, B. Lay, B. Cochener, C. Trone, P. Gain, R. Ordonez, P. Massin, A. Erginay, et al. Feedback on a publicly distributed image database: the messidor database. *Image Analysis & Stereology*, 33(3):231–234, 2014.
30. W. Denk, J. H. Strickler, and W. W. Webb. Two-photon laser scanning fluorescence microscopy. *Science*, 248(4951):73–76, 1990.
31. S. K. Devalla, P. K. Renukanand, B. K. Sreedhar, G. Subramanian, L. Zhang, S. Perera, J.-M. Mari, K. S. Chin, T. A. Tun, N. G. Strouthidis, T. Aung, A. H. Thiéry, and M. J. A. Girard. Drunet: a dilated-residual u-net deep learning network to segment optic nerve head tissues in optical coherence tomography images. *Biomed. Opt. Express*, 9(7):3244–3265, Jul 2018.
32. J. Donahue, Y. Jia, O. Vinyals, J. Hoffman, N. Zhang, E. Tzeng, and T. Darrell. Decaf: A deep convolutional activation feature for generic visual recognition. In *International conference on machine learning*, pages 647–655, 2014.
33. W. Drexler and J. G. Fujimoto. *Optical coherence tomography: technology and applications*. Springer Science & Business Media, 2008.
34. T. Eto, H. Suzuki, A. Honda, and Y. Nagashima. The changes of the stromal elastotic framework in the growth of peripheral lung adenocarcinomas. *Cancer: Interdisciplinary International Journal of the American Cancer Society*, 77(4):646–656, 1996.
35. C. L. Evans, E. O. Potma, M. Puoris' haag, D. Côté, C. P. Lin, and X. S. Xie. Chemical imaging of tissue in vivo with video-rate coherent anti-stokes raman scattering microscopy. *Proceedings of the national academy of sciences*, 102(46):16807–16812, 2005.
36. C. L. Evans and X. S. Xie. Coherent anti-stokes raman scattering microscopy: chemical imaging for biology and medicine. *Annu. Rev. Anal. Chem.*, 1:883–909, 2008.
37. C. L. Evans, X. Xu, S. Kesari, X. S. Xie, S. T. Wong, and G. S. Young. Chemically-selective imaging of brain structures with cars microscopy. *Optics express*, 15(19):12076–12087, 2007.
38. J. A. Evans, B. E. Bouma, J. Bressner, M. Shishkov, G. Y. Lauwers, M. Mino-Kenudson, N. S. Nishioka, and G. J. Tearney. Identifying intestinal metaplasia at the squamocolumnar junction by using optical coherence tomography. *Gastrointestinal endoscopy*, 65(1):50–56, 2007.
39. J. A. Evans, J. M. Poneros, B. E. Bouma, J. Bressner, E. F. Halpern, M. Shishkov, G. Y. Lauwers, M. Mino-Kenudson, N. S. Nishioka, and G. J. Tearney. Optical coherence tomography to identify intramucosal carcinoma and high-grade dysplasia in barrett's esophagus. *Clinical Gastroenterology and Hepatology*, 4(1):38–43, 2006.

40. S. Farsiu, S. J. Chiu, R. V. O'Connell, F. A. Folgar, E. Yuan, J. A. Izatt, C. A. Toth, A.-R. E. D. S. . A. S. D. O. C. T. S. Group, et al. Quantitative classification of eyes with and without intermediate age-related macular degeneration using optical coherence tomography. *Ophthalmology*, 121(1):162–172, 2014.
41. A. F. Fercher, W. Drexler, C. K. Hitzenberger, and T. Lasser. Optical coherence tomography-principles and applications. *Reports on progress in physics*, 66(2):239, 2003.
42. D. C. Fernández, H. M. Salinas, and C. A. Puliafito. Automated detection of retinal layer structures on optical coherence tomography images. *Opt. Express*, 13(25):10200–10216, Dec 2005.
43. e. P. Franken, A. E. Hill, C. e. Peters, and G. Weinreich. Generation of optical harmonics. *Physical Review Letters*, 7(4):118, 1961.
44. C. W. Freudiger, W. Min, B. G. Saar, S. Lu, G. R. Holtom, C. He, J. C. Tsai, J. X. Kang, and X. S. Xie. Label-free biomedical imaging with high sensitivity by stimulated raman scattering microscopy. *Science*, 322(5909):1857–1861, 2008.
45. C. W. Freudiger, R. Pfannl, D. A. Orringer, B. G. Saar, M. Ji, Q. Zeng, L. Ottoboni, W. Ying, C. Waeber, J. R. Sims, et al. Multicolored stain-free histopathology with coherent raman imaging. *Laboratory investigation*, 92(10):1492, 2012.
46. Y. Fu, T. B. Huff, H.-W. Wang, H. Wang, and J.-X. Cheng. Ex vivo and in vivo imaging of myelin fibers in mouse brain by coherent anti-stokes raman scattering microscopy. *Optics express*, 16(24):19396–19409, 2008.
47. J. G. Fujimoto, C. Pitris, S. A. Boppart, and M. E. Brezinski. Optical coherence tomography: an emerging technology for biomedical imaging and optical biopsy. *Neoplasia (New York, NY)*, 2(1–2):9, 2000.
48. G. González and C. L. Evans. Biomedical image processing with containers and deep learning: An automated analysis pipeline: Data architecture, artificial intelligence, automated processing, containerization, and clusters orchestration ease the transition from data acquisition to insights in medium-to-large datasets. *BioEssays*, page 1900004, 2019.
49. I. Goodfellow, Y. Bengio, and A. Courville. *Deep Learning*. The MIT Press, 2016.
50. A. Gulli and S. Pal. *Deep learning with Keras*. Packt Publishing Ltd, 2017.
51. V. Gulshan, L. Peng, M. Coram, M. C. Stumpe, D. Wu, A. Narayanaswamy, S. Venugopalan, K. Widner, T. Madams, J. Cuadros, R. Kim, R. Raman, P. C. Nelson, J. L. Mega, and D. R. Webster. Development and validation of a deep learning algorithm for detection of diabetic retinopathy in retinal fundus photographs. *JAMA - Journal of the American Medical Association*, 316(22):2402–2410, 2016.
52. V. Gulshan, R. P. Rajan, K. Widner, D. Wu, P. Wubbels, T. Rhodes, K. Whitehouse, M. Coram, G. Corrado, K. Ramasamy, R. Raman, L. Peng, and D. R. Webster. Performance of a Deep-Learning Algorithm vs Manual Grading for Detecting Diabetic Retinopathy in India. *JAMA Ophthalmology*, 137(9):987–993, 2019.
53. L. Hamers et al. Similarity measures in scientometric research: The jaccard index versus salton's cosine formula. *Information Processing and Management*, 25(3):315–18, 1989.
54. F. Hausdorff. Dimension und äußeres maß. *Mathematische Annalen*, 79(1–2):157–179, 1918.
55. K. He, X. Zhang, S. Ren, and J. Sun. Deep residual learning for image recognition. *CoRR*, abs/1512.03385, 2015.
56. J. I. Hoffman. *Biostatistics for medical and biomedical practitioners*. Academic press, 2015.
57. K. Hornik. Approximation capabilities of multilayer feedforward networks. *Neural networks*, 4(2):251–257, 1991.
58. K. Hornik, M. Stinchcombe, and H. White. Multilayer feedforward networks are universal approximators. *Neural Networks*, 2(5):359–366, 1989.
59. N. G. Horton, K. Wang, D. Kobat, C. G. Clark, F. W. Wise, C. B. Schaffer, and C. Xu. In vivo three-photon microscopy of subcortical structures within an intact mouse brain. *Nature photonics*, 7(3):205, 2013.
60. M. Huttunen, A. Hassan, C. McCloskey, S. Fasih, J. Upham, B. Venderhyden, R. Boyd, and S. Murugkar. Automated classification of multiphoton microscopy images of ovarian tissue using deep learning. *Journal of Biomedical Optics*, 23(6), 2018.

61. H. Iqbal. Plot neural net, 2018. Latex code for drawing neural networks for reports and presentation.
62. A. Işın, C. Direkoğlu, and M. Şah. Review of mri-based brain tumor image segmentation using deep learning methods. *Procedia Computer Science*, 102:317–324, 2016.
63. M. Ji, D. A. Orringer, C. W. Freudiger, S. Ramkissoon, X. Liu, D. Lau, A. J. Golby, I. Norton, M. Hayashi, N. Y. Agar, et al. Rapid, label-free detection of brain tumors with stimulated raman scattering microscopy. *Science translational medicine*, 5(201):201ra119–201ra119, 2013.
64. Y. Jia, E. Shelhamer, J. Donahue, S. Karayev, J. Long, R. Girshick, S. Guadarrama, and T. Darrell. Caffe: Convolutional architecture for fast feature embedding, 2014.
65. K. Kawaguchi, L. P. Kaelbling, and Y. Bengio. Generalization in deep learning. *arXiv preprint arXiv:1710.05468*, 2017
66. T. Kepp, C. Droigk, M. Casper, M. Evers, G. Hüttmann, N. Salma, D. Manstein, M. P. Heinrich, and H. Handels. Segmentation of mouse skin layers in optical coherence tomography image data using deep convolutional neural networks. *Biomed. Opt. Express*, 10(7):3484–3496, Jul 2019.
67. W. Kirch. Pearson's correlation coefficient. *Encyclopedia of Public Health; Springer: Dordrecht, The Netherlands*, pages 1090–2013, 2008.
68. N. D. Kirkpatrick, M. A. Brewer, and U. Utzinger. Endogenous optical biomarkers of ovarian cancer evaluated with multiphoton microscopy. *Cancer Epidemiology and Prevention Biomarkers*, 16(10):2048–2057, 2007.
69. A. Krizhevsky, I. Sutskever, and G. E. Hinton. Imagenet classification with deep convolutional neural networks. In *Proceedings of the 25th International Conference on Neural Information Processing Systems - Volume 1*, NIPS'12, pages 1097–1105, USA, 2012. Curran Associates Inc.
70. N. Kumar, R. Verma, S. Sharma, S. Bhargava, A. Vahadane, and A. Sethi. A dataset and a technique for generalized nuclear segmentation for computational pathology. *IEEE transactions on medical imaging*, 36(7):1550–1560, 2017.
71. Y. LeCun, P. Haffner, L. Bottou, and Y. Bengio. Object recognition with gradient-based learning. *Feature Grouping*, 1999.
72. K. Lee, M. Garvin, S. Russell, M. Sonka, and M. Abràmoff. Automated intraretinal layer segmentation of 3-d macular oct scans using a multiscale graph search. *Investigative Ophthalmology & Visual Science*, 51(13):1767–1767, 2010.
73. J. Li, F. Luisier, and T. Blu. Pure-let image deconvolution. *IEEE Transactions on Image Processing*, 27(1):92–105, 2017.
74. H. Lin, F. Deng, K. Huang, H. J. Lee, and J. Cheng. High-speed, high-sensitivity spectroscopic stimulated raman scattering microscopy by ultrafast delay-line tuning and deep learning. In *2019 Conference on Lasers and Electro-Optics (CLEO)*, pages 1–2, May 2019.
75. S. Lundberg and S. Lee. A unified approach to interpreting model predictions. *CoRR*, abs/1705.07874, 2017.
76. F. Mahmood, D. Borders, R. J. Chen, G. N. McKay, K. J. Salimian, A. S. Baras, and N. J. Durr. Deep adversarial training for multi-organ nuclei segmentation in histopathology images. *CoRR*, abs/1810.00236, 2018.
77. B. Manifold, E. Thomas, A. T. Francis, A. H. Hill, and D. Fu. Denoising of stimulated raman scattering microscopy images via deep learning. *Biomed. Opt. Express*, 10(8):3860–3874, Aug 2019.
78. S. Mannor, D. Peleg, and R. Rubinstein. The cross entropy method for classification. In *Proceedings of the 22nd international conference on Machine learning*, pages 561–568, 2005.
79. B. Masters, P. So, and E. Gratton. Optical biopsy of in vivo human skin: multi-photon excitation microscopy. *Lasers in Medical Science*, 13(3):196–203, 1998.
80. M. Mirza and S. Osindero. Conditional generative adversarial nets. *CoRR*, abs/1411.1784, 2014.
81. A. Mishra, A. Wong, K. Bizheva, and D. A. Clausi. Intra-retinal layer segmentation in optical coherence tomography images. *Opt. Express*, 17(26):23719–23728, Dec 2009.

82. D. Mojahed, P. Chang, Y. Gan, X. Yao, B. Angelini, H. Hibshoosh, R. Ha, and C. P. Hendon. Convolutional neural network (cnn) classification of breast cancer in optical coherence tomography (oct) images. In *Optical Coherence Tomography and Coherence Domain Optical Methods in Biomedicine XXIII*, volume 10867, page 108671N. International Society for Optics and Photonics, 2019.
83. W. J. Murdoch, C. Singh, K. Kumbier, R. Abbasi-Asl, and B. Yu. Definitions, methods, and applications in interpretable machine learning. *Proceedings of the National Academy of Sciences*, 116(44):22071–22080, 2019.
84. O. Nadiarnykh, R. B. LaComb, M. A. Brewer, and P. J. Campagnola. Alterations of the extracellular matrix in ovarian cancer studied by second harmonic generation imaging microscopy. *BMC cancer*, 10(1):94, 2010.
85. N. Nassif, B. Cense, B. Park, M. Pierce, S. Yun, B. Bouma, G. Tearney, T. Chen, and J. De Boer. In vivo high-resolution video-rate spectral-domain optical coherence tomography of the human retina and optic nerve. *Optics express*, 12(3):367–376, 2004.
86. A. Y. Ng. Feature selection, l 1 vs. l 2 regularization, and rotational invariance. In *Proceedings of the twenty-first international conference on Machine learning*, page 78, 2004.
87. S. J. Pan and Q. Yang. A survey on transfer learning. *IEEE Transactions on knowledge and data engineering*, 22(10):1345–1359, 2009.
88. A. Paszke, S. Gross, S. Chintala, G. Chanan, E. Yang, Z. DeVito, Z. Lin, A. Desmaison, L. Antiga, and A. Lerer. Automatic differentiation in PyTorch. In *NIPS Autodiff Workshop*, 2017.
89. S. W. Perry, R. M. Burke, and E. B. Brown. Two-photon and second harmonic microscopy in clinical and translational cancer research. *Annals of biomedical engineering*, 40(2):277–291, 2012.
90. P. Porwal, S. Pachade, R. Kamble, M. Kokare, G. Deshmukh, V. Sahasrabuddhe, and F. Meriaudeau. Indian diabetic retinopathy image dataset (idrid): A database for diabetic retinopathy screening research. *Data*, 3(3):25, 2018.
91. P. P. Provenzano, D. R. Inman, K. W. Eliceiri, J. G. Knittel, L. Yan, C. T. Rueden, J. G. White, and P. J. Keely. Collagen density promotes mammary tumor initiation and progression. *BMC medicine*, 6(1):11, 2008.
92. K. P. Quinn, G. V. Sridharan, R. S. Hayden, D. L. Kaplan, K. Lee, and I. Georgakoudi. Quantitative metabolic imaging using endogenous fluorescence to detect stem cell differentiation. *Scientific reports*, 3:3432, 2013.
93. M. T. Ribeiro, S. Singh, and C. Guestrin. "why should I trust you?": Explaining the predictions of any classifier. *CoRR*, abs/1602.04938, 2016.
94. T. W. Rogers, N. Jaccard, F. Carbonaro, H. G. Lemij, K. A. Vermeer, N. J. Reus, and S. Trikha. Evaluation of an ai system for the automated detection of glaucoma from stereoscopic optic disc photographs: the european optic disc assessment study. *Eye*, 33(11):1791–1797, 2019.
95. O. Ronneberger, P. Fischer, and T. Brox. U-net: Convolutional networks for biomedical image segmentation. *CoRR*, abs/1505.04597, 2015.
96. C. Ross and I. Swetlitz. Ibm's watson supercomputer recommended 'unsafe and incorrect'cancer treatments, internal documents show, 2018.
97. C. Rudin. Stop explaining black box machine learning models for high stakes decisions and use interpretable models instead. *Nature Machine Intelligence*, 1(5):206–215, 2019.
98. D. Rumelhart and J. McClelland. *Parallel Distributed Processing: Explorations in the Microstructure of Cognition, Vol. 1: Foundations*. Cambridge, MA: Bradford Books/MIT Press, 1985.
99. O. Russakovsky, J. Deng, H. Su, J. Krause, S. Satheesh, S. Ma, Z. Huang, A. Karpathy, A. Khosla, M. Bernstein, et al. Imagenet large scale visual recognition challenge. *International Journal of Computer Vision*, 115(3):211–252, 2015.
100. B. G. Saar, C. W. Freudiger, J. Reichman, C. M. Stanley, G. R. Holtom, and X. S. Xie. Video-rate molecular imaging in vivo with stimulated raman scattering. *science*, 330(6009):1368–1370, 2010.

101. J. M. Schmitt. Optical coherence tomography (oct): a review. *IEEE Journal of selected topics in quantum electronics*, 5(4):1205–1215, 1999.
102. A. Shah, L. Zhou, M. D. Abrámoff, and X. Wu. Multiple surface segmentation using convolution neural nets: application to retinal layer segmentation in oct images. *Biomed. Opt. Express*, 9(9):4509–4526, Sep 2018.
103. W. Shi and S. Dustdar. The promise of edge computing. *Computer*, 49(5):78–81, 2016.
104. K. Sies, J. K. Winkler, C. Fink, F. Bardehle, F. Toberer, T. Buhl, A. Enk, A. Blum, A. Rosenberger, and H. A. Haenssle. Past and present of computer-assisted dermoscopic diagnosis: performance of a conventional image analyser versus a convolutional neural network in a prospective data set of 1,981 skin lesions. *European Journal of Cancer*, 135:39–46, 2020.
105. K. Simonyan, A. Vedaldi, and A. Zisserman. Deep inside convolutional networks: Visualising image classification models and saliency maps. *CoRR*, abs/1312.6034, 2013.
106. K. Simonyan and A. Zisserman. Very deep convolutional networks for large-scale image recognition. *arXiv preprint arXiv:1409.1556*, 2014.
107. K. Simonyan and A. Zisserman. Very deep convolutional networks for large-scale image recognition. *arXiv preprint arXiv:1409.1556*, 2014.
108. S. W. Smith et al. *The scientist and engineer's guide to digital signal processing*. California Technical Pub. San Diego, 1997.
109. C. Szegedy, V. Vanhoucke, S. Ioffe, J. Shlens, and Z. Wojna. Rethinking the inception architecture for computer vision. In *Proceedings of IEEE Conference on Computer Vision and Pattern Recognition*, 2016.
110. C. Szegedy, Wei Liu, Yangqing Jia, P. Sermanet, S. Reed, D. Anguelov, D. Erhan, V. Vanhoucke, and A. Rabinovich. Going deeper with convolutions. In *2015 IEEE Conference on Computer Vision and Pattern Recognition (CVPR)*, pages 1–9, June 2015.
111. L. Torrey and J. Shavlik. Transfer learning. In *Handbook of research on machine learning applications and trends: algorithms, methods, and techniques*, pages 242–264. IGI Global, 2010.
112. B. van Ginneken, S. Kerkstra, and J. Meakin. Grand challenges in biomedical image analysis, 2015.
113. D. A. Van Valen, T. Kudo, K. M. Lane, D. N. Macklin, N. T. Quach, M. M. DeFelice, I. Maayan, Y. Tanouchi, E. A. Ashley, and M. W. Covert. Deep learning automates the quantitative analysis of individual cells in live-cell imaging experiments. *PLOS Computational Biology*, 12:1–24, 11 2016.
114. D. Wang, A. Khosla, R. Gargeya, H. Irshad, and A. H. Beck. Deep learning for identifying metastatic breast cancer. *arXiv preprint arXiv:1606.05718*, 2016.
115. H. Wang, S. Osseiran, V. Igras, A. J. Nichols, E. M. Roider, J. Pruessner, H. Tsao, D. E. Fisher, and C. L. Evans. In vivo coherent raman imaging of the melanomagenesis-associated pigment pheomelanin. *Scientific reports*, 6:37986, 2016.
116. J. Welzel. Optical coherence tomography in dermatology: a review. *Skin Research and Technology: Review article*, 7(1):1–9, 2001.
117. S. Weng, X. Xu, J. Li, and S. T. C. Wong. Combining deep learning and coherent anti-Stokes Raman scattering imaging for automated differential diagnosis of lung cancer. *Journal of Biomedical Optics*, 22(10):1–10, 2017.
118. R. M. Williams, A. Flesken-Nikitin, L. H. Ellenson, D. C. Connolly, T. C. Hamilton, A. Y. Nikitin, and W. R. Zipfel. Strategies for high-resolution imaging of epithelial ovarian cancer by laparoscopic nonlinear microscopy. *Translational Oncology*, 3(3):181, 2010.
119. J. K. Winkler, C. Fink, F. Toberer, A. Enk, T. Deinlein, R. Hofmann-Wellenhof, L. Thomas, A. Lallas, A. Blum, W. Stolz, and H. A. Haenssle. Association between Surgical Skin Markings in Dermoscopic Images and Diagnostic Performance of a Deep Learning Convolutional Neural Network for Melanoma Recognition. *JAMA Dermatology*, 155(10):1135–1141, 2019.
120. R. Yamashita, M. Nishio, R. K. G. Do, and K. Togashi. Convolutional neural networks: an overview and application in radiology. *Insights into imaging*, 9(4):611–629, 2018.

121. C. Zhang, D. Zhang, and J.-X. Cheng. Coherent raman scattering microscopy in biology and medicine. *Annual review of biomedical engineering*, 17:415–445, 2015.
122. A. P. Zijdenbos, B. M. Dawant, R. A. Margolin, and A. C. Palmer. Morphometric analysis of white matter lesions in mr images: method and validation. *IEEE transactions on medical imaging*, 13(4):716–724, 1994.

Part III
Deep Learning for Medical Diagnostics

Deep Learning for Lung Disease Detection from Chest X-Rays Images

Ebenezer Jangam, Chandra Sekhara Rao Annavarapu, and Mourad Elloumi

Abstract Chest x-rays images are the primary source to detect tuberculosis and different thoracic diseases. Lung nodules also can be identified from the chest x-rays images. In the past decade, the concept of automatic disease detection from the datasets of chest x-rays images has gained importance and researchers have proposed a variety of techniques for tuberculosis screening, thoracic disease detection and lung nodule detection. With the availability of massive public datasets of chest x-rays images, *Deep Learning* (DL) techniques played a key role in lung disease detection. This book chapter tries to cover the DL techniques used to detect lung diseases from chest x-rays datasets. It contains the description of the public datasets of chest x-rays images available for thoracic disease detection, tuberculosis screening and lung nodule detection. It also lists most commonly used performance metrics for the evaluation of disease detection techniques.

Keywords Deep learning · Disease detection · Tuberculosis detection · Lung nodule detection · Chest x-rays datasets · Performance metrics

1 Introduction

In 1895, Wilhelm Conrad Rontgen discovered x-rays which revolutionised the field of diagnostic medicine [46]. With the invention of the modern digital computer in the late 1940s, attempts were made to make computers perform intelligent tasks.

E. Jangam
Department of IT, Velagapudi Ramakrishna Siddhartha Engineering College, Vijayawada, Andhra Pradesh, India

C. S. R. Annavarapu (✉)
Department of Computer Science Engineering, Indian Institute of Technology, Dhanbad, Jharkhand, India
e-mail: acsrao@iitism.ac.in

M. Elloumi
Faculty of Computing and Information Technology, University of Bisha, Bisha, Saudi Arabia

M. Elloumi (ed.), *Deep Learning for Biomedical Data Analysis*,
https://doi.org/10.1007/978-3-030-71676-9_10

During 1960s, researchers published articles about chest x-rays image analysis using computers [17]. During 1970s, the focus was upon the detection of abnormalities in chest x-rays images using a computer. During 1980s, chest x-rays images were the primary source to diagnose pulmonary diseases like pneumonia, tuberculosis, emphysema and lung cancer [64]. The focus of research was to develop algorithms to automatically screen a specific disease from the chest x-rays images. During the past decade, automatic tuberculosis screening techniques and thoracic disease detection techniques have received more attention.

One of the major problems faced by the researchers was the lack of public datasets of chest x-rays images, which can act as a benchmark for the comparison of the performance of different techniques proposed. If public datasets are available, they can be used for evaluation of any proposed technique. Usage of public datasets has two advantages. The first advantage is the time and resources can be saved as new chest x-rays images set need not be obtained and researchers can spend their efforts on the development of their algorithms and implementations. The second advantage is the use of common datasets enables comparison of the performance of different approaches proposed for a given task [7, 51].

For about three decades from the 1970s to late 1990s, there were no public datasets available. All the techniques proposed for chest x-rays processing were evaluated on the customary datasets. In 2000, a public dataset from *Japanese Society of Radiological Technology* (JSRT) [59] was made available to researchers. JSRT database is primarily made available for lung nodule detection. Some of the techniques evaluated using complete JSRT dataset are deep feature fusion [68], lifetime approach [63], virtual dual energy approach [8], two stage nodule enhancement using *Support Vector Machine* (SVM) [9], *multi scale nodule detection* [52], *Artificial Neural Network* (ANN) [12]. A few researchers [5, 43] have addressed segmentation of ribs from the partial public datasets of chest x-rays images.

In 2006, *Segmented Chest Radiograph* (SCR) dataset [65] is made available to evaluate the techniques proposed for segmentation of anatomical structures. It is the modified version of JSRT dataset [59] where manually segmented anatomical structures are available for comparison. Most of the research is focused on lung segmentation and SCR dataset is most commonly used dataset for evaluation. Some of the techniques which resulted in high accuracy using SCR dataset are described in [6, 11, 13, 25, 31, 36, 54, 55, 65, 69, 72, 73, 75]. SCR dataset can be used for heart and clavicle segmentation [6, 48, 69]. *Montgomery* (MC) dataset [27] and *Chest Radiograph Anatomical Structure Segmentation* (CRASS) dataset [22] were recently made public for segmentation of anatomical structures.

In the past decade, a lot of attention was given to tuberculosis screening and thoracic disease detection. Researchers have dedicated public datasets [14, 22, 27, 70] of chest x-rays images. Some datasets are explicitly available for tuberculosis screening [27, 47]. A few large datasets are dedicated for thoracic disease detection [14, 70]. DL techniques [1, 2] are used for lung disease detection on a customised dataset of 93 images and 433 images respectively.

As there are large public datasets [14, 22, 27, 70], researchers focused more on DL techniques [10, 30, 34, 41, 45, 71] to process the large datasets of chest x-rays images.

This book chapter provides details of DL techniques, performance metrics and public datasets of chest x-rays images for thoracic disease detection, tuberculosis screening and lung nodule detection. Section 2 contains the classification of DL techniques. Section 3 gives details about the datasets, performance metrics and DL techniques for thoracic disease detection. Section 4 gives insight about the datasets, performance metrics and DL techniques for tuberculosis screening. Section 5 contain information about the datasets, performance metrics and DL techniques for lung nodule detection. Finally, Sect. 6 concludes the chapter by giving future directions.

2 Classification of Deep Learning Techniques for Lung Disease Detection from Chest X-Rays Images

Deep Learning (DL) techniques for lung disease detection from chest x-rays images can be classified into three categories based on the type of disease.

- DL for the thoracic disease detection [10, 34, 41, 70, 71]
- DL for tuberculosis screening [38, 44, 61]
- DL for the lung nodule detection [18, 68]

3 Deep Learning for the Thoracic Disease Detection

There are multiple thoracic diseases that can be detected using chest x-rays images [70]. Some of them are atelectasis cardiomegaly, effusion, infiltration, mass, nodule, pneumonia, pneumothorax, consolidation, edema, emphysema, fibrosis, pleural thickening and hernia [70]. This section contains details about public datasets, performance metrics and DL techniques for thoracic disease detection from chest x-rays images.

3.1 Datasets for the Thoracic Disease Detection

There are two massive datasets which are publicly available for thoracic disease detection, *ChestX-ray8* dataset [70] and *Indiana* dataset [14].

3.1.1 *ChestX-ray8* Dataset

ChestX-ray8 Dataset [70] is a massive dataset released publicly in 2017 for the detection of multiple diseases. Initially, it was a dataset of chest x-rays images with eight different disease labels and no finding label. Later the dataset is updated to have 112,120 frontal-view x-rays images of 30,805 unique patients with fourteen disease labels. Each image can have multi labels which are mined from the associated radiological reports using natural language processing. There are 880 unique images with bounding boxes to use them for training models for localization. Along with images and labels, there was other information about patients visit, gender, etc.

3.1.2 *Indiana* Dataset

Indiana Dataset [14] consists of 3996 radiology reports and corresponding 8300 *Posterior-Anterior* (PA) and lateral chest x-rays images. *U.S. National Library of Medicine* (USNLM) maintains the dataset and the images were acquired from Indiana University School of Medicine and its affiliated hospitals. Size of the images vary from 1024 to 4248 pixels along each dimension and are available in *Digital Imaging and Communications in Medicine* (DICOM) format. The images in the dataset have a wide range of abnormalities. USNLM researchers use this dataset for chest x-rays images view classification, detecting rotations in chest x-rays images, and foreign object detection [45].

Before *National Institute of Health* (NIH) released the *ChestX-ray14* dataset, this was the most extensive dataset available with 7470 frontal-view and lateral-view x-rays images (where each image can have multi-labels). Since the labels were not having a defined set of disease categories, and occasionally also had more information, e.g. calcified granuloma/lung/upper lobe/left, this dataset was used initially to perform binary classification of images. The images are categorised into two classes: normal (taking all the images marked normal, in total 2706 images) and abnormal (rest of the images, in total 4764 images).

3.2 Performance Metrics

Two kinds of tasks are performed on *ChestX-ray8* dataset. They are multi-label classification and pathology location.

For multi-label classification, *Area Under Curve* (AUC) of ROC curve is used [70] as a performance metric. AUC is calculated for each disease label. Higher the AUC, better is the performance of the approach. Mean AUC is used for evaluating performance of each model.

For pathology localisation, the performance metrics are accuracy and average false positive number [70].

Performance metrics used with *Indiana* dataset [14] are *Bilingual Evaluation Understudy* (BLEU) [49], *Metric for Evaluation for Translation with Explicit Ordering* (METEOR) [39], *Recall Oriented Understudy for Gisting Evaluation* (ROUGE) [42] and *Consensus-based Image Description Evaluation* CIDER [66]. All the four metrics define the similarity over words or n-grams of reference and candidate descriptions by considering different formulas. BLEU, METEOR and ROGUE are used to measure the similarity between two sentences. BLEU is a machine translation metric that measures the similarity between two sentences. METEOR measures similarity between two sentences but it additionally makes use of synonyms and paraphrase matching. ROGUE was proposed for document summarization. CIDER measures the consensus between candidate image description C_i and the reference sentences, which is a set $S_i = \{s_{i1}, \ldots, s_{im}\}$ provided by human annotators.

Table 2 gives the summary of comparison among different techniques [16, 32, 57, 67, 74] evaluated on *Indiana* dataset.

3.3 Deep Learning Techniques for Thoracic Disease Detection

DL techniques for thoracic disease detection can be classified into two categories based on the dataset used.

First category comprises DL techniques which are evaluated on *ChestX-ray8* dataset [70] which are listed in Table 1.

Second category comprises DL techniques evaluated on *Indiana* dataset [14] which are listed in Table 2.

Among DL techniques in the first category, *Attention Guided CNN* (AG-CNN) [19] has yeilded high AUC. AG-CNN is the DL technique which learns from disease-specific regions to avoid noise and improve alignment. It also integrates a global branch to compensate the lost discriminative cues by local branch. The comprehensive experiment is conducted on the *ChestX-ray14* dataset [70]. When *ResNet50* is used as backbone, a strong global baseline producing an average AUC of 0.841 is recorded. After combining the local cues with the global information,

Table 1 Comparison of performances for thoracic disease classification on the *ChestX-ray14* dataset [70]

Model	CNN	Mean AUC
AG-CNN [19]	D-121	87.1
AG-CNN [19]	R-50	86.8
ResNet-50 [56]	Fine tuned	84.32
ChexNet [50]	D-121	84.2
TieNet DenseNet LSTM [76]	D-121	80.3
Boosted Cascaded Convnets [37]	D-161	79.5
DCNN [70]	R-50	73.8

Table 2 Comparison of performances for thoracic disease labeling on the *Indiana* dataset [14]

Model	BLEU-1/-2/-3/-4	METEOR	ROUGE	CIDER
CNN-RNN [67]	0.316/0.211/0.140/0.095	0.159	0.267	1.111
LRCN [16]	0.369/0.229/0.149/0.099	0.155	0.278	0.190
Soft ATT [74]	0.399/0.251/0.168/0.118	0.167	0.323	0.302
ATT-RK [77]	0.369/0.226/0.151/0.108	0.171	0.323	0.155
Coattention [32]	0.577/0.386/0.306/0.247	0.217	0.447	0.327
Recurrent Neural Cascade Model [57]	0.785/0.144/0.047/0.0	–	–	–

AG-CNN improves the average AUC to 0.868. When *DenseNet-121* [23] is used, the average AUC achieved is 0.871.

Another model tested on *ChestX-ray14* dataset [70] is *ChexNet* [50]. It is a 121-layer *Deep Convolutional Neural Network* (DCNN) which is trained to detect pneumonia from the chest x-rays images. *ChexNet* takes a chest x-rays image as input and gives the probability of pneumonia and heatmap as output. For the pneumonia detection task, the *ChestX-ray8* [70] dataset is split into training (28,744 patients, 98,637 images), validation (1672 patients, 6351 images), and test (389 patients, 420 images). The performance of *ChexNet* is satisfactory with an AUC of 0.842 [50].

An alternative DL technique using LSTMs [76] is to leverage interdependencies among target labels in predicting 14 different patterns from chest x-rays images without pre-training. A two stage ANN model [76] which combines a densely connected image encoder with a *Recurrent Neural Network* (RNN) decoder [76] was proposed.

One more model is built using DCNN architectures, which are tailored for weakly-supervised object localisation. A unified weakly-supervised multi-label image classification and pathology localisation framework was applied on datasets of chest x-rays images to identify multiple pathologies and to generate bounding boxes around the corresponding pathologies [70]. The pathologies are located using activation and weights extracted from the DCNN. DCNN framework was initialized with four different training models of *AlexNet* [35], *GoogleNet* [62], VGG [60] and *ResNet-50* [21]. The model based on *ResNet-50* [70] had given best result.

Another DL model [37] was trained from scratch for multi label classification of *ChestX-ray14* dataset [70] and achieved better results from baseline which used *Transfer Learning* (TL) [58]. Two approaches to calculate loss for multi label classification: *Binary Relevance* (BR) and *PairWise Error* (PWE) loss [37] were implemented. A boosted cascade architecture [37] was designed which is specifically tailored for multi label classification task of type of *ChestX-ray14* dataset [70]. Complex dependencies between class labels was modelled using boosting methods and the resulting performance is improved as compared to single classifiers trained using cross-entropy and PWE loss [37].

As a summary, DL techniques play a key role in the detection of lung diseases from massive datasets[70] of chest x-rays images.

Let us turn our attention on DL techniques evaluated on *Indiana* dataset [14] which are listed in Table 2.

Among DL techniques evaluated on *Indiana* dataset [14], *Neural Image Caption Generator* (NIC) [67] has shown promising results. It is an automatic end-to-end ANN system that can generate a reasonable English description of an input image. NIC is built using a *Convolutional Neural Network* (CNN) [40] which encodes an image into compact representation followed by a RNN that generates a sentence. Metrics like BLEU [49] were used to establish the quality of sentences generated by NIC.

Another DL model called *Long-term Recurrent Convolutional Network* (LRCN) [15] model combines a deep hierarchical visual feature extractor (such as a CNN) with a model that can learn to recognize and synthesize temporal dynamics. This is applicable for the tasks involving sequential data (inputs or outputs), visual, linguistic, or otherwise. An algorithm for image captioning based on a semantic attention model was proposed in [77]. The attention model naturally combines the visual information in both top-down and bottom-up approaches in the framework of RNNs.

A multi-task learning framework which jointly predicts tags and generates descriptions was proposed in [32]. A co-attention mechanism [32] was proposed which can simultaneously explore visual and semantic information to accurately localize and describe abnormal regions. Moreover, a hierarchical LSTM network that can more effectively capture long-range semantics and produce high quality long texts was developed. A DL model was employed to detect a disease from an image and annotate its contexts (e.g., location, severity and the affected organs) [57]. *ChestX-ray14* dataset [70], reports and its image annotations, were used to mine disease names to train CNNs. RNNs were then trained to describe the contexts of a detected disease, based on the DCNN features.

Indiana dataset is a large dataset and hence DL techniques [16, 32, 57, 67, 74] play key role in disease labeling.

4 Deep Learning for Tuberculosis Detection

As we mentioned earlier, tuberculosis detection can be done via DL from chest x-rays images. This section gives an overview about the datasets available and the techniques used for tuberculosis detection.

4.1 Datasets for Tuberculosis Detection

There are three datasets of chest x-rays images, *Montgomery County* [27], *Shenzen* [27] and *Belarus* [47] datasets. Techniques used for tuberculosis detection are listed in Tables 3 and 4. Some of them are DL techniques [38, 44, 61].

4.1.1 The *Montgomery County* Dataset

U.S. National Library of Medicine (USNLM) has collected the *Montgomery County* (MC) dataset [27] in collaboration with the Department of Health and Human Services, MC, MD, USA. There are 138 chest x-rays images in this dataset, which are collected under the tuberculosis control program. 80 chest x-rays images are considered to be normal and 58 are abnormal with manifestations of tuberculosis.

Patient's Age: 031Y Cavitary nodular infiltrate in RUL: active tuberculosis.

Manual segmentation on images of MC dataset was performed under the supervision of a radiologist, and binary lung masks were generated. Mask images for left and right lungs are stored separately with ".png" extension and are included in separate folders in the dataset [27].

4.1.2 The *Shenzen* Dataset

USNLM and Shenzhen No. 3 People's Hospital, Guangdong Medical College, Shenzhen, China [27] have collaborated to create the *Shenzen* dataset. The chest x-rays images were collected from outpatients hospitals within a time frame of one month. The dataset contains 662 chest x-rays images, of which 326 are normal while 336 have manifestations of tuberculosis. The uniqueness of the dataset lies in the fact that it also includes some paediatric *AnteroPosterior* (AP) chest x-rays images [27]. However, researchers have used customised datasets [4] for the segmentation of lungs in pediatric chest x-rays images.

As the dataset contains images of different sizes, the width of the images vary from 1130 to 3001 pixels, and the image height varies from 948 to 3001 pixels. However, most of the images have a size of 3000×3000 pixels. The images are provided in the ".png" format and have a pixel size of 0.353mm. All image file names follow the standard template "CHNCXR_####_X.png", where the value of X is 0 for normal x-rays images and 1 for abnormal x-rays images [27].

For each chest x-rays image in *Shenzen* dataset, clinical readings are available in a text file with ".txt" extension. Clinical reading comprises of the details like the patient age, gender and abnormality seen in the lung. Unlike the MC dataset [27], the *Shenzen* dataset [27] is not suitable for segmentation of anatomical structures because segmented anatomical structures are not available with this dataset.

4.1.3 The *Belarus* Dataset

The Ministry of Health, the Republic of Belarus created the *Belarus* dataset [47] by collecting chest x-rays images, and *Computed Tomography* (CT) scans from various institutes under its supervision. The original purpose is to study drug-resistant tuberculosis. There are 420 sets of chest x-rays images in the dataset. Each chest x-rays image in the dataset has a resolution of 2248 by 2248 pixels. Clinical data such as patient's gender, age, height, weight, and social information is present in the dataset along with the medical images. This dataset is available for researchers at the portal [47].

4.2 Performance Metrics for Tuberculosis Detection

For multi-label classification, AUC of *Receiver Operating Charecteristic* (ROC) curve is used [70] as a performance metric. AUC is calculated for each disease label. Higher the AUC, better is the performance of the approach. Mean AUCs for the different methods are listed in Table 1. For pathology localisation, the performance metrics are *accuracy* and the average number of *False Positives* (FPs) [70].

4.3 DL Techniques for Tuberculosis Detection

Techniques for tuberculosis detection are listed in Tables 3 and 4. Table 3 compares techniques evaluated on MC dataset [27]. Table 4 compares techniques evaluated on *Shenzen* dataset [27].

Among the techniques evaluated on MC dataset, pretrained CNN [44] has given high accuracy for tuberculosis detection. Pretrained CNN [44] used three different CNN architectures to extract features from a resized chest x-rays image or sub regions of the chest x-rays image. Using CNN for tuberculosis detection has

Table 3 Comparison of performances for tuberculosis detection techniques on the MC dataset [27]

Model	AUC	*Accuracy*
Feature extraction using pretrained CNN [44]	92.6	82.6
Automatic tuberculosis detection [29]	87.0	78.3
Tuberculosis detection using DCNN [24]	86.9	78.3
Combined lung masks [28]	83.12	75.0

Table 4 Comparison of performances for tuberculosis detection techniques on the *Shenzen* dataset [27]

Model	AUC	*Accuracy*
Abnormality detection using DCNN [26]	94.0	90.0
Feature extraction using pretrained CNN [44]	92.6	84.7
Automatic tuberculosis detection [29]	90.0	84.0
Aircavity segmentation using texture and shape features [33]	91.0	93.4
Lung segmentation using texture and shape features [33]	90.7	90.8

yeilded high AUC and accuracy when compared to other *Machine Learning* (ML) techniques [28] as shown in Table 3 when MC dataset is used [27].

When the *Shenzen* dataset [27] is used, DCNN [26] and pretrained CNN [44] have given high accuracy and AUC values when compared to other ML techniques as shown in Table 4.

As the datasets of chest x-rays images used for tuberculosis detection are of moderate size, DL techniques [26, 29, 44] have superior performance when compared to other techniques [28, 33].

5 Deep Learning for Lung Nodule Detection

Detection of lung nodules from chest x-rays images is possible using ML and DL techniques. This section contains the overview about the public datasets, ML and DL techniques used for lung nodule detection.

5.1 Datasets for Lung Nodule Detection

JSRT [59] and *Japanese Radiological Society* (JRS) [59] has developed a chest x-rays images database of 247 chest radiographs with and without nodule. The images are collected from thirteen distinct institutions in Japan and one in the USA in 1988 and made it as a public dataset [59]. Out of 247 images, 154 chest x-rays images have lung nodules, while 93 are normal with no nodules. Researchers have used JSRT dataset for a variety of applications such as image processing, image compression, evaluation of image display, *Computer Aided Diagnosis* (CAD), *Picture Archiving and Communication System* (PACS), and for training and testing. Accuracy in the detection of lung nodules can be increased using deep learning techniques [18].

The chest x-rays images have high resolution with matrix size 2048×2048 and 0.175mm pixel size. The chest x-rays images are stored in universal image format with ".img" extension. Information like the patient's age, gender, X and Y coordinates of the lung nodule, along with a basic diagram of nodule area is available in the JSRT dataset [45]. JSRT was used in numerous studies where researchers have used it for lung nodule detection from chest x-rays images as shown in the Table 5.

Three radiologists manually grouped the chest x-rays images into five levels based on the degree of subtlety. Images which are extremely difficult to detect are designated by *level 1*. Due to low contrast, small size, or overlap with a normal structure, detection of these images is highly tricky. Images which are very difficult to detect are designated by *level 2*. Images which are difficult to detect are designated by *level 3*. Images where nodule can be detected relatively easily are designated by *level 4*. Finally, images where nodule can be detected easily are designated by *level 5*. The number of images in *level 1* is 25, *level 2* contains 29, *level 3* contains 50, *level 4* contains 38, and *level 5* contains 12 images.

5.2 Performance Metrics for Lung Nodule Detection

The primary purpose of JSRT dataset [59] is lung nodule detection. Sensitivity (*sens*), specificity (*spec*) and the average number of false positives per image (FP*average*) are the three most commonly used metrics to measure the performance of a proposed lung nodule detection method [68].

$$sens = \frac{\text{TP}}{\text{TP} + \text{FN}} \tag{1}$$

$$spec = \frac{\text{TN}}{\text{TN} + \text{FP}} \tag{2}$$

$$\text{FP}avearge = \frac{\text{FP}}{N} \tag{3}$$

Where TP is the number of true positives, FN is the number of false negatives, TN is the number of true negatives, and FP is the number of false positives. FP*average* is the average number of false positives per image and *N* is the total number of images.

JSRT was the most common dataset used in the performance evaluation of a proposed technique for lung nodule detection. DL technique [68] for lung nodule detection has given high sensitivity as shown in Table 5.

5.3 Deep Learning Techniques for Lung Nodule Detection

Both ML [9, 12, 20, 52, 53] and DL techniques [3, 68] are applied on chest x-rays images for lung nodule detection as shown in Table 5.

Among DL techniques [3, 68], *Deep feature fusion* [68] has given high sensitivity value. This is because *Deep feature fusion* used a deep model obtained from the non medical training and hand-crafted features to reduce the FP results.

ML techniques [9, 12, 20, 52, 53] achieved better sensitivity when compared to DL techniques [68]. Among ML techniques, SVM [9] achieved high sensitivity.

We can't conclude that ML techniques [9, 12, 20, 52, 53] are better than DL techniques [3, 68] for lung nodule detection from chest x-rays images because large datasets of chest x-rays images are not available.

Table 5 Comparison of performances for lung nodule detection techniques on JSRT dataset [59]

Model	Sensitivity	FP per image
Deep feature fusion [68]	69.27	1.19
CAD for lung nodule detection [53]	67.0	4.0
Lifetime approach [63]	82.0	7
Virtual dual energy approach [8]	85.0	5
Two stage nodule enhancement using SVM [9]	79.0	5
Multi scale nodule detection [52]	67.0	4
CAD system [20]	80.0	5
DL [68]	65.92	1.3
Handcrafted [68]	62.01	1.43
Fully automated [3]	78.0	2.5
ANN [12]	70.0	7.6

6 Conclusion and Future Scope

In this chapter, we explored DL techniques for chest x-rays images processing with respect to thoracic disease detection, tuberculosis and lung nodule detection. Details about public datasets of chest x-rays images and performance metrics are also discussed for each category.

Although researchers have proposed massive datasets with respect to thoracic disease detection from chest x-rays images, accuracy of DL techniques need to be improved. Existing DL models consider only captions of the image to perform multi label classification. Instead of that, a DL model is supposed to consider both the chest x-rays label and the content of the chest x-rays image in order to perform multi label classification.

Massive datasets are needed for tuberculosis and lung nodule detection tasks from chest x-rays images. Moreover, DL techniques which yield high accuracy for lung nodule detection from chest x-rays images are needed.

References

1. Bar, Y., Diamant, I., Wolf, L., Greenspan, H.: Deep learning with non-medical training used for chest pathology identification. In: Medical Imaging 2015: Computer-Aided Diagnosis. vol. 9414, p. 94140V. International Society for Optics and Photonics (2015)
2. Bar, Y., Diamant, I., Wolf, L., Lieberman, S., Konen, E., Greenspan, H.: Chest pathology detection using deep learning with non-medical training. In: ISBI. pp. 294–297. Citeseer (2015)
3. Campadelli, P., Casiraghi, E., Artioli, D.: A fully automated method for lung nodule detection from postero-anterior chest radiographs. IEEE transactions on medical imaging **25**(12), 1588–1603 (2006)
4. Candemir, S., Antani, S., Jaeger, S., Browning, R., Thoma, G.R.: Lung boundary detection in pediatric chest x-rays. In: Medical Imaging 2015: PACS and Imaging Informatics: Next Generation and Innovations. vol. 9418, p. 94180Q. International Society for Optics and Photonics (2015)

5. Candemir, S., Jaeger, S., Antani, S., Bagci, U., Folio, L.R., Xu, Z., Thoma, G.: Atlas-based rib-bone detection in chest x-rays. Computerized Medical Imaging and Graphics **51**, 32–39 (2016)
6. Candemir, S., Jaeger, S., Palaniappan, K., Musco, J.P., Singh, R.K., Xue, Z., Karargyris, A., Antani, S., Thoma, G., McDonald, C.J.: Lung segmentation in chest radiographs using anatomical atlases with nonrigid registration. IEEE transactions on medical imaging **33**(2), 577–590 (2014)
7. Chaquet, J.M., Carmona, E.J., Fernández-Caballero, A.: A survey of video datasets for human action and activity recognition. Computer Vision and Image Understanding **117**(6), 633–659 (2013)
8. Chen, S., Suzuki, K.: Computerized detection of lung nodules by means of virtual dual-energy radiography. IEEE Transactions on Biomedical Engineering **60**(2), 369–378 (2013)
9. Chen, S., Suzuki, K., MacMahon, H.: Development and evaluation of a computer-aided diagnostic scheme for lung nodule detection in chest radiographs by means of two-stage nodule enhancement with support vector classification. Medical physics **38**(4), 1844–1858 (2011)
10. Ching, T., Himmelstein, D.S., Beaulieu-Jones, B.K., Kalinin, A.A., Do, B.T., Way, G.P., Ferrero, E., Agapow, P.M., Zietz, M., Hoffman, M.M., *et al.*: Opportunities and obstacles for deep learning in biology and medicine. Journal of The Royal Society Interface **15**(141), 20170387 (2018)
11. Chondro, P., Yao, C.Y., Ruan, S.J., Chien, L.C.: Low order adaptive region growing for lung segmentation on plain chest radiographs. Neurocomputing **275**, 1002–1011 (2018)
12. Coppini, G., Diciotti, S., Falchini, M., Villari, N., Valli, G.: Neural networks for computer-aided diagnosis: detection of lung nodules in chest radiograms. IEEE Transactions on Information Technology in Biomedicine **7**(4), 344–357 (2003)
13. Dai, W., Doyle, J., Liang, X., Zhang, H., Dong, N., Li, Y., Xing, E.P.: Scan: Structure correcting adversarial network for organ segmentation in chest x-rays. arXiv preprint arXiv:1703.08770 (2017)
14. Demner-Fushman, D., Kohli, M.D., Rosenman, M.B., Shooshan, S.E., Rodriguez, L., Antani, S., Thoma, G.R., McDonald, C.J.: Preparing a collection of radiology examinations for distribution and retrieval. Journal of the American Medical Informatics Association **23**(2), 304–310 (2015)
15. Donahue, J., Hendricks, L.A., Guadarrama, S., Rohrbach, M., Venugopalan, S., Saenko, K., Darrell, T.: Long-term recurrent convolutional networks for visual recognition and description. CoRR **abs/1411.4389** (2014), http://arxiv.org/abs/1411.4389
16. Donahue, J., Anne Hendricks, L., Guadarrama, S., Rohrbach, M., Venugopalan, S., Saenko, K., Darrell, T.: Long-term recurrent convolutional networks for visual recognition and description. In: Proceedings of the IEEE conference on computer vision and pattern recognition. pp. 2625–2634 (2015)
17. Giger, M.L., Chan, H.P., Boone, J.: Anniversary paper: History and status of cad and quantitative image analysis: the role of medical physics and aapm. Medical physics **35**(12), 5799–5820 (2008)
18. Gordienko, Y., Gang, P., Hui, J., Zeng, W., Kochura, Y., Alienin, O., Rokovyi, O., Stirenko, S.: Deep learning with lung segmentation and bone shadow exclusion techniques for chest x-rays analysis of lung cancer. In: International Conference on Theory and Applications of Fuzzy Systems and Soft Computing. pp. 638–647. Springer (2018)
19. Guan, Q., Huang, Y., Zhong, Z., Zheng, Z., Zheng, L., Yang, Y.: Diagnose like a radiologist: Attention guided convolutional neural network for thorax disease classification. arXiv preprint arXiv:1801.09927 (2018)
20. Hardie, R.C., Rogers, S.K., Wilson, T., Rogers, A.: Performance analysis of a new computer aided detection system for identifying lung nodules on chest radiographs. Medical Image Analysis **12**(3), 240–258 (2008)
21. He, K., Zhang, X., Ren, S., Sun, J.: Deep residual learning for image recognition. In: Proceedings of the IEEE conference on computer vision and pattern recognition. pp. 770–778 (2016)

22. Hogeweg, L., Sánchez, C.I., de Jong, P.A., Maduskar, P., van Ginneken, B.: Clavicle segmentation in chest radiographs. Medical image analysis **16**(8), 1490–1502 (2012)
23. Huang, G., Liu, Z., Weinberger, K.Q.: Densely connected convolutional networks. CoRR **abs/1608.06993** (2016), http://arxiv.org/abs/1608.06993
24. Hwang, S., Kim, H.E., Jeong, J., Kim, H.J.: A novel approach for tuberculosis screening based on deep convolutional neural networks. In: Medical Imaging 2016: Computer-Aided Diagnosis. vol. 9785, p. 97852W. International Society for Optics and Photonics (2016)
25. Ibragimov, B., Likar, B., Pernuš, F., Vrtovec, T.: Accurate landmark-based segmentation by incorporating landmark misdetections. In: Biomedical Imaging (ISBI), 2016 IEEE 13th International Symposium on. pp. 1072–1075. IEEE (2016)
26. Islam, M.T., Aowal, M.A., Minhaz, A.T., Ashraf, K.: Abnormality detection and localization in chest x-rays using deep convolutional neural networks. arXiv preprint arXiv:1705.09850 (2017)
27. Jaeger, S., Candemir, S., Antani, S., Wáng, Y.X.J., Lu, P.X., Thoma, G.: Two public chest x-rays datasets for computer-aided screening of pulmonary diseases. Quantitative imaging in medicine and surgery **4**(6), 475 (2014)
28. Jaeger, S., Karargyris, A., Antani, S., Thoma, G.: Detecting tuberculosis in radiographs using combined lung masks. In: Engineering in Medicine and Biology Society (EMBC), 2012 Annual International Conference of the IEEE. pp. 4978–4981. IEEE (2012)
29. Jaeger, S., Karargyris, A., Candemir, S., Folio, L., Siegelman, J., Callaghan, F.M., Xue, Z., Palaniappan, K., Singh, R.K., Antani, S.K., *et al.*: Automatic tuberculosis screening using chest radiographs. IEEE Trans. Med. Imaging **33**(2), 233–245 (2014)
30. Jaeger, S., Karargyris, A., Candemir, S., Siegelman, J., Folio, L., Antani, S., Thoma, G.: Automatic screening for tuberculosis in chest radiographs: a survey. Quantitative imaging in medicine and surgery **3**(2), 89 (2013)
31. Jangam, E., Rao, A.: Segmentation of lungs from chest x-rays using firefly optimized fuzzy c-means and level set algorithm. In: International Conference on Recent Trends in Image Processing and Pattern Recognition. pp. 303–311. Springer (2018)
32. Jing, B., Xie, P., Xing, E.: On the automatic generation of medical imaging reports. arXiv preprint arXiv:1711.08195 (2017)
33. Karargyris, A., Siegelman, J., Tzortzis, D., Jaeger, S., Candemir, S., Xue, Z., Santosh, K., Vajda, S., Antani, S., Folio, L.: Combination of texture and shape features to detect pulmonary abnormalities in digital chest x-rays. International journal of computer assisted radiology and surgery **11**(1), 99–106 (2016)
34. Ker, J., Wang, L., Rao, J., Lim, T.: Deep learning applications in medical image analysis. IEEE Access **6**, 9375–9389 (2018)
35. Krizhevsky, A., Sutskever, I., Hinton, G.E.: Imagenet classification with deep convolutional neural networks. In: Advances in neural information processing systems. pp. 1097–1105 (2012)
36. Kumar, K., Rao, A.C.S.: Breast cancer classification of image using convolutional neural network. In: 2018 4th International Conference on Recent Advances in Information Technology (RAIT). pp. 1–6 (2018)
37. Kumar, P., Grewal, M., Srivastava, M.M.: Boosted cascaded convnets for multilabel classification of thoracic diseases in chest radiographs. In: International Conference Image Analysis and Recognition. pp. 546–552. Springer (2018)
38. Lakhani, P., Sundaram, B.: Deep learning at chest radiography: automated classification of pulmonary tuberculosis by using convolutional neural networks. Radiology **284**(2), 574–582 (2017)
39. Lavie, A., Denkowski, M.J.: The meteor metric for automatic evaluation of machine translation. Machine translation **23**(2–3), 105–115 (2009)
40. Lecun, Y., Bottou, L., Bengio, Y., Haffner, P.: Gradient-based learning applied to document recognition. Proceedings of the IEEE **86**(11), 2278–2324 (1998)
41. Li, Z., Wang, C., Han, M., Xue, Y., Wei, W., Li, L.J., Li, F.: Thoracic disease identification and localization with limited supervision. arXiv preprint arXiv:1711.06373 (2017)

42. Lin, C.Y.: Rouge: A package for automatic evaluation of summaries. In: Text summarization branches out. pp. 74–81 (2004)
43. Loog, M., Ginneken, B.: Segmentation of the posterior ribs in chest radiographs using iterated contextual pixel classification. IEEE Transactions on Medical Imaging **25**(5), 602–611 (2006)
44. Lopes, U., Valiati, J.: Pre-trained convolutional neural networks as feature extractors for tuberculosis detection. Computers in biology and medicine **89**, 135–143 (2017)
45. Mittal, A., Hooda, R., Sofat, S.: Lung field segmentation in chest radiographs: a historical review, current status, and expectations from deep learning. IET Image Processing **11**(11), 937–952 (2017)
46. Mould, R.F.: A century of x-rays and radioactivity in medicine: with emphasis on photographic records of the early years. CRC Press (1993)
47. NA: Belarus dataset, http://www.tuberculosis.by
48. Novikov Alexey ALenis, Dimitrios, M.J.H.M.W.K.B.: Fully convolutional architectures for multi-class segmentation in chest radiographs. IEEE Transactions on Medical Imaging (2018)
49. Papineni, K., Roukos, S., Ward, T., Zhu, W.J.: Bleu: a method for automatic evaluation of machine translation. In: Proceedings of the 40th annual meeting of the Association for Computational Linguistics. pp. 311–318 (2002)
50. Rajpurkar, P., Irvin, J., Zhu, K., Yang, B., Mehta, H., Duan, T., Ding, D., Bagul, A., Langlotz, C., Shpanskaya, K., *et al.*: Chexnet: Radiologist-level pneumonia detection on chest x-rays with deep learning. arXiv preprint arXiv:1711.05225 (2017)
51. Rao, A.C.S., Somayajulu, D., Banka, H., Chaturvedi, R.: Outlier detection in microarray data using hybrid evolutionary algorithm. Procedia Technology **6**, 291–298 (2012)
52. Schilham, A.M., van Ginneken, B., Loog, M.: Multi-scale nodule detection in chest radiographs. In: International Conference on Medical Image Computing and Computer-Assisted Intervention. pp. 602–609. Springer (2003)
53. Schilham, A.M., Van Ginneken, B., Loog, M.: A computer-aided diagnosis system for detection of lung nodules in chest radiographs with an evaluation on a public database. Medical Image Analysis **10**(2), 247–258 (2006)
54. Seghers, D., Loeckx, D., Maes, F., Vandermeulen, D., Suetens, P.: Minimal shape and intensity cost path segmentation. IEEE Transactions on Medical Imaging **26**(8), 1115–1129 (2007)
55. Shao, Y., Gao, Y., Guo, Y., Shi, Y., Yang, X., Shen, D.: Hierarchical lung field segmentation with joint shape and appearance sparse learning. IEEE transactions on medical imaging **33**(9), 1761–1780 (2014)
56. Shen, L., Song, R.: Semi-supervised learning for multi-label classification. Reconstruction **1**, 1
57. Shin, H.C., Roberts, K., Lu, L., Demner-Fushman, D., Yao, J., Summers, R.M.: Learning to read chest x-rays: Recurrent neural cascade model for automated image annotation. In: Proceedings of the IEEE conference on computer vision and pattern recognition. pp. 2497–2506 (2016)
58. Shin, H.C., Roth, H.R., Gao, M., Lu, L., Xu, Z., Nogues, I., Yao, J., Mollura, D., Summers, R.M.: Deep convolutional neural networks for computer-aided detection: Cnn architectures, dataset characteristics and transfer learning. IEEE transactions on medical imaging **35**(5), 1285–1298 (2016)
59. Shiraishi, J., Katsuragawa, S., Ikezoe, J., Matsumoto, T., Kobayashi, T., Komatsu, K.i., Matsui, M., Fujita, H., Kodera, Y., Doi, K.: Development of a digital image database for chest radiographs with and without a lung nodule: receiver operating characteristic analysis of radiologists' detection of pulmonary nodules. American Journal of Roentgenology **174**(1), 71–74 (2000)
60. Simonyan, K., Zisserman, A.: Very deep convolutional networks for large-scale image recognition. arXiv preprint arXiv:1409.1556 (2014)
61. Suzuki, K.: Overview of deep learning in medical imaging. Radiological physics and technology **10**(3), 257–273 (2017)
62. Szegedy, C., Liu, W., Jia, Y., Sermanet, P., Reed, S.E., Anguelov, D., Erhan, D., Vanhoucke, V., Rabinovich, A.: Going deeper with convolutions. CoRR **abs/1409.4842** (2014), http://arxiv.org/abs/1409.4842

63. Ture, H., Kayikcioglu, T.: Detection and segmentation of nodules in chest radiographs based on lifetime approach. In: CMBEBIH 2017, pp. 539–547. Springer (2017)
64. Van Ginneken, B., Romeny, B.T.H., Viergever, M.A.: Computer-aided diagnosis in chest radiography: a survey. IEEE Transactions on medical imaging **20**(12), 1228–1241 (2001)
65. Van Ginneken, B., Stegmann, M.B., Loog, M.: Segmentation of anatomical structures in chest radiographs using supervised methods: a comparative study on a public database. Medical image analysis **10**(1), 19–40 (2006)
66. Vedantam, R., Lawrence Zitnick, C., Parikh, D.: Cider: Consensus-based image description evaluation. In: Proceedings of the IEEE conference on computer vision and pattern recognition. pp. 4566–4575 (2015)
67. Vinyals, O., Toshev, A., Bengio, S., Erhan, D.: Show and tell: A neural image caption generator. In: Proceedings of the IEEE conference on computer vision and pattern recognition. pp. 3156–3164 (2015)
68. Wang, C., Elazab, A., Wu, J., Hu, Q.: Lung nodule classification using deep feature fusion in chest radiography. Computerized Medical Imaging and Graphics **57**, 10–18 (2017)
69. Wang, C.: Segmentation of multiple structures in chest radiographs using multi-task fully convolutional networks. In: Scandinavian Conference on Image Analysis. pp. 282–289. Springer (2017)
70. Wang, X., Peng, Y., Lu, L., Lu, Z., Bagheri, M., Summers, R.M.: Chestx-ray8: Hospital-scale chest x-rays database and benchmarks on weakly-supervised classification and localization of common thorax diseases. In: Computer Vision and Pattern Recognition (CVPR), 2017 IEEE Conference on. pp. 3462–3471. IEEE (2017)
71. Wang, X., Peng, Y., Lu, L., Lu, Z., Summers, R.M.: Tienet: Text-image embedding network for common thorax disease classification and reporting in chest x-rays. In: Proceedings of the IEEE Conference on Computer Vision and Pattern Recognition. pp. 9049–9058 (2018)
72. Wu, G., Zhang, X., Luo, S., Hu, Q.: Lung segmentation based on customized active shape model from digital radiography chest images. Journal of Medical Imaging and Health Informatics **5**(2), 184–191 (2015)
73. Xiong, J., Shao, Y., Ma, J., Ren, Y., Wang, Q., Zhao, J.: Lung field segmentation using weighted sparse shape composition with robust initialization. Medical physics **44**(11), 5916–5929 (2017)
74. Xu, K., Ba, J., Kiros, R., Cho, K., Courville, A., Salakhudinov, R., Zemel, R., Bengio, Y.: Show, attend and tell: Neural image caption generation with visual attention. In: International conference on machine learning. pp. 2048–2057 (2015)
75. Yang, W., Liu, Y., Lin, L., Yun, Z., Lu, Z., Feng, Q., Chen, W.: Lung field segmentation in chest radiographs from boundary maps by a structured edge detector. IEEE journal of biomedical and health informatics **22**(3), 842–851 (2018)
76. Yao, L., Poblenz, E., Dagunts, D., Covington, B., Bernard, D., Lyman, K.: Learning to diagnose from scratch by exploiting dependencies among labels. arXiv preprint arXiv:1710.10501 (2017)
77. You, Q., Jin, H., Wang, Z., Fang, C., Luo, J.: Image captioning with semantic attention. In: Proceedings of the IEEE conference on computer vision and pattern recognition. pp. 4651–4659 (2016)

Deep Learning in Multi-Omics Data Integration in Cancer Diagnostic

Abedalrhman Alkhateeb, Ashraf Abou Tabl, and Luis Rueda

Abstract Human cell is a complex of interacting small molecules which work together to perform daily tasks of the cell. The reading and the measurements of this different molecules are called *omics*, where any dysfunction among these omics may cause different diseases, and cancer is not any exception. The advances in biomedical technology in general and in software development help to speed up, and reduce the cost of reading these omics. However, the challenge is how to integrate the big amount of data from these omics to help the predictions systems in cancer outcomes, where these outcomes include the diagnosis, development, and treatment of cancer.

Due to the heterogeneity nature of the tumor tissue, different types of cancers such as breast, liver, prostate, and other cancers are considered to be different disease. Nevertheless, the subtypes of a certain cancer such as breast cancer are considered to be a different disease. That means the genomic activities varies among these different diseases and the normal tissue as well. Thanks to the power of computing, *Deep Learning* (DL) techniques have become feasible to integrate multi-omics data generated from the cells/tissue to study the outcomes of cancer as discussed in this chapter.

Keywords Multi-Omics data integration · Deep learning · Cancer diagnostic · Next-generation sequencing

---

A. Alkhateeb (✉) · L. Rueda
School of Computer Science, University of Windsor, Windsor, ON, Canada
e-mail: alkhate@uwindsor.ca; lrueda@uwindsor.ca

A. Abou Tabl
Zekelman School of Business & Information Technology, St. Clair College, Windsor, ON, Canada
e-mail: aaboutabl@stclaircollege.ca

M. Elloumi (ed.), *Deep Learning for Biomedical Data Analysis*,
https://doi.org/10.1007/978-3-030-71676-9_11

1 Introduction

Omics consists of integrating the analysis of cell and molecules that contain quantification, events, changes, images, text, and other genomic-related data in the aim of understanding how cells molecules communicate with each others and the surrounding environment. Omics are variety of disciplines including but not restricted to *Genomics*, *Epigenomics*, *Proteomics*, *Foodomics*, *Transcriptomics*, and *Metabolomics*, among others. [1]. Human cells divide to generate new cells when old cells die or become broken. If the old cell survive and stay in the tissue, or new cells are formed when they are not needed, they form a tumor. Tumor cells are *malignant* (cancerous) if they invade nearby tissues or travel to distant locations. Unlike malignant, benign tumor cells do not spread, and they are usually treated by taken away surgically and do not grow back. A benign tumor is not life-threatening unless it grows in the brain [2].

Cancer develops when pathways controlling cell survival, cell fate, or genome maintenance are disrupted by the somatic alteration of key driver genes. Study cancer driver genes and their genomics, epigenomics, and transcriptomics is essential for accurate characterization of cancer biology and identification of therapeutic targets [3]. The central dogma of biology viewed RNA as an intermediate molecular between DNA and proteins.

Transcriptomes studies analyze the quantitative aspect of the RNA such as how much genes transcripts to protein, and the qualitative aspects such as RNA editing sites, events, and novel splicing sites. Many of the studies have focus on the non-coding (unknown protein-coding) parts of the DNA, these identified elements can be correspondence to sequence variants linked to human disease [4]. Thousands of long non-coding RNAs transcribed in mammalian cells play essential roles in many physiological processes of cancer [5]. On the other hand, small non-coding RNAs (miRNA) were reported to regulate the expression of various oncogenes or tumor suppressor genes [6].

Transcriptomics examines RNA levels genome-wide, both qualitatively (which transcripts are present, identification of novel splice sites, RNA editing sites) and quantitatively (how much of each transcript is expressed). The central dogma of biology viewed RNA as a molecular intermediate between DNA and proteins. Sequencing RNA comes with artifacts that require intensive preprocessing [7].

Proteomics is used to quantify peptide abundance, modification, and interaction. The analysis and quantification of proteins has been revolutionized by mass spectrometry MS-based methods and, recently, these have been adapted for high-throughput analyses of thousands of proteins in cells or body fluids [1]. Chromatin immunoprecipitation with Sequencing (*Chip-Seq*) has been used to examine global interactions between proteins and nucleic acids [8].

Metabolomics simultaneously quantifies multiple small molecule types, such as amino acids, fatty acids, carbohydrates, or other products of cellular metabolic functions [1]. While DNA methylation is an epigenetic mark that has suspected regulatory roles in a broad range of biological processes and diseases [9]. Several tools are now available for processing data obtained with all common methods for genome-wide DNA methylation mapping such as MeDIP-Seq [10] and meDIP-Chip [11].

Microbiomics is analyzing all the microorganisms of a given community together. Human skin, mucosal surfaces, and the gut are colonized by microorganisms, including bacteria, viruses, and fungi, collectively known as the microbiota (and their genes constituting the microbiome). The human microbiome is enormously complex; for example, the gut contains roughly 100 trillion bacteria from 1000 different species. Several analytic tools have been developed for analyzing *Next-Generation Sequencing* (NGS) data from targeted 16S or metagenomics analysis, such as *Quantitative Insights Into Microbial Ecology* (QIIME) [12]. These allow accurate quantitative determination of taxa that can be correlated with disease or other phenotypes of interest [13]. Associated technologies include NGS application for 16S ribosomal abundance and metagenomics quantification [14].

Clinical data can be integrated with multi-omics data as another dimension of the data. Features such as disease family history, age, race, gender, allergies, etc. may provide an interesting prior knowledge to the learning system. pathological information about tumor tissue can help to determine the prognosis and treatment of the disease, for example, *Gleason score* which is a pathological grading score based on the visual attributes of the prostate tumor tissue is found to have different genomic activities among different scores(patterns) [15].

Each type of omics will provide unique characteristics of cancer on it is own. collecting all these pieces of details will unveil the overall disease development. By integrating the different omics in the learning and prediction phases, the *Computer-Aided Diagnosis* (CADs) systems [16] can generalize the global features and can also individualize different cases based on their omics. personalize the treatment and medication for the patient will be more precise.

The main aims of integrating multi-omic data can be summarized as follows:

- Understand the structure of the cell: multi-omics approaches work together to learn how daily cell functions work. The main questions here are how the omics communicate to perform a specific function, which omics participate in a specific task, and the amount of impact of omics in a specific task such as how a specific mutation can change the produced protein from a certain gene [17].
- Build the smaller networks in specific organs to understand the relationships among different cells in such organ, and possibly different organs [18].
- Understand the development of a specific disease: Current research consider the flow of information among a diseased cell and the molecules involved, and compare to an adjacent normal cell. The main idea is to isolate a panel of molecules that serve as potential biomarkers for the disease [1].
- Understand the reactions of the cell molecules after applying different drugs and treatments. These kinds of studies can relate a set of omics to specific drugs or clinical trials, which serve well in finding the precise treatment and avoid

unnecessary reactions or diminishing the patient's quality of life. Nevertheless, it can help in prescribing a more accurate dosage, and in drug re-purposing [19].

- Understand the reactions of the cell molecules to a different type of foods, finding omics as biomarkers associated with reaction to food can help in personalizing nutrition plans, and reduce obesity and improve the patient's quality of life [20].
- Study the effect of the environment such as pollution and radiation on different cell's omes [21].
- Analyze the effect of different choices of life including smoking and drinking alcohol on different cells' omics [22].

Due to the development of more powerful computing and big data analysis techniques, combined with the advancements in technologies such as NGS, researchers nowadays tend to utilize a more comprehensive and integrative approach for analysis, targeting *multi-omics data*. The current approaches, also, aim to link their findings to publicly available datasets in order to validate the results and gain more information about the outcomes of the study. Multi-omics data is a heterogeneous multi-source data that has data views from different omics which are varying in volume, value, and structure.

The rest of the chapter is organized as follows: in Sect. 2, we review the categories of *Deep Learning* (DL) methods for data integration based on which level in the *Artificial Neural Network* (ANN), where the data from multi-sources integrate. In Sects. 3, 4 and 5, we discuss the application of these DL methods in breast cancer, colorectal cancer, and bladder cancer in order. Finally, in the last Sect. 6, we conclude the contribution of this chapter and the importance of applying DL methods to integrate multi-omics data for better diagnosis and treatment of the disease.

2 Deep Learning for Multi-Omics Data Integration Methods Categories

Due to the evolving technology in computing powers and parallel computing, *Machine Learning* (ML) was utilized on big data analysis and multi-source data. The literature shows many categories for ML methods for multi omics data integration. Li et al. [23] categorized Implementations of machine learning methods for multi-view data analysis based on the type of the ML method into six categories, which are feature concatenation, Bayesian models or networks, Ensemble learning, multiple kernel learning, network-based learning, multi-view matrix or tensor factorization [23]. Lin et al. [24] divided the ML implementations based on how the data is integrated, the three categories include *Model-Based Integration* (MBI), *Concatenation-Based Integration* (CBI), and *Transformation-Based Integration*.

For DL approaches, the structural category is considered based on which layer the data integration has happened, the implementations were divided into early layer integration, middle layer integration, and late layer integration [25]. In this chapter, we classify the models based on these layers categories as the following:

2.1 Early Layer Integration

Multiple data views are concatenated/merged to form one matrix of data, the merged data matrix is fed to the input layer of the DL approach as seen in Fig. 1. The integration of the omics happens before the input layer of the learning process.

Different omics data requires different preprocessing, where some of the preprocessing may use learning systems, including DL methods [26]. The learning we are referring to in this category is the main learning system, In other words, if the

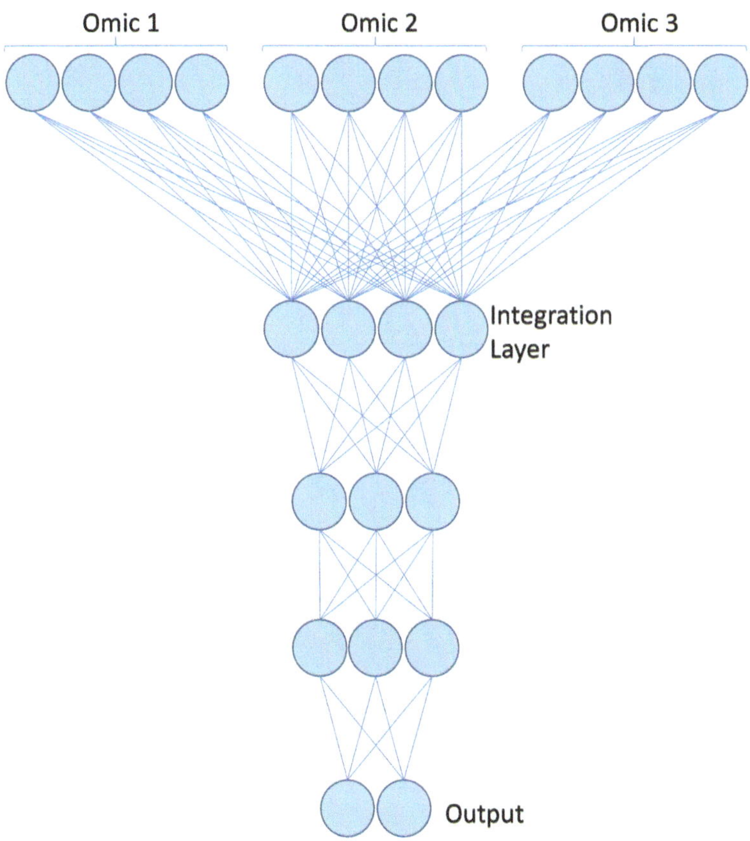

Fig. 1 ELI model

data sources have merged together in the early stage of the main learning systems, it considers as *Early Layer Integration* (ELI) model. Chaudhary et al. [27] normalized the omics datasets before concatenating the data into the DL method.

2.2 Middle Layer Integration

The different omics views are integrated in the hidden layers of the main learning system as seen in Fig. 2, it is the *Middle Layer Integration* (MLI) [28] model. Ibrahim et al. [29] have proposed DL approach to integrate gene expression data and miRNA data in the hidden layers of the ANN, then predict the breast cancer subtype as the output in the last layer.

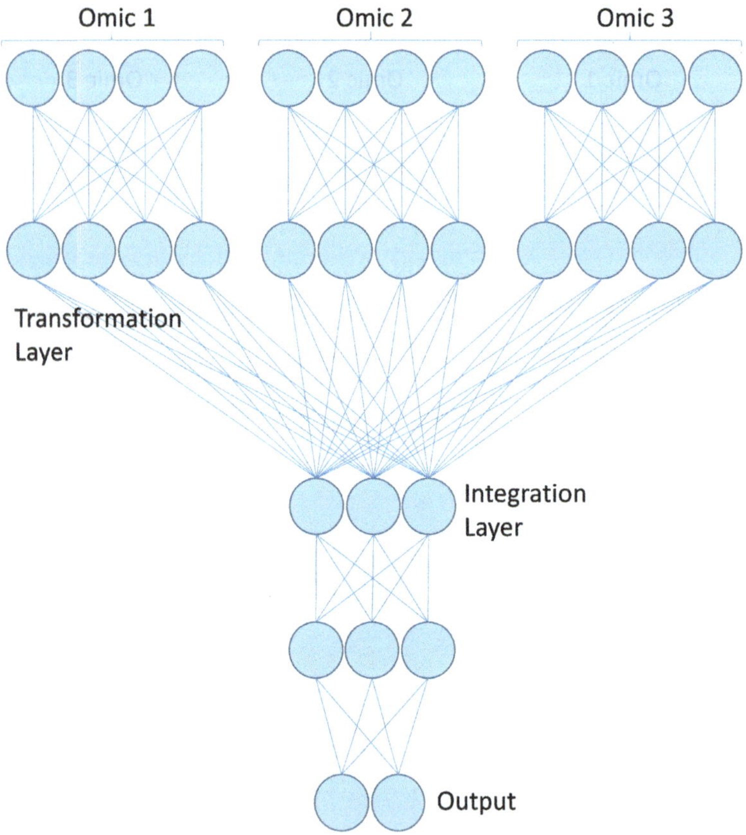

Fig. 2 MLI model

2.3 Late Layer Integration

Multiple predictive models that work on different omics. Each model is built on a specific source of data (omic) independently from the other models, the results from the independent models are integrated to make the final decision on each patient sample as shown in Fig. 3. In this category, The integration happens right before the output layer using fusion layer [30], or ensembles the models by majority voting criteria to select the class [31].

Khoshghalbvash et al. [30] presented a multimodal deep *Artificial Neural Network* (ANN) to exploit the mutual information between three different modalities to predict the survival rates of breast cancer. Each model is a deep ANN that is built on a specific omic, the value in each layer is calculated as Eq. (1):

$$h^{(i)} = \sigma(W_{mj}^{(i)} \times h_{mj}^{(i-1)} + b_{mj}^{(i)}) \tag{1}$$

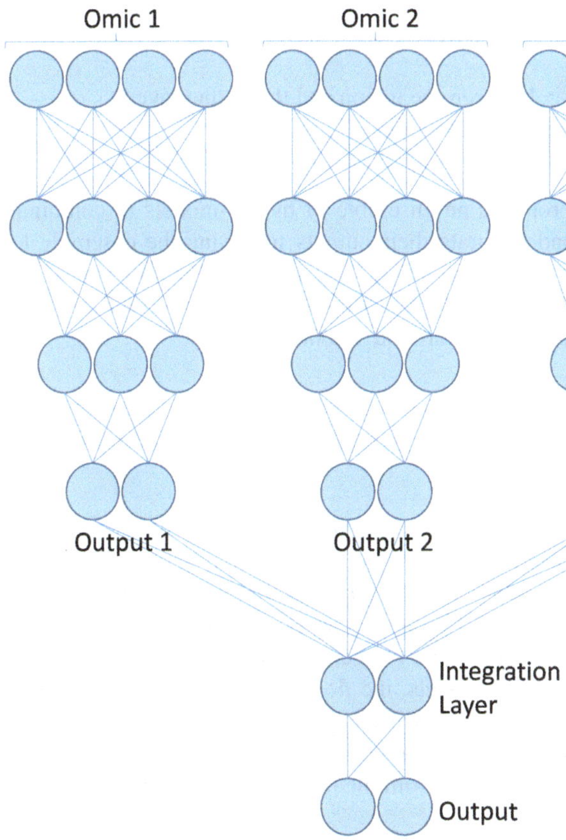

Fig. 3 LLI model

where:

- σ is a *Rectified Linear Unit* (ReLu) activation function [32]
- $W_{mj}^{(i)}$ is the weight matrix between $i-1$ to i hidden layer
- $b_{mj}^{(i)}$ is the bias for ith hidden layer for modality mi The fusion layer is integrating the different modalities m as Eq. (2):

$$h^{(f)} = \sigma(W_{m1}^{(f)} \times h_{m1}^{(f-1)} + W_{m2}^{(f)} \times h_{m2}^{(f-1)} + W_{m3}^{(f)} \times h_{m3}^{(f-1)} + b^{(f)}) \quad (2)$$

$m1$ is the modality that is built on *Copy Number Alteration* (CNA) [33] data, $m2$ on gene expression, while $m3$ on clinical data. The output layer is predicting the outcome which is the survival rate by connecting to the fusion layer as the following:

$$predicted = \sigma'(W^{(o)} \times h^{(f)} + b^{(o)}) \quad (3)$$

Where :

$W^{(o)}$ is the weight matrix between the fusion and the output layers
σ' is the *sigmoid* [34] activation function
$b^{(o)}$ is the bias for the output layer

Poirion et al. [31] proposed an ensemble of hybrid-models (a combination of DL and ML models) and integrate their outputs following the ensemble learning paradigm. The model, called *DeepProg* [31], is built on multi-omics including RNASeq, Methylation, and miRNA data from 32 cancers in *The Cancer Genome Atlas* (TCGA) [35], with a total of around 10,000 samples.

3 Breast Cancer

Breast cancer is the most commonly occurring cancer in women and the second most common cancer overall. There were over 2 million new cases in 2018 [36]. Sun et al. [37] proposed a *Multimodal Deep Neural Network by integrating Multi-dimensional Data* (MDNNMD) for the prognosis prediction of breast cancer. The method's architecture is LLI category where the integration happens by fusion the output of the independent models, the flowchart of the method is depicted in Fig. 4. The results show that MDNNMD achieves a better performance than the prediction methods with single-dimensional data and other existing approaches [37]. The multidimensional data is taken from the heterogeneous infamous breast cancer dataset, called METABRIC [38].

Multi-Level Feature Selection (MLFS) [29] approach was proposed for selecting genes/miRNAs based on expression profiles. The approach is based on both DL

Fig. 4 MDNNMD model

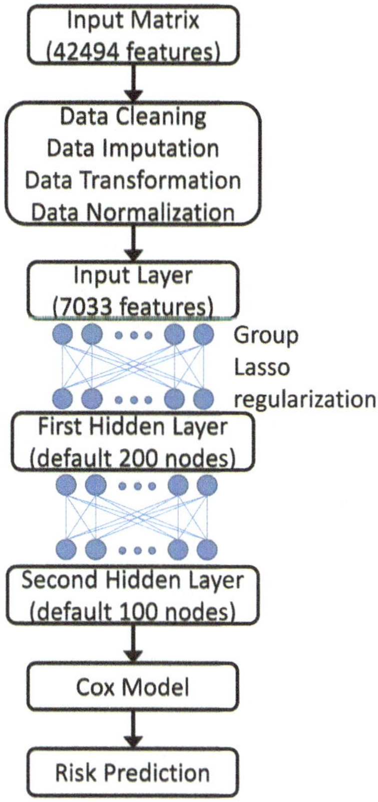

and *Active Learning* (AL). The method reduced the number of discriminative genes that related to a specific cancer using *Deep Belief Net* (DBN) [39], The model is extended by extracting miRNAS of the target genes using the well known *miRanda* database [40], and finally apply DBN on the selected miRNAs to detect the cancer outcomes. MLFS was applied to a breast cancer dataset to predict the subtype.

Middle Integration Deep Neural Network (MI-DNN) [25] is an example of LLI category. The model predicts breast cancer subtypes based on multi-omics TCGA-BRCA dataset [35], where gene expression data and miRNA data has been downloaded using TCGA-Assembler tool [41], and lncRNA from TANRIC [42]. Despite the fact that the authors have called the model *Middle Integration*, MI-DNN constructs the hidden layers prior to integration, which happened to be right before the output layer. The hidden layer number k_{th} has a modality m, where m is mathematically calculated similar to Eq. (1).

Xu et al. [26] proposed *Hierarchical Integration Deep Flexible Neural Forest* (HI-DFNForest) framework. The framework integrates multi-omics data for breast cancer subtypes prediction. A *Stacked AutoEncoder* (SAE) is built on each omic data to transform the omic matrix into a high-complex presentation. Then all the presentations from all omics are concatenated to form one input matrix for the final

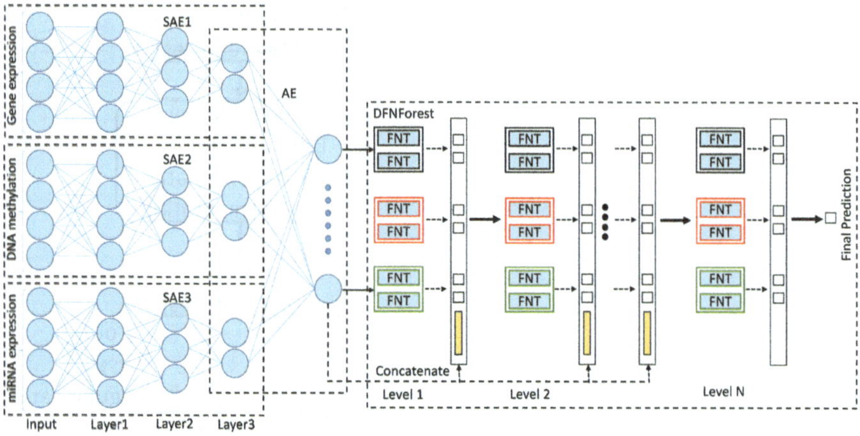

Fig. 5 A hierarchical integration deep flexible neural forest framework for cancer subtype classification by integrating multi-omics data

classifier that is *Deep Flexible Neural Forest* (DFNForest) model as seen in Fig. 5. The framework was applied to TCGA-BRCA dataset. Table 1 shows the applied DL methods on different breast cancers.

4 Colorectal Cancer

ColoRectal Cancer (CRC) is the third most common cancer diagnosed in both men and women each year in the United States, excluding skin cancer [43]. In 2004, ANN was introduced to detect the tumor in CRC [44]. 182 samples were detected by using *Surface-Enhanced Laser Desorption/Ionization Mass Spectrometry* (SELDI-MS) [45]. These samples include 55 samples from CRC patients, 35 samples from *ColoRectal Adenoma* (CRA) patients, and 92 samples from *Healthy Persons* (HP). The proteomic approach achieved high sensitivity and specificity for the diagnosis of cancer, and it as able to find potential biomarkers for CRC.

Many DL methods for CRC analysis have been proposed [46, 47]. Hu et al. [46] proposed a ML approach to distinguish the left side CRC from the right side CRC as they are different diseases. The multi-omics dataset, called *COlon ADenocarcinoma* (COAD) [48], was downloaded from TCGA portal [35]. It contains somatic mutations, DNA methylation, gene expression, and miRNA. Bychkov et al. [47] structured convolutional and recurrent architectures to train a deep ANN to predict colorectal cancer five-year survivability based on images of tumour tissue samples. The images are *Hematoxylin and Eosin* (H&E) stained histopathology images of the tumour *Tissue MicroArrays* (TMAs), which is a technique that is used by pathologists to identify what is known as tissue entities such mitoses, nuclear pleomorphism, infiltrating immune cells, and tumour budding. The model only

Table 1 DL model for breast cancer multi-omics data integration

Type	Method	Classes	Dataset	Acc	Pre	Sn	ROC	Platform	Hardware
LLI	MDNNMD	Binary - Tumor vs Normal	METABRIC	82.6	0.749	0.45	0.846	Tensorflow1.0	2 Nvidia GTX Titan Z graphics cards
LLI	MI-DNN	Multi-class: Normal, Stage I , Stage II breast carcinoma	TCGA-BRCA	83	0.83	0.83	0.97	Sklearn	–
MLF	MLFS	Breast cancer subtypes			0.7815	0.6047	–	Theano	
ELI	HI-DFNForest	Breast cancer subtypes	TCGA-BRCA	84.6	80.9	84	–		

works on the TMAs images without prior knowledge of the tissue entities, which simplifies the diagnosis of the disease. However, the visual attributes of the TMAs images will not reveal the intrinsic functional molecules features in the tumor tissue. The lack of DL on multi-omics data—up to date—to predict the outcome of CRC makes it an interesting area to research.

5 Bladder Cancer

The general 5-year survival rate for people with bladder cancer is 77%. However, if bladder cancer is invasive which means that it has spread to distant parts of the body, the 5-year survival rate is 5% [49]. Poirion et al. [50] proposed a DL autoencoder algorithm, to predict the survival subtypes in bladder cancer. The ELI model is built on a multi-omics TCGA dataset which has mRNA, miRNA, and methylation [50]. The samples were downloaded and assembled using TCGA-Assembler. The model is an extension of a previously proposed predictive model by the authors. The model is an autoencoder that transforms multi-omics features followed by integration of survival information to identify subtypes linked with survival [27]. For each samples omic vector $x = (x_1, x_2, \ldots x_n)$ of dimension n, x through successive hidden layers. For a given layer i using tanh as an activation function between input layer x and output layer y. denoted as:

$$y = f(x) = tanh(W_i x + bi) \tag{4}$$

Considering having k layers, the model has k functions which forms the autoencoder that transforms the vector x into the reduced dimensional vector x' as the following:

$$x' = f_{1\to k}(x) = F_1 \circ ..F_{k-1} \circ F_k(x) \tag{5}$$

Where $F_{k-1} \circ F_k(x)$ is the composition function between the 2 functions as $F_{k-1}(F_k(x))$. The model obtains the weight matrix W_i which minimizes the objective loss function between the input x and output x' as:

$$logloss(x, x') = \sum_k^n (x_k log(x'_k) + (1 - x_k) log(1 - x'_k)) \tag{6}$$

The model controls overfitting using L_1 regularization penalty α_w on the weight vector W_i, and a L_2 regularization penalty α on the nodes activities: $F_{1\to k}(x)$. Thus the objective function above becomes:

$$L(x, x') = logloss(x, x') + \sum_{i=1}^k (\alpha_w ||W_i||_1 + \alpha_a ||F_{1\to i}(x)||_2^2) \tag{7}$$

The values α_a and α_w were set to 0.0001 and 0.001, and the model selected 100 reduced features x' for each omic, then they applied *Cox Proportional Hazard* (Cox-PH) model [51] to select the features that increases the survivablity of bladder cancer. The selected features from each omic were concatenated together before applying *K-means* clustering [52] on combined features from all omics. The original model was proposed to identify robust survival subgroups of *HepatoCellular Carcinoma* (HCC) to improve patient care. The authors presented a DL based model on HCC that robustly differentiates survival sub-populations of patients in six cohorts.

A recent model to analyze bladder cancer survivals from multi-omics TCGA dataset is *Group lass regularized Deep learning for cancer Prognosis* (GDP). The omics used in GDP were pre-processed first; gene expression data using *RNA-Seq by Expectation-Maximization* (RSEM) [53], *Copy Number Variation* (CNV) data by *Genomic Identification of Significant Targets in Cancer* (GISTIC 2.0) [54], normalized RPPA protein expression data, *Variant Call Format* (VCF) [55] cancer DNA somatic mutation data, and part of the clinical data from *Broad* GDAC *Firehose* [56]. The method integrated a DL framework and Cox-ph together, and applied group lasso regularization [57] to incorporate gene-level group prior knowledge into the model training process . The overall optimization process for group lasso and DL method to reduce the cost as defined as:

$$C = -l(\beta) + scale \times (alpha \times \sum_{l=1}^{L} \sqrt{p1}||W_{1l}||_2 + (1-\alpha) \times ||W_o||_1) \tag{8}$$

6 Conclusion

CADs have been evolved in the diagnosis of treatment of cancer patients. The revolutionary NGS technologies allowed us to read different omics measurements from the tumor cells. In this chapter, we have introduced DL algorithms to integrate these multi-sources data. The researchers have used well-known biological databases to process and validate the outcomes of their models.

The future direction of multi-omics integration methods is to use the *Transfer Learning* (TL) [58] systems to utilize the proposed CADs in health care systems. TL is an ML method where a model developed for a task is reused as the starting point for a model on a second task. The reusable TL systems can be integrated with the omics reads by NGS systems to predict the outcome of cancer. This kind of genomics profiling system can personalize the treatment of the patients and can be integrated into the right drugs with the proper dosage to reduce the side effects of these chemical molecules. This can improve the efficiency in the health care system, and cut the high cost of redundant unnecessary health practices.

Another direction in this area is to create gene panels based on the outcomes of the DL methods. Many current cancer gene panels are meant for cancer diagnosis and treatment in general for research purposes, such as *AmpliSeq* for *Illumina Focus Panel* [59]. More gene panels based on the outcomes of recent computer methods for specific cancer problem can be utilized for a better quality of life, and to improve the overall survivability.

References

1. Yehudit Hasin, Marcus Seldin, and Aldons Lusis. Multi-omics approaches to disease. *Genome Biology*, 18(1):83, 2017.
2. H Lodish, A Berk, SL Zipursky, P Matsudaira, D Baltimore, and J Darnell. Tumor cells and the onset of cancer. *Molecular cell biology*, 2000.
3. H Robert Frost and Christopher I Amos. A multi-omics approach for identifying important pathways and genes in human cancer. *BMC bioinformatics*, 19(1):479, 2018.
4. ENCODE Project Consortium et al. An integrated encyclopedia of DNA elements in the human genome. *Nature*, 489(7414):57, 2012.
5. Rajnish A Gupta, Nilay Shah, Kevin C Wang, Jeewon Kim, Hugo M Horlings, David J Wong, Miao-Chih Tsai, Tiffany Hung, Pedram Argani, John L Rinn, et al. Long non-coding rna hotair reprograms chromatin state to promote cancer metastasis. *Nature*, 464(7291):1071, 2010.
6. Kaladhar B Reddy. Microrna (miRNA) in cancer. *Cancer cell international*, 15(1):38, 2015.
7. Abedalrhman Alkhateeb and Luis Rueda. Zseq: an approach for preprocessing next-generation sequencing data. *Journal of Computational Biology*, 24(8):746–755, 2017.
8. Yong Zhang, Tao Liu, Clifford A Meyer, Jérôme Eeckhoute, David S Johnson, Bradley E Bernstein, Chad Nusbaum, Richard M Myers, Myles Brown, Wei Li, et al. Model-based analysis of chip-seq (macs). *Genome biology*, 9(9):R137, 2008.
9. Christoph Bock. Analysing and interpreting dna methylation data. *Nature Reviews Genetics*, 13(10):705, 2012.
10. Oluwatosin Taiwo, Gareth A Wilson, Tiffany Morris, Stefanie Seisenberger, Wolf Reik, Daniel Pearce, Stephan Beck, and Lee M Butcher. Methylome analysis using medip-seq with low dna concentrations. *Nature protocols*, 7(4):617, 2012.
11. Nina Pälmke, Diana Santacruz, and Jörn Walter. Comprehensive analysis of dna-methylation in mammalian tissues using medip-chip. *Methods*, 53(2):175–184, 2011.
12. J Gregory Caporaso, Justin Kuczynski, Jesse Stombaugh, Kyle Bittinger, Frederic D Bushman, Elizabeth K Costello, Noah Fierer, Antonio Gonzalez Pena, Julia K Goodrich, Jeffrey I Gordon, et al. Qiime allows analysis of high-throughput community sequencing data. *Nature methods*, 7(5):335, 2010.
13. Elin Org, Margarete Mehrabian, and Aldons J Lusis. Unraveling the environmental and genetic interactions in atherosclerosis: central role of the gut microbiota. *Atherosclerosis*, 241(2):387–399, 2015.
14. Anna Klindworth, Elmar Pruesse, Timmy Schweer, Jörg Peplies, Christian Quast, Matthias Horn, and Frank Oliver Glöckner. Evaluation of general 16s ribosomal RNA gene pcr primers for classical and next-generation sequencing-based diversity studies. *Nucleic Acids Research*, 41(1):e1–e1, 2013.
15. Osama Hamzeh, Abedalrhman Alkhateeb, Julia Zhuoran Zheng, Srinath Kandalam, Crystal Leung, Govindaraja Atikukke, Dora Cavallo-Medved, Nallasivam Palanisamy, and Luis Rueda. A hierarchical machine learning model to discover gleason grade-specific biomarkers in prostate cancer. *Diagnostics*, 9(4):219, 2019.
16. Computer-aided Diagnosis: The Tipping Point for Digital Pathology, 2020. https://digitalpathologyassociation.org/blog/computer-aided-diagnosis-the-tipping-point-for-digital-pathology/ [Online; Last accessed June 2020].

17. Harald Hampel, Nicola Toschi, Claudio Babiloni, Filippo Baldacci, Keith L Black, Arun LW Bokde, René S Bun, Francesco Cacciola, Enrica Cavedo, Patrizia A Chiesa, et al. Revolution of alzheimer precision neurology. passageway of systems biology and neurophysiology. *Journal of Alzheimer's Disease*, 64(s1):S47–S105, 2018.
18. Oliver Fiehn. Combining genomics, metabolome analysis, and biochemical modelling to understand metabolic networks. *International Journal of Genomics*, 2(3):155–168, 2001.
19. Craig Knox, Vivian Law, Timothy Jewison, Philip Liu, Son Ly, Alex Frolkis, Allison Pon, Kelly Banco, Christine Mak, Vanessa Neveu, et al. Drugbank 3.0: a comprehensive resource for 'omics' research on drugs. *Nucleic acids research*, 39(suppl_1):D1035–D1041, 2010.
20. Martin Kussmann, Frédéric Raymond, and Michael Affolter. Omics-driven biomarker discovery in nutrition and health. *Journal of biotechnology*, 124(4):758–787, 2006.
21. Panagiotis Georgiadis, Dennie G Hebels, Ioannis Valavanis, Irene Liampa, Ingvar A Bergdahl, Anders Johansson, Domenico Palli, Marc Chadeau-Hyam, Aristotelis Chatziioannou, Danyel GJ Jennen, et al. Omics for prediction of environmental health effects: Blood leukocyte-based cross-omic profiling reliably predicts diseases associated with tobacco smoking. *Scientific Reports*, 6:20544, 2016.
22. Charles D Parry, Jayadeep Patra, and Jürgen Rehm. Alcohol consumption and noncommunicable diseases: epidemiology and policy implications. *Addiction*, 106(10):1718–1724, 2011.
23. Yifeng Li, Fang-Xiang Wu, and Alioune Ngom. A review on machine learning principles for multi-view biological data integration. *Briefings in bioinformatics*, 19(2):325–340, 2016.
24. Eugene Lin and Hsien-Yuan Lane. Machine learning and systems genomics approaches for multi-omics data. *Biomarker research*, 5(1):2, 2017.
25. Fariba Khoshghalbvash and Jean X Gao. Integrative feature ranking by applying deep learning on multi source genomic data. In *Proceedings of the 10th ACM International Conference on Bioinformatics, Computational Biology and Health Informatics*, pages 207–216. ACM, 2019.
26. Jing Xu, Peng Wu, Yuehui Chen, Qingfang Meng, Hussain Dawood, and Hassan Dawood. A hierarchical integration deep flexible neural forest framework for cancer subtype classification by integrating multi-omics data. *BMC bioinformatics*, 20(1):1–11, 2019.
27. Kumardeep Chaudhary, Olivier B Poirion, Liangqun Lu, and Lana Garmire. Deep learning based multi-omics integration robustly predicts survival in liver cancer, biorxiv, 2017.
28. I. Bica, P. Velickovic, H. Xiao, and P. Li, "Multi-omics data integration using cross-modal neural networks.," in *ESANN*, 2018.
29. Rania Ibrahim, Noha A Yousri, Mohamed A Ismail, and Nagwa M El-Makky. Multi-level gene/mirna feature selection using deep belief nets and active learning. In *2014 36th Annual International Conference of the IEEE Engineering in Medicine and Biology Society*, pages 3957–3960. IEEE, 2014.
30. Fariba Khoshghalbvash and Jean X Gao. Integrating heterogeneous datasets by using multimodal deep learning. In *International Conference in Communications, Signal Processing, and Systems*, pages 279–285. Springer, 2018.
31. Olivier B Poirion, Kumardeep Chaudhary, Sijia Huang, and Lana X Garmire. Multi-omics-based pan-cancer prognosis prediction using an ensemble of deep-learning and machine-learning models. *medRxiv*, page 19010082, 2019.
32. R. H. Hahnloser, R. Sarpeshkar, M. A. Mahowald, R. J. Douglas, and H. S. Seung, "Digital selection and analogue amplification coexist in a cortex-inspired silicon circuit," *Nature*, vol. 405, no. 6789, pp. 947–951, 2000.
33. W. Li, A. Lee, and P. K. Gregersen, "Copy-number-variation and copy-number-alteration region detection by cumulative plots," *BMC bioinformatics*, vol. 10, no. 1, pp. 1–11, 2009.
34. P.-F. Verhulst, "Notice sur la loi que la population suit dans son accroissement," *Corresp. Math. Phys.*, vol. 10, pp. 113–126, 1838.
35. The Cancer Genome Atlas Program, 2020. https://www.cancer.gov/tcga [Online; Last accessed June 2020].

36. World Cancer Research Fund. Breast cancer statistics Breast cancer is the most common cancer in women worldwide, 2020. https://www.wcrf.org/dietandcancer/cancer-trends/breast-cancer-statistics [Online; Last accessed January 2020].
37. Dongdong Sun, Minghui Wang, and Ao Li. A multimodal deep neural network for human breast cancer prognosis prediction by integrating multi-dimensional data. *IEEE/ACM Transactions on Computational Biology and Bioinformatics (TCBB)*, 16(3):841–850, 2019.
38. Christina Curtis, Sohrab P Shah, Suet-Feung Chin, Gulisa Turashvili, Oscar M Rueda, Mark J Dunning, Doug Speed, Andy G Lynch, Shamith Samarajiwa, Yinyin Yuan, et al. The genomic and transcriptomic architecture of 2,000 breast tumours reveals novel subgroups. *Nature*, 486(7403):346–352.
39. Geoffrey E Hinton and Ruslan R Salakhutdinov. Reducing the dimensionality of data with neural networks. *science*, 313(5786):504–507, 2006.
40. Bino John, Anton J Enright, Alexei Aravin, Thomas Tuschl, Chris Sander, and Debora S Marks. Human microrna targets. *PLoS biology*, 2(11):e363, 2004.
41. Yitan Zhu, Peng Qiu, and Yuan Ji. Tcga-assembler: open-source software for retrieving and processing tcga data. *Nature methods*, 11(6):599, 2014.
42. Jun Li, Leng Han, Paul Roebuck, Lixia Diao, Lingxiang Liu, Yuan Yuan, John N Weinstein, and Han Liang. Tanric: an interactive open platform to explore the function of lncrnas in cancer. *Cancer research*, 75(18):3728–3737, 2015.
43. American Society of Clinical Oncology, Cancers.NET, Colorectal Cancer: Statistics, 2019. https://www.cancer.net/cancer-types/colorectal-cancer/statistics [Online; Last accessed December 2019].
44. Jie-Kai Yu, Yi-Ding Chen, and Shu Zheng. An integrated approach to the detection of colorectal cancer utilizing proteomics and bioinformatics. *World Journal of Gastroenterology: WJG*, 10(21):3127, 2004.
45. N. Tang, P. Tornatore, and S. R. Weinberger, "Current developments in seldi affinity technology," *Mass spectrometry reviews*, vol. 23, no. 1, pp. 34–44, 2004.
46. Wangxiong Hu, Yanmei Yang, Xiaofen Li, Minran Huang, Fei Xu, Weiting Ge, Suzhan Zhang, and Shu Zheng. Multi-omics approach reveals distinct differences in left-and right-sided colon cancer. *Molecular Cancer Research*, 16(3):476–485, 2018.
47. Dmitrii Bychkov, Nina Linder, Riku Turkki, Stig Nordling, Panu E Kovanen, Clare Verrill, Margarita Walliander, Mikael Lundin, Caj Haglund, and Johan Lundin. Deep learning based tissue analysis predicts outcome in colorectal cancer. *Scientific reports*, 8(1):3395, 2018.
48. C. G. A. Network *et al.*, "Comprehensive molecular characterization of human colon and rectal cancer," *Nature*, vol. 487, no. 7407, p. 330, 2012.
49. American Society of Clinical Oncology, Cancers.NET, Bladder Cancer: Statistics, 2019. https://www.cancer.net/cancer-types/bladder-cancer/statistics [Online; Last accessed December 2019].
50. Olivier B Poirion, Kumardeep Chaudhary, and Lana X Garmire. Deep learning data integration for better risk stratification models of bladder cancer. *AMIA Summits on Translational Science Proceedings*, 2018:197, 2018.
51. David R Cox. Regression models and life-tables. *Journal of the Royal Statistical Society: Series B (Methodological)*, 34(2):187–202, 1972.
52. J. MacQueen *et al.*, "Some methods for classification and analysis of multivariate observations," in *Proceedings of the Fifth Berkeley Symposium on mathematical statistics and Probability*, vol. 1, pp. 281–297, Oakland, CA, USA, 1967.
53. Bo Li and Colin N Dewey. Rsem: accurate transcript quantification from RNA-Seq data with or without a reference genome. *BMC bioinformatics*, 12(1):323, 2011.
54. Craig H Mermel, Steven E Schumacher, Barbara Hill, Matthew L Meyerson, Rameen Beroukhim, and Gad Getz. Gistic2. 0 facilitates sensitive and confident localization of the targets of focal somatic copy-number alteration in human cancers. *Genome Biology*, 12(4):R41, 2011.

55. Petr Danecek, Adam Auton, Goncalo Abecasis, Cornelis A Albers, Eric Banks, Mark A DePristo, Robert E Handsaker, Gerton Lunter, Gabor T Marth, Stephen T Sherry, et al. The variant call format and vcftools. *Bioinformatics*, 27(15):2156–2158, 2011.
56. Broad GDAC Firehose, 2020. https://confluence.broadinstitute.org/display/GDAC/fbget [Online; Last accessed June 2020].
57. Gangcai Xie, Chengliang Dong, Yinfei Kong, Jiang F Zhong, Mingyao Li, and Kai Wang. Group lasso regularized deep learning for cancer prognosis from multi-omics and clinical features. *Genes*, 10(3):240, 2019.
58. J. West, D. Ventura, and S. Warnick, "Spring research presentation: A theoretical foundation for inductive transfer," *Brigham Young University, College of Physical and Mathematical Sciences*, vol. 1, no. 08, 2007.
59. Illumina focused panel, 2020. https://www.illumina.com/products/by-type/sequencing-kits/library-prep-kits/ampliseq-focus-panel.html [Online; Last accessed June 2020].

Using Deep Learning with Canadian Primary Care Data for Disease Diagnosis

Hasan Zafari, Leanne Kosowan, Jason T. Lam, William Peeler, Mohammad Gasmallah, Farhana Zulkernine, and Alexander Singer

Abstract The majority of Canadian primary care systems record patient data in the form of *Electronic Medical Records* (EMR). EMRs hold structured, semi-structured and unstructured demographic and health care data about patients. The value of EMR data for research, health surveillance and quality improvement continues to be explored. Data analytics such as *Machine Learning* (ML) and statistical modeling techniques have been applied to de-identified EMR data repositories to advance our understanding of different health conditions and patient care. More recently, the application of *Deep Learning* (DL) approaches to structured, semi-structured and unstructured data of the EMRs is being investigated as an avenue for improved identification of health conditions. Supervised ML methods have dominated disease classification for more prevalent diseases. A large cohort of labeled data is required to train ML models using supervised learning methods. For less common diseases, the amount of available labeled data is often insufficient, and a variety of strategies are being explored to deal with inadequate, noisy and missing data. This chapter describes the benefits of using DL models with EMR data for research to improve provisioning of health care in primary care settings. A few prominent DL models such as *Multi-Layered Perceptron* (MLP), *Convolutional Neural Network* (CNN) and *Recurrent Neural Network* (RNN) are discussed with example scenarios that demonstrate application of some of these predictive analytics models to both structured and unstructured EMR data using regular and weak supervision methods for diagnosing both prevalent and non-prevalent diseases.

Keywords Deep learning · Machine learning · Medical data analysis · Electronic medical record · Disease diagnosis · Natural language processing

H. Zafari · J. T. Lam · M. Gasmallah · F. Zulkernine (✉)
School of Computing, Queen's University, Kingston, ON, Canada
e-mail: farhana.zulkernine@queensu.ca

L. Kosowan · W. Peeler · A. Singer
Department of Family Medicine, Max Rady College of Medicine, Rady Faculty of Health Sciences, University of Manitoba, Winnipeg, MB, Canada

M. Elloumi (ed.), *Deep Learning for Biomedical Data Analysis*,
https://doi.org/10.1007/978-3-030-71676-9_12

1 Introduction

Health care in Canada is a provincial/territorial responsibility. Federal funding for health care is available to provinces and territories that meet the standards of the Canadian Health Act. Administrative health records include information submitted for billing and registry purposes. They are stored and maintained by a provincial/territorial governing body and provide a large population-based data repository that can be used for research and disease surveillance. However, these records do not have the same depth of clinical details that can be found in primary care *Electronic Medical Records* (EMR).

Primary health care is typically the first contact an individual has with the health system. Primary care providers include family physicians, nurse practitioners, and community pediatricians that provide health services within the community, including primary prevention, secondary prevention, and disease management. In Canada, the use of EMRs has become a common place in the last 20 years. In primary care, the shift from traditional paper records to EMRs has produced a variety of improvements including legibility, accessibility, and structure, as well as introducing efficiencies designed to reduce the overall cost. EMRs were intended to improve the provision of health care [1]. EMRs are now being used in disease surveillance, advance analytics of disease symptoms, quality improvement, pragmatic clinical trials, and research on the health care continuum through linkage of administrative health records and primary care EMRs [2–6]. The type of information typically captured in a modern EMR includes demographic information, biometric information, such as height, weight and blood pressure, problem lists, health/lifestyle behaviors, personal and family medical histories, billing information, prescribed medications, laboratory test results and referrals to inter-professional health care services. Primary care EMRs provide longitudinal data on a patients' health including details of disease progression.

In Canada, the *Canadian Primary Care Sentinel Surveillance Network* (CPC-SSN) [8] provides a national data repository of primary care EMR data [6, 8]. CPCSSN holds de-identified EMR data from over 2 million Canadians being cared for by over 1800 primary care providers in nearly 500 clinics spread across 8 Provinces and 1 Territory in a centrally maintained data repository [8]. Therefore, it provides a great source of data that is ready to be analyzed and explored through research [9]. With ethical approval from research institutions and CPCSSN, researchers in Canada can get access to selected subsets of patients' anonymized data for research [10, 11]. CPCSSN dataset contains data available in tables such as encounter table, medication table, laboratory, and exam result tables. Validation studies have been published outlining case definition algorithms for the identification of a number of chronic diseases including hypertension [5], type 2 diabetes mellitus [4, 5], depression [5], Parkinson disease [5], osteoarthritis [5], *Chronic Obstructive Pulmonary Disease* (COPD) [5], epilepsy [5], dementia [5], pediatric asthma [3], multiple sclerosis [2], dyslipidemia [12], chronic kidney disease [13], and herpes zoster [14]. Documentation is also available in the EMR for other acute and chronic conditions for which case definitions are yet to be validated.

More recently, there have been efforts to understand the value of health care provider EMR encounter notes for health research. One of our ongoing research projects focuses on using de-identified free text encounter notes from consenting primary care providers to study the symptoms, evolution and health care provision for patients with *Post-Traumatic Stress Disorder* (PTSD) in both the general and military/veteran populations [15]. Some diseases such as PTSD are difficult to diagnose and predict based only on the structured EMR data. Free text data from a provider's notes can offer valuable information about the diagnosis and progression of PTSD.

Although the primary care data from the EMRs have proven invaluable for many research studies, there are several challenges in curating this data. Different EMR systems (e.g., TELUS [16], QHR [17], Oscar [18]) are used at primary care clinics across Canada. Each EMR stores the health care information differently and is supported by different *Information Technology* (IT) companies. Data extraction systems must be carefully designed and customized for each EMR and clinic with help from the IT support company to filter, extract and transform a predefined set of selected data fields before loading and appending them to the existing CPCSSN data. Furthermore, the information entered into the EMR is subject to human error and physician's bias. None the less, CPCSSN and primary care EMR data in Canada serve as a valuable resource for research studies specifically with the advancements in ML techniques [19].

The history of DL trace back to 1943, when Walter Pitts and Warren McCulloch created a computational cognitive model based on the neural networks [20]. The application of ML really flourished in 2010 with the parallel advancements in fast computing platforms, availability of digital information, and advancement in the state-of-the-art ML algorithms [21–23]. Traditional data exploration algorithms relied on historical data and carefully concocted techniques by human domain experts knowledgeable about the data who could select important aspects of the data and design algorithms to transform the data and extract knowledge [25–27]. In contrast, modern ML algorithms are designed to automatically learn important aspects or features of the data and categorize the data based on the common features [28, 29]. Therefore, as the data changes, models can be retrained or designed to automatically adapt to new features influencing the outcome. DL models are designed to execute the learning in multiple hierarchically organized layers. At each layer, the information is refined and filtered to extract the key aspects from the information to create the final categorization. DL approaches have been applied to complex data such as unstructured text, voice, signal, and image data to categorize text, generate replies, identify objects in images, translate voice to text and text to voice, and transform and extract important features from complex hybrid data [30].

The rest of the chapter is organized as follows. In Sect. 2, we will provide further information about the extraction and processing of the EMR data. In Sect. 3, different types of ML models with a focus on DL models are reviewed. Section 4 surveys the state-of-the-art DL research in healthcare and Sect. 5 demonstrates some of our recent work on the application of DL in EMR data for disease diagnosis.

2 Extraction and Processing of the EMR Data at CPCSSN for Research

The design of ML approaches depends heavily on: (1) the layout, (2) type of data and (3) the objective of the study. In this section we describe the data extraction workflow for the EMRs and explain the structure and content of the CPCSSN data bank, which we use for developing, training, and validating the DL models.

2.1 EMR Data Types

There are three basic types of health data stored in all EMRs: *structured*, *semi-structured* and *unstructured data*. *Structured data* in EMRs can be described as having a pre-defined data structure or reference sets, which may be developed and maintained by external medical or government organizations. Examples of the data include patients' biographical data such as name, date of birth, and health card number; health data such as height, weight, *Blood Pressure* (BP), and *Body Mass Index* (BMI), and standardized billing or service codes (provincially mandated), diagnosis codes (i.e., ICD9/10 or SNOMED), medication codes (i.e., DIN or ATC), and lab tests (i.e., HL-7). *Semi-structured data* items in the EMR contain information with a mixed data type such as numbers and text but follow some structure such as risk factor behavior. *Unstructured data* items often contain mixed data types and have no structure or format in representation such as narrative text in physicians' chart notes, which record the observations of the medical doctor during patients' visits. Often such records follow a problem-oriented record model. For example, a *SOAP note* includes the following headings to structure the note namely *subjective*, *objective*, *assessment,* and *plan*. Free form entries may also be found in other parts of the EMR where entry is permitted without any content restrictions.

2.2 Challenges

A key challenge in extracting information from many different EMR systems for research is the absence of standards in the representation and storage structure of the data. While detailed data transfer standards such as *Fast Healthcare Interoperability Resources* (FHIR) [33] have been adopted in some places, these have not yet been routinely adopted as standards for Canadian primary care EMR systems. However, EMR databases do share common features in terms of the types of data they store. It is worth noting that most semi-structured data elements often relate to data elements with defined parameters that could be mapped [33].

In Canada, access and use of patients' medical information is governed by provincial legislation and regional policies that define specific requirements designed to protect an individual's privacy and prohibit inappropriate disclosure of the data. Additionally, use of the data by an institution in all jurisdictions requires approval from corresponding Research Ethics boards (e.g., university, research centers) which follow the Tri-Council policies [10, 11].

2.3 *Data Extraction, Anonymization, Transfer and Load*

In 2008, CPCSSN, funded by the *Public Health Agency of Canada*, developed a national network to collect health information on patients across Canada. Since its conception, CPCSSN has developed and refined its processes for collecting and storing primary care EMR data. CPCSSN is comprised of regional networks; with each regional CPCSSN network employing a *Data Manager* who is responsible for the semi-annual data extraction from the EMRs of the participating physicians. The regional EMR data goes through a cycle of *Extraction, Transfer* and *Load* (ETL) where it is curated in order to fit into a common database structure. Prior to any use for secondary purposes, the data goes through a de-identification process whereby patient identifiers are removed, and the data is anonymized. The anonymized data sets derived from the regional networks are pass through a set of nationally standardized processes before being combined into one national centralized CPCSSN research database. Some of CPCSSN's regional networks such as the *Manitoba Primary Care Research Network* (MaPCReN) [34] have initiated efforts to extract, de-identify and process unstructured data from the narrative notes of the primary care providers. Our current research initiative leverages notes derived from over 56,000 patients which represent nearly one million patient-physician encounters. Narrative physician notes contain valuable information that may be useful for health research in many domains. However, significant work remains to address the inherent challenges that make individually derived narratives difficult to process such as, but not limited to, misspelled words, atypical or irregular sentence structures, liberal use of parenthetical expression, unpredictable repetition of sentences or phrases, dramatic local variation of terms and idiosyncratic use of jargon/abbreviations/acronyms. Figure 1 shows how the data flow occurs in collecting and building the CPCSSN data repository.

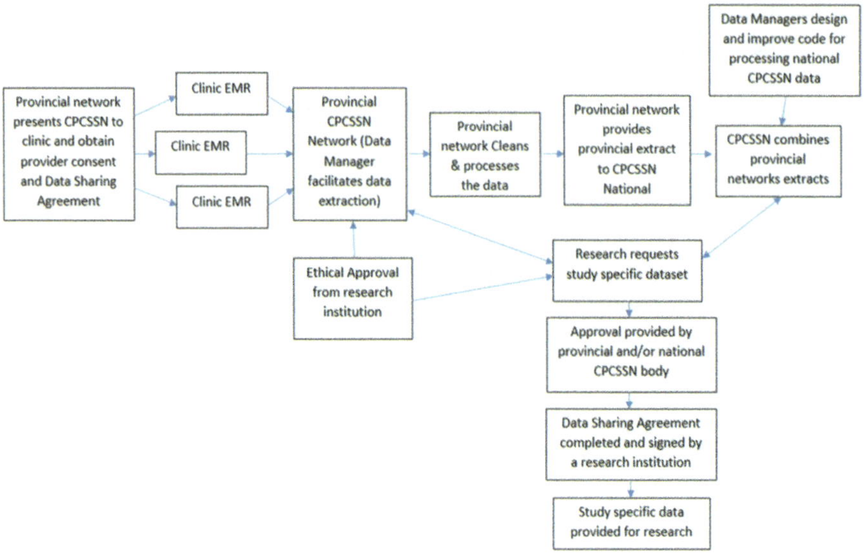

Fig. 1 CPCSSN Data Flow Diagram

3 Artificial Intelligence in Medical Decision Support Systems

Four experimental systems are generally regarded as having started the research on application of *Artificial Intelligence* (AI) algorithms in medicine [35]. These were *MYCIN*, a program to advise physicians on antimicrobial selection for patients with bacteremia or meningitis [38]; the *Present Illness Program* (PIP), a system that gathered data and generated hypotheses about disease processes in patients with renal disease [37]; *INTERNIST-1*, a large system to assist with diagnosing complex problems in general internal medicine [36], and *CASNET*, an ophthalmology advisor designed to assess disease states and recommend management for patients with glaucoma [39]. All four systems drew on AI techniques, emphasizing the usefulness of encoding large amounts of specialized medical knowledge acquired from the clinical literature and domain experts. However, rule-based expert systems are challenged with the extraction of both explicit and tacit expert knowledge and the continual update of algorithms for inclusion of new knowledge and information. Therefore, other AI techniques are being explored.

Classic statistical techniques have been applied on medical data, for example, learning the distribution of patient demography for a critical disease outbreak [40–42]. With the availability of large amounts of electronic and digital medical information, data driven ML algorithms evolved over the years. These algorithms are built and trained based on patients' data banks. The AI models automatically learn the optimum parameter values and the decision rules from the data. As a result, when the data changes, the models can be adapted to these changes to continue to provide the desired results. The risk lies in models being fed *bad data*, which can deteriorate the accuracy and performance of the models.

Recent successes of DL models and the increasing availability of digital EMRs have inspired researchers to apply DL to medical data to analyze medical text reports [29, 43] radiology images [44] and signals from instruments monitoring patients' vital signs in hospitals [45]. DL models are structured into multiple layers where each subsequent layer extracts, filters, refines, transforms and selects important features from a variety of data items that help categorize the data or learn the mapping of the features to obtain a final decision outcome such as a diagnosis. A supervised learning approach can train ML models with labeled data such as patients' health data. The training occurs iteratively until the models are capable of producing satisfactory results. If the label (e.g., the diagnosis) is not given then an unsupervised learning approach can be applied to train the models to extract and group similar features from the data such as disease symptoms. Some of the commonly used DL models are discussed later in the context of specific use case scenarios of a *Medical Decision Support System* (MDSS).

MDSS have shown immense value in extracting important aspects of patients' records from years of data, reducing diagnosis time and human error, and exploring new knowledge about diseases, changing disease symptoms, drugs, and treatment approaches [4]. Reappearance of an old symptom or co-occurrence of several trivial symptoms can be detected by a MDSS and indicate serious health issues. Outcomes can be improved with diagnostic information linked from multiple data sources such as primary care, hospitals, and long-term care institutions. However, in Canada due to strong privacy and confidentiality policies, patient records are anonymized and held outside of the health care setting for access by researchers. This process prevents the re-identification of patients by researchers or within research results [10, 11].

3.1 Types of ML Models

Traditionally two types of ML models have been prevalent in the literature: the traditional ML models and the DL based models. These traditional models apply statistical algorithms to the data to extract knowledge, discover patterns, and learn parameters to optimize the model's objective as new data comes in. The DL models, on the other hand, are based on *Artificial Neural Networks* (ANNs) which are inspired by biological neural networks defined originally to model a human brain (i.e., biological neural structure and function). ANNs are designed as a network of connected artificial neurons as indicated by the circular nodes in Fig. 2, which are simple computational units that process data in parallel. Data flows from the input layer to the output layer through one or more hidden layers.

The links between the artificial neurons have associated weights that change as the ANN extracts and combines features from the data or classify the data based on the existence of specific features in the data. For example, a model can learn to

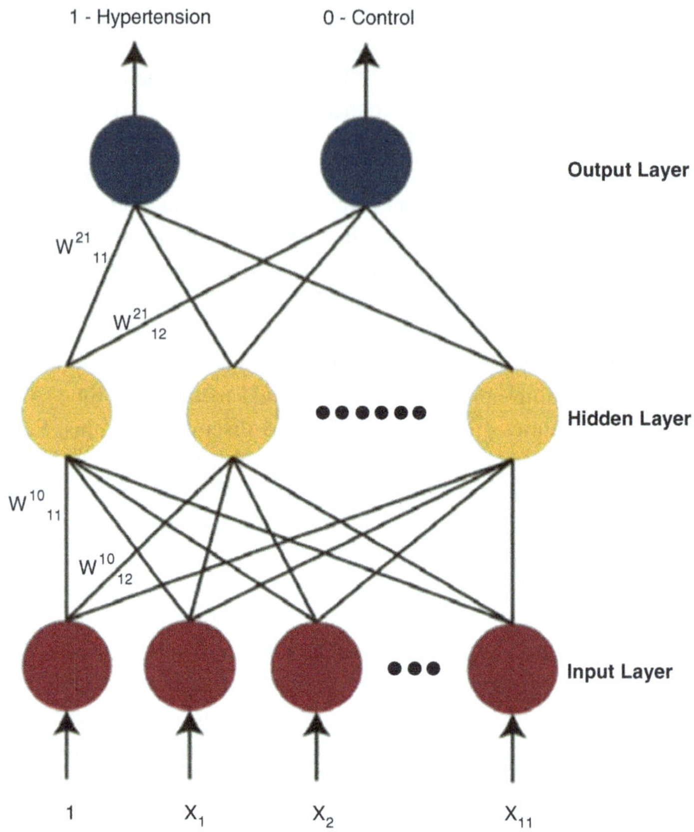

Fig. 2 A one hidden layer ANN – Multilayered Perceptron [19]

cluster all types of mental diseases into one group based on the similarity of the symptoms. Alternatively, a model can be trained to *classify* or predict specific types of back pain after learning to map symptoms and medications to specific diagnosis. Labels (e.g., a diagnosis) are not required to learn similarity within a feature set. Given a large number of sample data, models can learn to extract similar patterns from the data and group them together. However, to train models to associate a class label to an extracted feature set such as influenza or pneumonia, a label must be provided in the training data. The presence of labels allows the model to adjust the weights iteratively until it can predict the class labels with a satisfactory level of accuracy. In the ML paradigm, regression, classification, and clustering are the three most common applications [46]. Other applications include feature extraction and reduction, data transformation, generation of data distribution and pattern recognition [47].

DL Models

DL models are ANNs with generally more than two hidden layers between the data input and output layers. As more and more layers are added to an ANN, training of the weight values becomes more complex and computationally expensive. However, with more layers, DL models can learn a variety of combinations of the extracted features and optimum mapping of these features to generate the correct output. For the same reason, DL models have shown great success in classifying features extracted from complex data such as image, voice or other hybrid multimodal data. To learn the optimum values of these weights, DL models require a large number of samples of training data without which the models fail to provide decent results.

A common property of these models is that they all have a data input layer, an output layer and one or more layers between these two layers called the hidden layers (Fig. 2). Due to the type of connections in the layers and the functions performed by each ANN node, some of the DL models perform particularly well at tracking and predicting temporal features from time series data (heart rate monitoring signals) or spatial features from image data (medical image data), or generally extracting and encoding key features from multimodal data for clustering or classification. Some of the common DL models are briefly described below.

- **Multi-Layered Perceptron:** *Multi-Layered Perceptrons* (MLP) are the simplest form of ANNs. A MLP with multiple hidden layers (similar to the model shown in Fig. 2 but having multiple hidden layers) can be considered as a DL model [48]. Each layer in a MLP extracts and composes features from the data coming from the previous layer and the neurons at the final output layer generates numerical values that can be transformed to fit the task at hand. Some of the common ML algorithms are explained in the next section.
- **Convolutional Neural Network:** A *Convolutional Neural Network* (CNN) applies multiple convolving filters at each layer to extract and transform spatial features from the input data as shown in Fig. 3 [49]. When convolutions are paired with max pooling, it is important to also include the batch normalization layer and the ReLU activation function in the typical CNN block, which are stacked in a CNN. These blocks allow the network to extract a concise set of spatial features which are then fed into a MLP type subnetwork to generate the final result. Although initial applications of the CNN were in the area of image processing and computer vision [50], the efficiency of CNN models in extracting spatial features have extended their applications to other types of data such as text data [51]. Figure 3 shows a CNN model used to identify handwritten characters in text data.
- **Recurrent Neural Network:** *Recurrent Neural Networks* (RNN) are good at modelling and learning contexts from a sequential or time series data and can predict the next data item when trained with sequences of data. RNNs implement memory units and feedback or backward data flow connections in each layer that allows correlating the recent data item with the previous data to learn the context and sequential patterns in the data. The stored context information is used as an additional input along with the recent input data to help predict

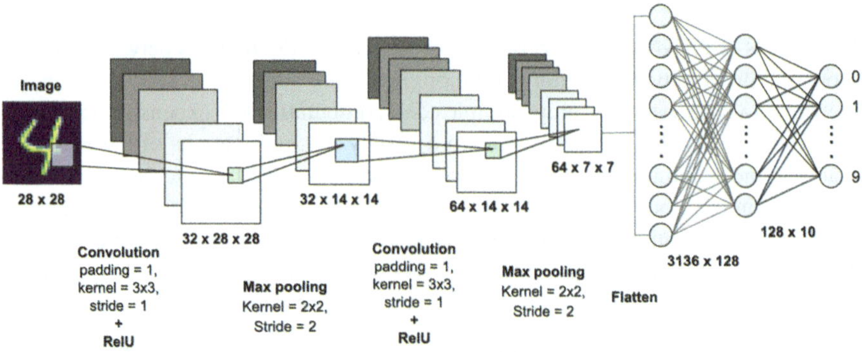

Fig. 3 A CNN model for handwritten character recognition [49]

the next item in the sequence Fig. 4a. Some of the popular applications of RNNs include sequential text processing to analyze disease trajectory from physicians' chart notes [52], general language modeling and translation [53], and prediction of stock prices [54], signal data prediction for monitoring heart rates [55] and processing audio or video data for online chat systems [56]. Many different variations and extensions of RNNs have been implemented over the years to reduce model training time (FastRNN, *Gated Recurrent Unit* (GRU)) [57], to increase or decrease the memory by deleting or forgetting the old and unnecessary data (*Long Short Term Memory* (LSTM)), and to enable bidirectional sequence learning (*Bidirectional RNN* (BRNN) or *GRU* (BiGRU)) [58] in both forward and backward directions.

- **Autoencoder**: An *autoencoder* is a type of ANN used to learn efficient data representations in an unsupervised manner [59]. It aims to learn an encoding by training a network to produce an output data that is very close to the input data by first extracting the key feature set through a series of encoder layers and then expanding the features to reproduce the original data through a series of decoder layers. The middle layer thus produces a concise representation of the input data by eliminating noise and unimportant features. Figure 4b shows a schema of a basic Autoencoder.

3.2 ML Algorithms

The key difference between the traditional rule-based and the ML models is the way parameters of the models are defined. Unlike the traditional models which have a static set of parameters that are determined computationally at the preset of the model, ML models can have a variable number of parameters whose values are learned by the model iteratively through optimizing a criterion based on a large sample of training data. Datasets are often split into three different sets that do not

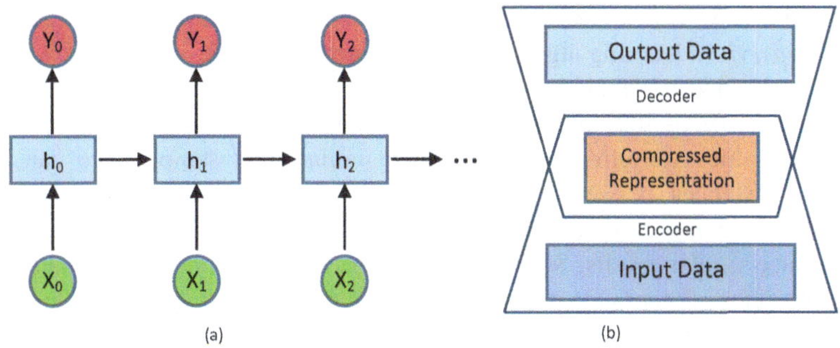

Fig. 4 (**a**) A RNN model, (**b**) An Autoencoder

overlap namely training, validation and testing data. Once the parameter values are learned using the training data, the model can be validated against a predefined criterion such as prediction accuracy. It is a common practice to use the validation set to select the best performing trained model. The training process is generally repeated for multiple *epochs* (an *epoch* is a full pass through the training dataset). Finally, the model is tested on a testing data set that the model has never seen, to assess its performance and generalizability. In the case of unavailability of a large enough data set, the validation step is often skipped.

There are three main types of ML algorithms. *Error correction learning* defines an approach to compute the error in model performance and gradually reduce the error by making some strategic adjustments to the parameter values [60]. *Reinforcement learning* algorithms aim to maximize a predefined goal or outcome and gradually adjust the parameter values to develop a policy for this objective [61]. *Correlation learning* algorithms allow models to learn strong correlations between different features in the training data that always co-occur to generate similar outcomes [62].

Error correction learning requires labeled data, where the labels represent the expected outcomes, to apply necessary corrections to the model parameters. This is also known as *supervised learning* approach where the labeled data is often referred to as the *ground truth* data [63]. *Backpropagation* is a popular algorithm which allows learning by computing the error at the output layer, propagating a gradient of the error backwards towards the input layer and applying proportionate adjustments to the relevant weight values at the different layers in the ANN [64]. Supervised learning is commonly applied to train ML models to classify data into predefined categories. Trained models can then be applied to predict the unknown class labels in new data. Supervised methods have been applied in sentiment analysis using social media data [67], criminal charge predictions [68–70], and disease prediction from symptoms [71]. A big difficulty in applying supervised learning is the requirement of large labeled data [71] that can be difficult to obtain especially when the labeling must be done by domain experts such as medical doctors. As a remedy, multiple strategies to augment and up-sample training data have been applied as well as other *semi-supervised* learning algorithms [65] that have evolved such as the *weakly supervised* learning algorithm as explained later [66].

In the absence of labeled training data, variations of *reinforcement* [61] and *correlation learning* [62] algorithms can be applied. For example, to learn the parameter values for effectively grouping similar data samples we can minimize the distances between the data samples in the same cluster. Similarly, *correlation learning* models assign greater weight values to links in the network that produce similar results. Such models can be used to fill in missing data values based on values of other data items in a sample data set and are also commonly used to model human memory. Since such algorithms do not require labeled data, they are called *unsupervised learning* algorithms. Since most of the real-world data are unlabeled, researchers are currently working on developing unsupervised learning algorithms for static and streaming data [72].

In the subsequent sections, we first describe the state-of-the-art research on applications of DL models in the paradigm of MDSS. Next a few example scenarios of applications of DL models to medical data are discussed. Due to privacy, security and reliability concerns, real life ML applications in MDSS are rare and all the examples given in this chapter are from research studies using structured, semi-structured and unstructured EMR data.

4 State-of-the-Art DL Research in Healthcare

Recent research in MDSS focuses on developing and validating ML models using medical data obtained through custom patient questionnaires, interviews with domain experts, *Electronic Health Records* (EHR) collected through designated authorities, hospital and clinical data, or other accessible data and knowledge sources such as online information, journals, and medical ontologies. The problem, and the type and the structure of the data such as numeric, signal, text, image, and hybrid structured, semi-structured or unstructured data influence the selection of the ML models and algorithms. Some of the recent research focusing on the development and application of DL models in MDSS are summarized in Table 1 followed by a brief discussion of the research work.

With widespread adaptation of EMRs and the success of DL in many domains, researchers have begun to use DL to process EMR data for various clinical informatics applications. These applications range from disease prediction, classification and inference [28, 29, 55, 74, 75, 77, 78, 80, 84, 86], summarizing and representing patient data [43], learning interpretable representations of medical concepts [73], estimating readmission possibility [74], extracting relations between medical concepts [76], finding optimal medication dosage [79], identifying disease outbreak [87], patient trajectory modelling [74, 89], and devising clinical decision support systems [82, 83].

Table 1 Existing research on the application of DL to medical text and signal data

References	Name	Data source	Goal	DL model	Learning
Choi et al. [73]	*Med2Vec*	Structured and free text EHR data	Learn interpretable representations of medical concepts without depending on expert knowledge	MLP	Supervised
Pham et al. [74]	*Deepcare*	Structured and free text EHR data	Disease prediction, recommending interventions, estimating unplanned re-admission	LSTM	Supervised
Choi et al. [28]	*Doctor AI*	Structured EHR data	Predict (all) the diagnosis and medication categories for a subsequent visit	GRU	Supervised
Wickramasinghe et al. [75]	*Deepr*	Structured and free text EHR data	Predictive analytics in healthcare services	CNN	Supervised
Liu et al. [29]	*Deep EHR*	Free-text notes + Structured HER data	Disease onset prediction	CNN, LSTM	Supervised
Lv et al. [76]	N/A	Clinical texts of EMR data	Relations extraction between medical concepts	*Autoencoder*, CRF	Supervised
Mallya et al. [77]	N/A	Structured and text data of EMR	Predicting the onset of Congestive Heart Failure	LSTM	Supervised
Nie et al. [78]	N/A	Question and answers pairs from HealthTap	A DL scheme to infer the possible diseases given the questions of heath seekers	MLP	Supervised
Nemati et al. [79]	N/A	Structured EMR data	Find optimal medication dosage	*Deep Reinforcement Learning*	Supervised
Choi et al. [80]	*RETAIN*	Structured EMR data	Disease predictive models	*Attention*, RNN	Supervised
Ong et al. [81]	N/A	Data obtained from sensors	Predict Fine particulate matter concentration, time-series prediction	*Deep RNN, autoencoder*	Supervised
Choi et al. [55]	N/A	Structured and text data of EMR	Predicting initial diagnosis of heart failure	GRU	Supervised

(continued)

Table 1 (continued)

References	Name	Data source	Goal	DL model	Learning
Che et al. [82]	*Interpretable mimic Learning*	Structured EHR data	Learn interpretable phenotype features, Clinical decision support system	*Stacked Denoising Autoencoder*, LSTM	Supervised
Jagannatha et al. [83]	N/A	Unstructured Text of EHR notes	Detecting medication, diagnosis, and adverse drug event	LSTM, GRU	Supervised
Bhatt et al. [84]	N/A	Structured EMR data	Predicting the onset of sepsis	*Feed-forwardmodels*	Supervised
Kinjo et al. [85]	N/A	R-R interval sensor data	Patient's stress estimation	*Deep neuralnetwork*	Supervised
Hu et al. [86]	N/A	Unstructured EMR data	Rare disease classification with EHR data	CNN	Supervised
Zhao et al. [87]	*Simnest*	Social media data	Disease outbreak identification	MLP	Semi-supervised
Banerjee et al. [88]	N/A	Unstructured EMR data	Assess the presence, absence, or risk discussion of patient-cantered outcomes	*Neural word embedding, multinomial logistic model*	Weakly supervised
Beaulieu-Jones et al. [89]	N/a	Structured and text data of EMR	Modelling longitudinal clinical data, Prediction of patient survival	*Autoencoder*, LSTM	Unsupervised, Supervised
Miotto et al [43]	*Deep Patient*	Structured and text clinical notes	Patient representation from EHR data that facilitates clinical predictive modelling	*Autoencoder*	Unsupervised

Both structured data and free-text notes have been used in these studies. With regard to EMR data, research results suggest that models that incorporate unstructured text data outperform models that assess structured data exclusively [29].

Although the majority of previous studies incorporated supervised DL algorithms [28, 29, 55, 74–77, 79–83, 85, 86], other types of learning including weakly supervised, semi-supervised and unsupervised learning methods have also been explored. Miotto et al. [43] and Beaulieu-Jones et al. [89] applied an unsupervised autoencoder to create a higher level representation from EMR data to facilitate clinical predictive modelling and enhance the performance of their prediction tasks. In an attempt to overcome the lack of labeled data, Zhao et al. [87] proposed a novel semi-supervised DL framework that integrates the strengths of computational epidemiology and social media mining techniques to identify influenza outbreak using the social media users' health states and intervention actions in real time. In another study, Banerjee et al. [88] used a weakly supervised method to assess the presence, absence, or risk discussion of patient-centered outcomes using free-text data of EMR.

In terms of temporal aspect of data, the prediction tasks can be divided into static and temporal prediction. The difference is that in the former, all features are processed using a MLP and/or a CNN model at once without considering the chronological order of the information [73, 75, 78, 84, 85, 86, 87] while in the latter, the data is associated with time-stamped information and is processed with RNNs [28, 29, 55, 74, 77, 80, 83, 89]. Temporal data can be used to look at trends and predict future outcomes. LSTMs have been researched heavily in this area and have yielded very good results in many medical and non-medical applications including stock price prediction [54, 90] and traffic flow prediction [91].

5 Example Scenarios of Applications of DL to EMRs

Some of our research on applications of DL models to EMRs are described in this section. DL models can be used in developing MDSS as well as extracting knowledge from a large amount of medical text data. Simple MLP models along with other statistical models have been developed using structured and semi-structured EMR data to predict a disease diagnosis such as hypertension [19], COPD [7] and PTSD [32]. DL models are particularly useful in processing complex text data. We have implemented different DL models to predict PTSD [32] and back pain [31] using both unstructured free-text note data and structured EMR data.

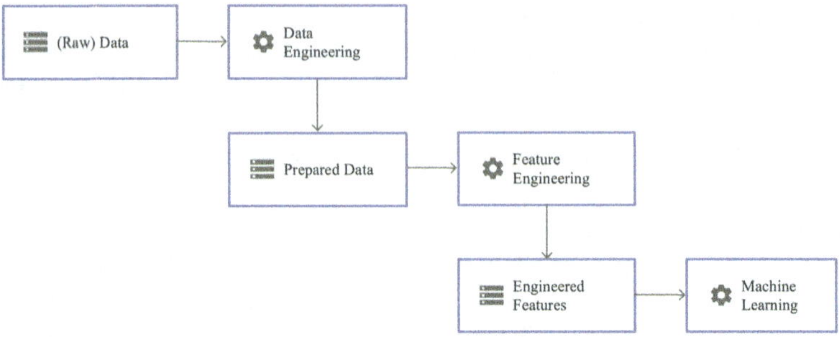

Fig. 5 The flow of data from raw data to ML [93]

5.1 Data Preprocessing

Preparing the data into a format that can be used to build and train a ML model is a critical part of AI. Use of bad data which may contain errors, noise or missing values may result in incorrect or ineffective models. The challenges of data preprocessing vary for different types of structured, semi-structured and unstructured data and the objectives of data modeling. We describe some data preprocessing techniques in this section that are commonly applied to process EMR data for developing MDSS, some of which have been applied in our research as illustrated in the next section.

Preprocessing the data for ML involves both *data engineering* and *feature engineering* as depicted in Fig. 5 [93]. *Data engineering* is the process of converting *raw data* into *prepared data. Raw data* refers to the data in its source form without any prior preparation. *Prepared data* refers to the dataset that have been parsed, joined, and put into a tabular form [93]. This data has been aggregated and summarized to the right granularity. For example, each row in the dataset represents a unique patient, irrelevant columns are removed, and invalid records are filtered out of the dataset. Feature engineering then further processes the prepared data to create the features to use as input into the ML model. DL models can perform feature engineering and extraction through multiple layers. *Engineered features* refer to the dataset with the tuned features expected by ML models as input, or extracted and transformed by ML models for generating the final outcome. Feature engineering operations can include performing certain ML-specific operations on the prepared datasets, including scaling numerical columns to a value between 0 and 1, clipping values, combining multiple values into one value and one-hot-encoding of categorical features.

5.1.1 Processing Structured and Semi-Structured EMRs

In CPCSSN and MaPCReN EMRs, raw data are organized in multiple tables in the database. Different tables contain logs of patient encounter data, patient demography, disease codes, information about prescribed medication, risk factors, health condition, and billing data. In this raw data, the number of patient records and the size of data vary widely. Some patients visit the physicians more often and have a lot of data, multiple risk factors, medications, and complex health situations while others can have very few records in the EMR. Prior to applying ML algorithms, data and feature engineering are applied to the data.

Processing Categorical Data

Categorical variables are those variables whose values are string or label. This type of variable is pretty common in medical data. For example, the *disease* variable in the MaPCReN data is a categorical variable representing a number of different health conditions. As a part of feature engineering, categorical variables are transformed into numerical variables. The most well-known method for this purpose is one-hot encoding. It converts each category value into a binary value and assigns 1 or 0 value to it to indicate if that feature exists or not. For example, one-hot encoding would transform the *disease* variable to a binary value, where different digits would indicate presence or absence of a disease condition by the value of 1 or 0 respectively.

Although one-hot encoding is a useful encoding method, it would be impractical to apply it to some variables in MaPCReN data due to the high number of unique values that could exist for those fields. For example, *medication* is a categorical variable with more than 1300 names. If we represent it using one-hot encoding the resulting feature set would have more than 1300 columns. Luckily, there are standard codes associated with some variables in MaPCReN data that facilitate the encoding process. For example, each medication name is associated with an *Anatomical Therapeutic Chemical* (ATC) classification system code [97]. Using the first three digits of the ATC code enables a researcher to categorize the medications which not only reduces the number of categories, but also accounts for changes in drug utilization by focusing on groups of similar medications.

General Data Cleaning and Balancing Techniques: Normalization and Balancing Data

There are often additional steps to help ML algorithms build better predictive models. For structured data, feature engineering operations include *feature tuning* which attempts to improve the quality of a feature for ML by scaling and normalizing numeric values, imputing missing values, clipping outliers, and adjusting values with skewed distributions [93].

A common problem in creating ML models, especially in medical domain is data imbalance, where only a small portion of patients have the diseases of interest; while most of the standard ML algorithms expect balanced class distribution. One solution to this problem is resampling the dataset to have the same number of positive and negative instances. This process includes techniques for repeated random sampling, oversampling of minority classes, and stratified partitioning.

5.1.2 Processing Unstructured Text Data

Natural language processing (NLP), a sub-field of AI, strives to understand and derive meaning from natural language text data. For medical data, the challenges in processing medical notes include frequent use of medical terms, frequent use of abbreviations, interpretation of negative or uncertain expressions, and subjective assessments of the health conditions. Interpreting medical terms and assigning a significance value to a term for diagnosis of a health condition requires the use of a dictionary or a medical ontology. For some MDSS structured and semi-structured health data such as blood pressure, weight, height, and medication, in addition to the unstructured EMR encounter notes, can provide important information for the detection of a medical condition, assessment of severity, or prediction of progression.

Text data from healthcare providers' encounter notes offer several challenges due to the subjective writing styles of different physicians. Some clinics may have an assistant who converts doctors' hand-written notes to electronic data and other providers may use tools such as speech-to-text generators. All of which introduce noise and error in the data; furthermore, encounter notes often contain expressions that are not always complete sentences, contain domain-specific medical terminology and abbreviations. Some of the commonly used text processing methods are described below.

When working with unstructured data, DL has enabled the domain knowledge-based feature engineering by folding it into the model architecture. Like other DL models, a convolutional layer is an automatic feature extractor [94]. The following three approaches are commonly used to transform text records into numerical vectors:

1. *Bag of words*: Involves a vocabulary of known words and a measure of the presence of a word in a document or a text corpus [95]. It could be created in two forms:
   (a) *Bag of word* (unigram)
   (b) *Bag of n-gram*
2. *Information Extraction tools*: Automatically extract structured information from text as concepts or named entities in the following ways [96]:
   (a) Named-entity linking associates different names referring to the same entity
   (b) Relation extraction identifies relations between concepts
   (c) Ontology mapping adds annotations by mapping biomedical text to the UMLS Meta-thesaurus
3. Word embeddings: Two famous methods for word embeddings are:
   (a) Glove [105]
   (b) Word2Vec [112]

Bag of words model involves a vocabulary of known words and a measure of the presence of known words in a document or text corpus, which is usually calculated as *Term Frequency-Inverted Document Frequency* (TF-IDF) scores [95]. The vocabulary can be created using all words or a subset of words (unigram or 1 word) or n-grams (e.g., bigram means 2 consecutive words) in the text. Then, by calculating scores for the individual words or n-grams for every document in the corpus, the vector representation can be calculated [98]. Usually, stemming and lemmatization are performed on the text tokens prior to creating the vocabulary and computing the scores [99].

Information Extraction (IE) is the task of automatically extracting and encoding clinical information from text data [100]. IE annotates extracted concepts with (i) named-entity linking that associates names referring to the same entity, (ii) relation extraction that identifies relations between concepts, and (iii) providing annotations by mapping biomedical text to the *Unified Medical Language System* (UMLS) Meta-thesaurus [106]. The most frequently used tools for IE in the clinical domain are *cTAKES* [101] and *MetaMap* [102]. *cTAKES*, developed by Mayo Clinic and later transitioned to an Apache project, is the most commonly used tool. It is built upon multiple Apache open-source projects, the Apache *Unstructured Information Management Architecture* (UIMA) framework [103] and the Apache OpenNLP toolkit [104]. It contains several analysis engines for various linguistics and clinical tasks, such as sentence detection, tokenization, part-of-speech tagging, concept detection, and normalization. It uses the UMLS Meta-thesaurus [106] for annotating medical concepts. Figure 6 shows an example sentence processed by cTAKES.

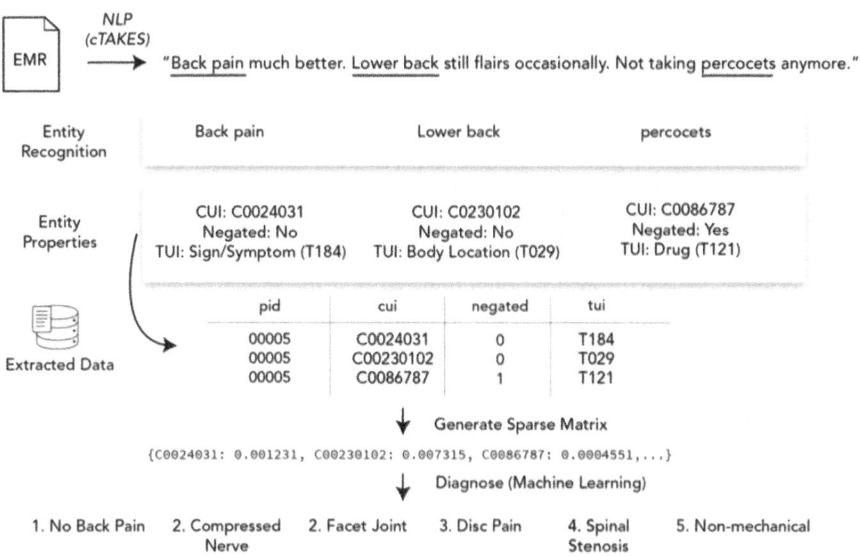

Fig. 6 An Example of Apache cTAKES output [31]

MetaMap was developed by the *National Library of Medicine* (NLM) with the goal of mapping biomedical text to the UMLS Meta-thesaurus, or vice versa. It was originally developed to improve biomedical text retrieval of MEDLINE/PubMed citations.

Word embeddings are a type of representation that allow words with similar meaning to have a similar vector representation. To do this, text is usually segmented into tokens where a single token is either a word, punctuation or special character. Tokens can then be transformed with the use of pre-trained word embeddings. Two widely used embeddings are Global Vectors (GloVe) for Word Representations [105] and Word2Vec [112]. Although the two embeddings are similar, there are a few key differences. GloVe attempts to leverage global context as it creates a large co-occurrence matrix of words that often appear around a target word with the final embeddings computed from a matrix reduction. While *word2vec* is a simple ANN that attempts to predict the target word by using the contextual words in a defined window, there are pre-trained vectors available online for both techniques. Since good embeddings are defined based on a large text corpus, training ML models to create word embeddings requires massive computational power. In practice both techniques yield similar results with more researchers preferring GloVe embeddings due to its propensity to capture the global context [105]. Following the embedding, the text segment is then encoded into vector representations.

Deep RNNs have shown great success in text encoding [109]. Each token from the text is treated as an input to the model at each time step, with the hidden state learning the context of the word token and the output layer generating the predicted next word in the sequence. Therefore, the hidden state at each time step is considered to represent the encoding of each word token. In a deep RNN, the final hidden layer represents an amalgamation of all previous states and hence, the context vector of the current word token. This method of encoding is considered as a *unidirectional* approach to embedding where representations only propagate forward in one direction.

Unfortunately, single direction encoding often fails to capture long-term dependencies. ANNs generate hidden representations through multiplication, as sequences get longer, the information from tokens earlier in the sequence begin to diminish. Naturally, this causes context vectors to only represent the latter portions of the sequence. As a remedy, Schuster et al. [58] proposed a *bidirectional* approach to help learn context from both directions so that the model can learn long term dependencies. In 2005 Graves et al. [108] extended the bidirectional approach to include LSTMs. The Bidirectional-LSTM concatenates representations from two separate LSTMs, where the text sequence is fed into the first LSTM in the forward direction and into the second LSTM in the reverse direction. Using the bidirectional approach, each time-step would now be informed by information from the past (i.e., from the beginning of the sequence to the current token) and the future (i.e., from the current token to the end of the sequence).

5.2 Model Validation Approaches

Common performance metrics that are used to evaluate the efficiency of classification models are *accuracy*, *sensitivity*, *specificity*, *precision*, *recall* and *Area Under the Curve* (AUC). The formulas to calculate these indicators are given below. Here TP indicates *true positive* i.e., data correctly classified by the model as having a positive label and TN indicates *true negative* i.e., data correctly classified by the model as having a *negative* label. FP indicates *false positive* i.e., data mistakenly classified by the model as having a positive label where it actually has a negative label. Similarly, FN indicates *false negative* i.e., data labeled as positive was identified by the model as negative. While *accuracy*, *precision* and *recall* are commonly used in ML research, *sensitivity* and *specificity* are more commonly used in medical research.

$$Accuracy = \frac{\mathrm{TP} + \mathrm{TN}}{\mathrm{TP} + \mathrm{FP} + \mathrm{FN} + \mathrm{TN}}$$

$$Precision = \frac{\mathrm{TP}}{\mathrm{TP} + \mathrm{FP}}$$

$$Recall = Sensitivity = \frac{\mathrm{TP}}{\mathrm{TP} + \mathrm{FN}}$$

$$Specificity = \frac{\mathrm{TN}}{\mathrm{TN} + \mathrm{FP}}$$

A *Receiver Operator Characteristic* (ROC) graph is a technique for visualizing the performance of classifiers [110]. The curve is created by plotting the sensitivity against (1–specificity) at various threshold settings. The *Area Under the Curve* (AUC) [92] is the portion of the area of the unit square under the ROC curve which always has a value between 0 and 1. The best possible prediction method would yield a point in the upper left corner or the coordinate (0,1) of the ROC space, representing 100% sensitivity (no FN) and 100% specificity (no FP).

5.3 Research Studies

In this section we present some of our research studies demonstrating the application of a variety of DL models to analyze EMR data for disease diagnosis as a MDSS. We also include some studies by other researchers to provide examples of other DL models in the medical domain.

5.3.1 Predicting Hypertension Using Structured EMR Data

LaFreniere et al. [19] utilized a very large sample size from the CPCSSN data set containing 185,371 primary care patients' diagnosed with hypertension and 193,656 controls i.e., individuals who did not have any chronic disease conditions. The goal was to develop a predictive model for hypertension, the most prevalent chronic disease in CPCSSN data. The EMR data included information about patients' current health conditions, medical records, and demographics. The factors of interest were exacted from multiple tables within the CPCSSN database. The data was aggregated into a master table for cleaning and preprocessing such as assigning class values (converting numeric values to categorical features). The final feature set comprised of age, gender, BMI, systolic and diastolic blood pressure, high and low density lipoproteins, triglycerides, cholesterol, micro albumin, and urine albumin creatinine ratio. As ANNs have proven to be promising in many data domains and have gained much success as incremental learners and predictors that mimic human intelligence, this method was used for predicting hypertension in this study. Because hypertension is the most prevalent chronic disease, sufficient number of data samples were available to apply supervised learning to train a MLP ANN. This method achieved 82.5% accuracy.

5.3.2 Diagnosing PTSD Using Structured EMR Data

Kaczmarek et al. [32] explored a small set of structured EMR data from MaPCReN to diagnose PTSD. This structured data came from several tables in the CPCSSN data repository including patient demographics, medical exam, disease, medication, and risk-factors. After applying data and feature engineering and converting these tables into a proper representation, 62 features were created for each patient from multiple years of doctors' visits or encounter data. This dataset was highly imbalanced as the number of patients without PTSD is significantly greater than the number of patients diagnosed with PTSD. Only 445 patients were found to have a PTSD positive diagnosis based on the ICD-9-CM code. To circumvent the unbalanced data problem, a random under sampling technique was applied on the dataset to create a balanced set of 890 *structured* data samples. To identify PTSD patients, three supervised ML models were developed namely a MLP ANN model, a *Support Vector Machine* (SVM), and a *Random Forest* (RF) classifier model. These methods obtained 0.79, 0.78, and 0.83 AUC values respectively. The best ANN structure, parameters and training algorithm were determined empirically through multiple trials. Similarly, five optimization algorithms were tested for backpropagation training and the *Bayesian Regularization* (BR) function was selected because it produced the best result. However, RF obtained better results than MLP. The reason might be related to the intrinsic noise in clinical and medical data and small data sample which is generally inadequate to train ANN models.

5.3.3 Classification of Low Back Pain Using EMR Chart Note Data

Judd et al. [31] developed a generic text-mining and decision support framework to diagnose chronic low back pain based on unstructured chart notes in the EMR data. The study used a very small sample data which contained only unstructured text chart note data from OSCAR EMR [24]. The framework utilized *Natural Language Processing* (NLP) and ML to classify low back pain into four groups. The authors applied open-source de-identification tool, De-id [111], to anonymize EMR chart note data, which often contains patients' names and other sensitive information. The system was designed in a pipeline, consisting of different modules including a sentence boundary detector, a tokenizer, a normalizer, a part-of-speech tagger, a shallow parser, and a named entity recognizer. Apache *cTAKES* [101] was used in the NLP step to extract relevant medical terminology and relevant negation from the free-text as depicted in Fig. 6. The pipeline used the *Unified Medical Language System* (UMLS) [106] for medical concept annotation including anatomical sites, signs/symptoms, procedures, diseases/disorders and medications. For every medical entity cTAKES recognizes, a data structure is returned containing a normalized UMLS Unique Concept Identifier, a Type Unique Identifier including values for negation, and a subject. Four ML models were implemented: Bernoulli Naïve-Bayes, Multinomial Naïve-Bayes, Linear Support Vector classifier, and MLP with stochastic gradient descent optimization, to classify low back pain. The results show a high accuracy for the limited thirty-four patients' data which were labeled by medical students as ground truth data for this pilot study. However, due to the small sample size used in this research, the perceptron did not get comparable results as the other ML models.

5.3.4 Predicting COPD from EMR Data

Zafari et al. [7] conducted a study with MapReCN patient EMR data to prediction *Chronic Obstructive Pulmonary Disease* (COPD). They explored seven tables from the MaPCReN namely *patient demographics*, *health condition*, *billing*, *disease case*, *examinations*, *medication*, and *risk factors*. They applied three supervised machine learning models, a MLP model, a SVM, and an *Extreme Gradient Boosting* (XGB), which the XGB model achieved an accuracy of 82%, compared to 81% accuracy of SVM and 80% of the MLP. Utilizing feature importance, they identified a set of key symptoms for diagnosing COPD within the EMR data that were mostly from medications, health condition and risk factors tables.

5.3.5 Weak Supervision Model for Text Embedding to Diagnose PTSD

Obtaining labeled data is time consuming and an expensive task that is often the problem and constraint in developing DL solutions. For this study, a medical student manually reviewed full patient chart notes spending approximately an hour to annotate each patient's data.

To remedy the lack of ground truth labels, Ratner et al. [113] introduced a method called *Data Programming*. It is a recently proposed programming paradigm to develop weakly-supervised ML models by defining rules called *Learning Functions* (LFs). Each learning function outputs a vote of '1' for positive, '-1' for negative or '0' for abstaining. This method of weak-supervision has shown success across a variety of domains like user recommendations and identifying trends [116], clinical text classification for smoking status and proximal femur (hip) fracture classification [114]. One approach for implementing weak-supervision is through the Snorkel platform [117], a system dedicated to leveraging Data Programming for a wide variety of analytics tasks. Ratner et al. [117] further demonstrated that Snorkel models achieved an average predictive performance without the use of labeled data, within 3.6% of models using large hand-annotated data sets.

To create well-defined LFs, a variety of taggers and processors can be used to augment existing data. For example, cTAKES [101] is an out-of-the-box tool that can perform *Named Entity Recognition* (NER), identify parts of speech, sentence segmentation, tokenization and much more. The cTAKES NER tags are *Concept Unique IDs* (CUIs) that are defined by the *Unified Medical Language System* (UMLS) Meta-thesaurus. The resulting cTAKES tags can be combined with the original text to be used as inputs in LFs. A sample of the tags generated by cTAKES can be seen in Fig. 6. CUI tags, word tokens, sentence boundaries and the original text were aligned to create a candidate for training the ML models where candidates refer to entities-of-interest. Currently, cTAKES seems to have the most comprehensive solution for biomedical NLP. While state-of-the-art systems exist for niche topic areas like *TaggerOne* [119] for chemicals and diseases, cTAKES supports decent functionality across numerous domains (e.g., drug names, diseases, signs and symptoms, anatomy, and test procedures).

Candidates generated from the preprocessing stage described above were embedded and used as input to generate probabilistic labels through Snorkel's [117] noise-aware generative model as shown in Fig. 7. A generative model was trained using LFs, which served as the source of weak-supervision. For simplicity, a rule-set could include all similar CUIs of a term and therefore prevent the need for word-sense disambiguation. To construct LFs for weak-supervision, a user would define semantic rules from a small hand-annotated dataset of true positive samples. For example, when assessing patients for PTSD, a common rule could be appearance of the word 'served' within five-word token distance from 'Afghanistan' or 'Iraq' (e.g., Recently served in Iraq). Resulting probabilistic labels which are the outcomes of the votes by the LFs, were used in-place of the traditional ground truth to train supervised discriminative models.

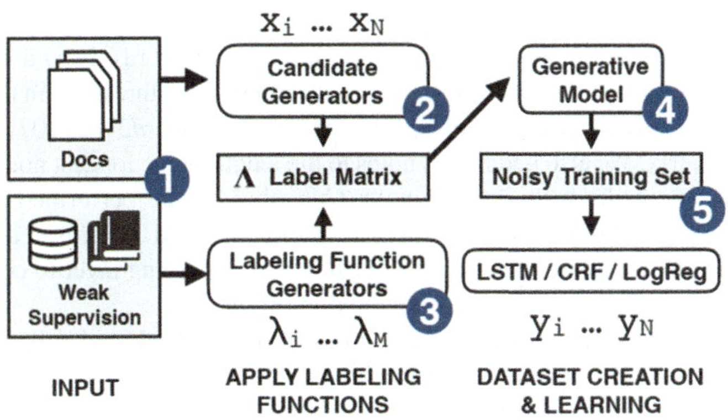

Fig. 7 Snorkel pipeline use case [115]

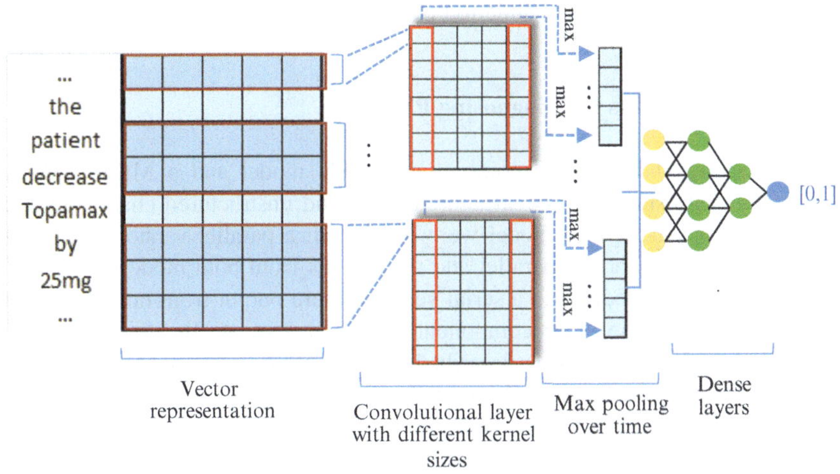

Fig. 8 Applying CNN model to physician notes for PTSD diagnosis

5.3.6 PTSD Diagnosis Using Encounter Notes and Deep CNN

In an ongoing research study, we used a CNN to identify PTSD using free text encounter notes of EMR. We processed unstructured encounter note using NLP and DL algorithms and specifically CNN.

As depicted in Fig. 8, the architecture of this DL model is composed of three parts: word embedding, convolutional model and fully connected model. *Word embedding* is a distributed representation of words where different words that have a similar meaning (based on their usage) also have a similar representation. The CNN model learns to extract salient features from documents encoded using a word embedding method. The fully connected part of the model maps the extracted features to a predictive output.

First, the narrative notes are preprocessed including tokenizing all input texts, removing all punctuation from words, and removing stop words. Next a vocabulary is created using words that occur for more than a predefined threshold in the dataset. Then sentences are mapped to embedding vectors using *Word2Vec* [107] algorithm to create a matrix. We also padded all notes to the length of the longest note to ensure that all documents have the same length. Convolutions are performed across the input words using different sized kernels, including 1, 2 and 3 words at a time. The resulting feature maps are then processed using a max-pooling layer to condense or summarize the extracted features as depicted in Fig. 8.

Finally, these extracted features are interpreted by a back-end fully connected model. It uses a sigmoid activation function to convert each note to a real value between 0 and 1. We fit the network on the training data using a binary cross-entropy loss function. The efficient *Adam* implementation of stochastic gradient descent is used. Additionally, we track the accuracy and loss during training. The accuracy obtained using a 5-fold cross validation is 0.87.

5.3.7 Hybrid Models for Diagnosing PTSD

We developed hybrid models combining a CNN model and a MLP model to diagnose PTSD using both the structured EMR and unstructured chart note data. Two versions of the hybrid model were developed: a parallel version where two component models work in parallel and the outputs from both models were combined using a third model, and a serial version where output from one component model was sent to the other component model to generate the final output. In the parallel version, a CNN as described in Sect. 5.3.6, was used to process the unstructured chart note data. A MLP model was used to processes the structured data in parallel. Extracted feature-sets from the two models were fed into a dense layer as shown in Fig. 9, which classified the data into PTSD or non-PTSD diagnosis. The model achieved 0.88 accuracy.

The second serial version of the hybrid model as shown in Fig. 10, implemented a CNN model which processed the unstructured chart note data and generated a probability value in the range of [0,1] indicating the probable diagnosis of PTSD.

A second component model of random forest (RF) classifier was implemented that accepted the output of the first component in addition to the structured EMR data to generate a diagnosis of PTSD. By stacking the two models that processed two different types of data, our hybrid model was able to process both structured EMR and unstructured chart note data in an innovative way to predict PTSD.

In this research the dataset was compiled from six tables in the MaPCReN dataset [34] namely patient demographics, encounter notes, disease case, exam, medication and risk factor using table join and merge operations. Various feature engineering techniques were used to create a proper representation of the data. Numerical exam values like blood pressure and BMI were described using descriptive statistics by computing the range and mean values of each exam as a separate feature. Categorical features like diseases and risk factors were represented as dichotomized

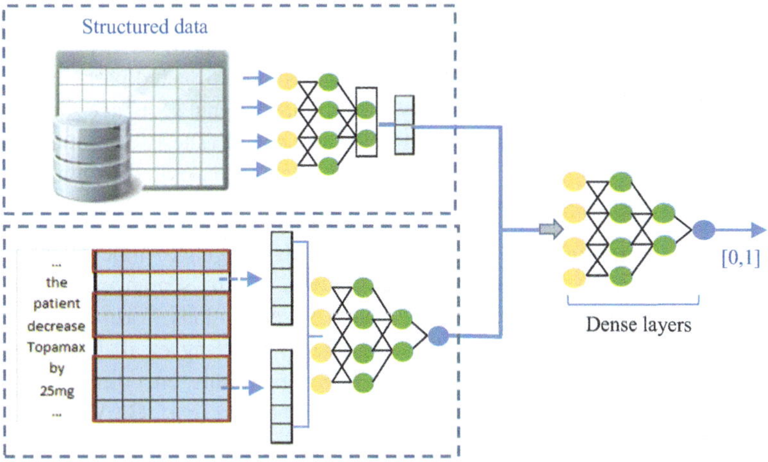

Fig. 9 Parallel architecture for processing free text and structured data

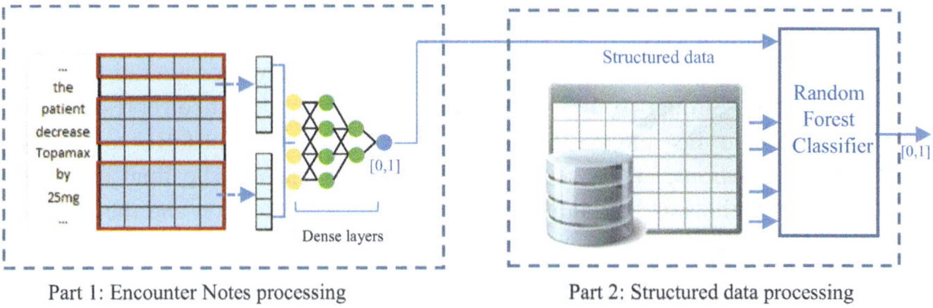

Fig. 10 Serial architecture for processing free text and structured data

variables. Prescribed medications were identified using the ATC codes of medications [97] and were represented and generalized by the first letter of the medication ATC code. Finally, a RF classifier was used to perform the final classification which resulted in 0.89 accuracy.

5.3.8 Other DL Models from State-of-the-Art Research on MDSS

Attention has been one of the most impactful contributions to the field of DL within the last few years. Bahdanau et al. [118] introduced the idea of attention in 2014 for a language translation model. The initial insight came from being able to align the translated text back to the original sequence. Language translation is rarely linear; the word orderings from language A is often different from the word ordering in language B. As ANNs translate a sentence one-word token at a time, attention allows the network to be able to look back at all the hidden representations of the original sequence to learn which parts of the original sequence best match the translated token. In addition to the benefits of aligning the text, it also expedites computation by limiting the processing to targeted parts of the data. Attention scores are calculated for every word in the original sequence representing its correspondence to the current word being translated. The use of attention has become so prevalent that a variation of it has been incorporated to almost every state-of-the-art NLP application.

Hierarchical Attention Network The English language inherently has a hierarchical structure where blocks of words form sentences, a number of sentences form paragraphs, and a set of paragraphs form papers. The *Hierarchical Attention Network* (HAN) introduced by Yang et al. [120] encodes smaller chunks of text that are then used to inform future encodings. Figure 11 shows the HAN architecture where words are embedded (e.g., GloVe, *Word2Vec*) and encoded using a bidirectional *Gated-Recurrent Unit* (GRU) type of RNN, as seen in the word encoder layer. A bidirectional LSTM type RNN can be used instead of a bidirectional GRU, as both architectures are similar and often yield analogous results. The encoded word-level representations are then sent through an attention layer to produce word-level attention weights. A weighted average of word level attention weights and the hidden word representations create sentence vectors. Similar to the word level, the resulting sentence representations of a document are then passed through a different bidirectional GRU to generate hidden representations, and a sentence-level attention layer. The final document representation is the weighted sum of the sentence level attention and the sentence encodings.

HAN in MDSS In recent applications, HANs has shown promises in analyzing EMRs for medical disease diagnosis. In 2017, Gao et al. [121] demonstrated that HAN largely outperformed traditional DL architectures in classifying free-text pathology notes. Mullenbach et al. [122] utilized a HAN inspired architecture in conjunction with a CNN to predict disease codes from physicians' notes. Baumel et al. [123] utilized a HAN in conjunction with a GRU to predict ICD-9 codes from patient notes.

In addition to the reported state-of-the art results from the aforementioned authors, the hierarchical approach has the added advantage of being able to provide transparency. Since physician notes typically comprise of an amalgamation of individual observations, the attention weights from the HAN can identify the specific sentence (or observation) and words that contribute the most towards the classification. Implementation on a custom dataset requires a more in-depth preprocessing step. To diagnose patients using physicians' notes, each patient object would be the amalgamation of notes in the EMR. *SpaCy* [124] is a light-weight tool that can be used on patients' notes to separate them into individual sentences and tokens. Although this requires two passes through a RNN, this implementation can run efficiently on GPU setups used for DL.

Self-Attention Several variants of the attention models have evolved over the years. *Self-attention* ANN models derive attention weights by attending to the

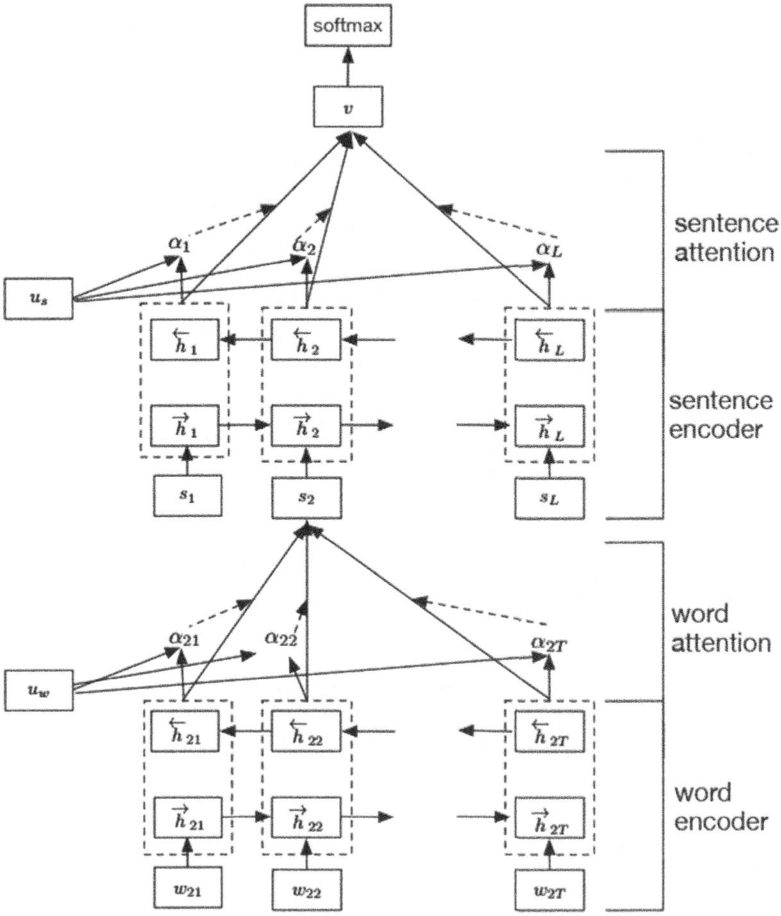

Fig. 11 HAN [120]

generated text embeddings and performing additional optimization to tune the attention weights to learn the important vectors that contribute more towards the correct classification. Self-attention proved easy to implement and showed significant benefits to down-stream classification tasks. The general self-attention mechanism forces multiple attention layers to attend to different areas of the text for document classification tasks and provides transparency by creating heat maps using the attention weights on the original text. Zhang et al. [125] utilized an attention penalty function in generating embeddings of patient chart notes and Sousa et al. [126] leveraged self-attention diagnose patients with diabetes using their medical records.

Transformer A lightweight non-recurrent Transformer ANN model was proposed by Vaswani et al. [127] for transforming the data using a two-phase encoder-

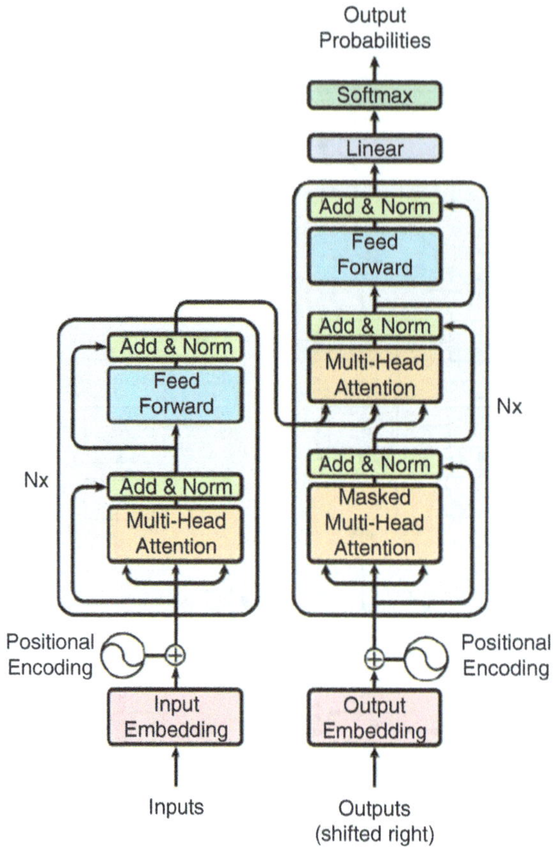

Fig. 12 Transformer Network [127]

decoder function which has led to almost every architecture that currently tops the NLP benchmarks. It is significantly faster to train and exploits parallelization in computing the attention weights using a single matrix multiplication. An encoding layer is used to first learn the positional information of tokens in fixed size sequences (typically consisting of a maximum of 512 tokens) which is summed with the embedded inputs and then encoded. The encoder is typically comprised of a stack of identical layers where each layer contains a multi-headed attention mechanism and a feed forward layer. The encodings are then passed to a decoder which also comprises a stack of identical layers including an additional multi-head attention over the encodings from the encoder. Masking is used in the decoder to prevent self-attention weights from attending to the same positions. A diagram of the encoder can be seen in Fig. 12.

Bidirectional Encoder Representations from Transformers The *Bidirectional Encoder Representations from Transformers* (BERT) [128] and BERT-inspired architectures have currently dominated NLP benchmarks (e.g., GLUE [129]). Although the architecture is not novel, its utilization of the Transformer combined with transfer learning and domain adaptations is a significant contribution to NLP. Devlin et al. [128] pre-trained a Transformer network using both a Book Corpus and the English Wikipedia comprising of 3300 million words, to create a model with a generalized understanding of the English language. The intuition behind this was to have a model that understood the nuances of a language prior to engaging in natural language problems. To achieve this generalized understanding, BERT was trained, unsupervised, using two language modelling objectives: *Next Sentence Prediction* (NSP) and *Masked Language Model* (MLM). As the relationship between sentences is essential for contextualized understanding of text, BERT utilized a binary sentence prediction objective by determining if a sentence B frequently followed a sentence A. To develop contextual understanding at a word level, a portion of the words were masked with a [MASK] token and the objective was to predict the masked token using the surrounding context [128]. As unsupervised learning was used to pre-train the model, it created an efficient methodology to understand the language in the underlying corpus. Once pre-trained, the model can be further fine-tuned on down-stream tasks. While training BERT on a specific language domain is possible, pre-training BERT-esque models takes a large amount of resources (BERTLARGE took 4 days using 16 Cloud TPUs). Companies like Facebook and Google have released public models that have been trained on a large corpus of data.

Due to its success, many variations of the model evolved from BERT (GLUE benchmarks [129]). Numerous versions of BERT have already been pre-trained for use in the medical domain such as SciBERT (trained on scientific text) [130], BioBERT [131] (trained on biomedical and life science articles from PubMed), and ClinicalBERT [132] (trained on chart notes from critical care patients).

6 Conclusion

This chapter describes the benefits of using DL models with EMR data. EMR data provides detailed information on a patient's health as it records each encounter of a patient with the physician over multiple years. We reviewed EMRs in Canada, and specifically explained the extraction and processing steps of the EMR data needed to build the CPCSSN database [8]. The value of EMR data for research, health surveillance and quality improvement as well as the existing challenges were explained. Considering the challenging nature of processing EMR data, especially the unstructured data, DL can be very effective in organizing this data and aid in decision-making. We explained different EMR data types, summarized the applications of AI and ML algorithms to EMR data, and described a variety of DL models that can be used to process the different types of EMR and medical data. Definitions of supervised, semi-supervised and unsupervised ML methods were provided. We also emphasized the need for a large amount of labeled data to develop and train DL models, which may be a challenge when identifying rare diseases. State-of-the-art research using DL model with medical data was presented. Finally, we discussed a number of research studies which focused on developing DL models for diagnosing a variety of diseases. The digitization of medical records has opened numerous opportunities to apply AI models to health data to learn disease symptoms, progression, comorbidity, patient care, diagnosis and effective treatments. However, the complex nature of health data (text, structured, codes, image, signal), jurisdictional differences in ownership, storage and access depending on the data type (e.g., administrative, EMR, clinical) and provincial regulations, availability of labeled data, and variation in disease progression among patients from different demography offer interesting challenges when developing a universal MDSS. Future work using AI within the medical data domain should aim to inform diagnoses, summarize and characterize data availability, assess prescribed treatments and innovate reliable models that can inform initiatives to improve patient care and health system outcomes.

References

1. Chang, F., & Gupta, N. (2015). Progress in electronic medical record adoption in Canada. Canadian Family Physician, 61(12), 1076-1084.
2. Marrie, R. A., Kosowan, L., Taylor, C., & Singer, A. (2019). Identifying people with multiple sclerosis in the Canadian primary care sentinel surveillance network. Multiple Sclerosis Journal–Experimental, Translational and Clinical
3. Cave, A. J., Davey, C., Ahmadi, E., Drummond, N., Fuentes, S., Kazemi-Bajestani, S. M. R., ... & Taylor, M. (2016). Development of a validated algorithm for the diagnosis of paediatric asthma in electronic medical records. NPJ primary care respiratory medicine, 26(1), 1-4.
4. Kosowan, L., Wicklow, B., Queenan, J., Yeung, R., Amed, S., & Singer, A. (2019). Enhancing Health Surveillance: Validation of a Novel Electronic Medical Records-Based Definition of Cases of Pediatric Type 1 and Type 2 Diabetes Mellitus. Canadian journal of diabetes, 43(6), 392-398.
5. Williamson, T., Green, M. E., Birtwhistle, R., Khan, S., Garies, S., Wong, S. T., ... & Drummond, N. (2014). Validating the 8 CPCSSN case definitions for chronic disease surveillance in a primary care database of electronic health records. The Annals of Family Medicine, 12(4), 367-372.
6. Singer, A., Kosowan, L., Katz, A., Ronksley, P., McBrien, K., Halas, G., & Williamson, T. (2020). Characterizing patients with high use of the primary and tertiary care systems: A retrospective cohort study. Health Policy, 124(3), 291-297.
7. Zafari, H.,Langlois, S.,Zulkernine, F., Kosowan, L., & Singer, A. (2020). Predicting Chronic Obstructive Pulmonary Disease from EMR data. International Conference on Computational Intelligence in Bioinformatics and Computational Biology.
8. Birtwhistle, R. V. (2011). Canadian Primary Care Sentinel Surveillance Network: A developing resource for family medicine and public health. Canadian Family Physician, 57(10), 1219-1220.
9. Queenan, J. A., Williamson, T., Khan, S., Drummond, N., Garies, S., Morkem, R., & Birtwhistle, R. (2016). Representativeness of patients and providers in the Canadian Primary Care Sentinel Surveillance Network: a cross-sectional study. CMAJ open, 4(1), E28.

10. TCPS-2. (2014). Canadian Institutes of Health Research, Natural Sciences and Engineering Research Council of Canada, and Social Sciences and Humanities Research Council of Canada. Tri-Council Policy Statement: Ethical Conduct for Research Involving Humans.
11. Kotecha, J. A., Manca, D., Lambert-Lanning, A., Keshavjee, K., Drummond, N., Godwin, M., ... & Birtwhistle, R. (2011). Ethics and privacy issues of a practice-based surveillance system: Need for a national-level institutional research ethics board and consent standards. Canadian Family Physician, 57(10), 1165-1173.
12. Oake, J., Aref-Eshghi, E., Godwin, M., Collins, K., Aubrey-Bassler, K., Duke, P., ... & Asghari, S. (2017). Using electronic medical record to identify patients with dyslipidemia in primary care settings: international classification of disease code matters from one region to a national database. Biomedical informatics insights, 9, 1178222616685880.
13. Bello, A. K., Ronksley, P. E., Tangri, N., Kurzawa, J., Osman, M. A., Singer, A., ... & Lindeman, C. (2019). Prevalence and demographics of CKD in Canadian primary care practices: a cross-sectional study. Kidney international reports, 4(4), 561-570.
14. Queenan, J. A., Farahani, P., Ehsani-Moghadam, B., & Birtwhistle, R. V. (2018). The prevalence and risk for herpes zoster infection in adult patients with diabetes mellitus in the Canadian Primary Care Sentinel Surveillance Network. Canadian journal of diabetes, 42(5), 465-469.
15. Zafari, H.,Zulkernine, Singer, A., & Kosowan, L. (2019). Weakly Supervised Text Classification for Assisting Patient Data Processing," in the 10th annual conference hosted by the Canadian Institute for Military and Veteran Health Research (CIMVHR).
16. Telus: https://www.telus.com, last accessed 2020/8/28
17. QHR Technologies: https://qhrtechnologies.com/, last accessed 2020/8/28
18. OSCAR EMR: https://oscar-emr.com/oscar/, last accessed 2020/8/28
19. LaFreniere, D., Zulkernine, F., Barber, D., & Martin, K. (2016, December). Using machine learning to predict hypertension from a clinical dataset. In 2016 IEEE Symposium Series on Computational Intelligence (SSCI) (pp. 1-7). IEEE.
20. McCulloch, W. S., & Pitts, W. (1943). A logical calculus of the ideas immanent in nervous activity. The bulletin of mathematical biophysics, 5(4), 115-133.
21. Schmidhuber, J. (2015). Deep learning in neural networks: An overview. Neural networks, 61, 85-117.
22. Oh, K. S., & Jung, K. (2004). GPU implementation of neural networks. Pattern Recognition, 37(6), 1311-1314.
23. Chellapilla, K., Puri, S., & Simard, P. (2006, October). High performance convolutional neural networks for document processing.
24. OSCAR Canada: About OSCAR, http://oscarcanada.org/about-oscar/brief-overview, last accessed 2020/8/28.
25. Xiao, L., Cousins, G., Fahey, T., Dimitrov, B. D., & Hederman, L. (2012, October). Developing a rule-driven clinical decision support system with an extensive and adaptative architecture. In 2012 IEEE 14th International Conference on e-Health Networking, Applications and Services (Healthcom) (pp. 250-254). IEEE.
26. Achour, S. L., Dojat, M., Rieux, C., Bierling, P., & Lepage, E. (2001). A UMLS-based knowledge acquisition tool for rule-based clinical decision support system development. Journal of the American Medical Informatics Association, 8(4), 351-360.
27. Kuo, K. L., & Fuh, C. S. (2011). A rule-based clinical decision model to support interpretation of multiple data in health examinations. Journal of medical systems, 35(6), 1359-1373.
28. Choi, E., Bahadori, M. T., Schuetz, A., Stewart, W. F., & Sun, J. (2016, December). Doctor ai: Predicting clinical events via recurrent neural networks. In Machine Learning for Healthcare Conference (pp. 301-318).
29. Liu, J., Zhang, Z., & Razavian, N. (2018). Deep ehr: Chronic disease prediction using medical notes. arXiv preprint arXiv:1808.04928..
30. Shickel, B., Tighe, P. J., Bihorac, A., & Rashidi, P. (2017). Deep EHR: a survey of recent advances in deep learning techniques for electronic health record (EHR) analysis. IEEE journal of biomedical and health informatics, 22(5), 1589-1604.

31. Judd, M., Zulkernine, F., Wolfrom, B., Barber, D., & Rajaram, A. (2018, September). Detecting low back pain from clinical narratives using machine learning approaches. In International Conference on Database and Expert Systems Applications (pp. 126-137). Springer, Cham.
32. Kaczmarek, E., Salgo, A., Zafari, H., Kosowan, L., Singer, A., & Zulkernine, F. (2019, December). Diagnosing PTSD using electronic medical records from canadian primary care data. In Proceedings of the 6th International Conference on Networking, Systems and Security (pp. 23-29).
33. Braunstein, M. L. (2015, June). Patient—Physician collaboration on FHIR (Fast Healthcare Interoperability Resources). In 2015 International Conference on Collaboration Technologies and Systems (CTS) (pp. 501-503). IEEE.
34. Coleman, N., Halas, G., Peeler, W., Casaclang, N., Williamson, T., & Katz, A. (2015). From patient care to research: a validation study examining the factors contributing to data quality in a primary care electronic medical record database. BMC family practice, 16(1), 11.
35. Shortliffe, E. H. (1986). Medical expert systems—knowledge tools for physicians. Western Journal of Medicine, 145(6), 830.
36. Miller, R. A., McNeil, M. A., Challinor, S. M., Masarie Jr, F. E., & Myers, J. D. (1986). The INTERNIST-1/quick medical REFERENCE project—Status report. Western Journal of Medicine, 145(6), 816.
37. Pauker, S. G., Gorry, G. A., Kassirer, J. P., & Schwartz, W. B. (1976). Towards the simulation of clinical cognition: taking a present illness by computer. The American journal of medicine, 60(7), 981-996.
38. MYCIN: https://web.archive.org/web/20120212093503/http://raa.ruby-lang.org/project/mycin/, last accessed 2020/8/28
39. Kulikowski, C. A., & Weiss, S. M. (1982). Representation of expert knowledge for consultation: the CASNET and EXPERT projects. Artificial Intelligence in medicine, 51, 21-55.
40. Kumar, A., Zarychanski, R., Pinto, R., Cook, D. J., Marshall, J., Lacroix, J., ... & Turgeon, A. F. (2009). Critically ill patients with 2009 influenza A (H1N1) infection in Canada. Jama, 302(17), 1872-1879.
41. Lewis, M. D., Pavlin, J. A., Mansfield, J. L., O'Brien, S., Boomsma, L. G., Elbert, Y., & Kelley, P. W. (2002). Disease outbreak detection system using syndromic data in the greater Washington DC area. American journal of preventive medicine, 23(3), 180-186.
42. Guthmann, J. P., Klovstad, H., Boccia, D., Hamid, N., Pinoges, L., Nizou, J. Y., ... & Ciglenecki, I. (2006). A large outbreak of hepatitis E among a displaced population in Darfur, Sudan, 2004: the role of water treatment methods. Clinical infectious diseases, 42(12), 1685-1691.
43. Miotto, R., Li, L., Kidd, B. A., & Dudley, J. T. (2016). Deep patient: an unsupervised representation to predict the future of patients from the electronic health records. Scientific reports, 6(1), 1-10.
44. Lakhani, P., & Sundaram, B. (2017). Deep learning at chest radiography: automated classification of pulmonary tuberculosis by using convolutional neural networks. Radiology, 284(2), 574-582.
45. Wang, N., Cui, L., Huang, X., Xiang, Y., & Xiao, J. (2018). EasiCSDeep: A deep learning model for Cervical Spondylosis Identification using surface electromyography signal. arXiv preprint arXiv:1812.04912.
46. Tomar, D., & Agarwal, S. (2013). A survey on Data Mining approaches for Healthcare. International Journal of Bio-Science and Bio-Technology, 5(5), 241-266.
47. Ding, S., Zhu, H., Jia, W., & Su, C. (2012). A survey on feature extraction for pattern recognition. Artificial Intelligence Review, 37(3), 169-180.
48. Reed, R., & MarksII, R. J. (1999). Neural smithing: supervised learning in feedforward artificial neural networks. Mit Press.
49. K. Patel, "MNIST Handwritten Digits Classification using a Convolutional Neural Network," 2020. [Online]. Available: https://towardsdatascience.com/mnist-handwritten-digits-classification-using-a-convolutional-neural-network-cnn-af5fafbc35e9.

50. Simard, P. Y., Steinkraus, D., & Platt, J. C. (2003, August). Best practices for convolutional neural networks applied to visual document analysis. In Icdar (Vol. 3, No. 2003).
51. Kim, Y. (2014). Convolutional neural networks for sentence classification. arXiv preprint arXiv:1408.5882.
52. Pham, T., Tran, T., Phung, D., & Venkatesh, S. (2017). Predicting healthcare trajectories from medical records: A deep learning approach. Journal of biomedical informatics, 69, 218-229.
53. Sutskever, I., Vinyals, O., & Le, Q. V. (2014). Sequence to sequence learning with neural networks. In Advances in neural information processing systems (pp. 3104-3112).
54. Chen, K., Zhou, Y., & Dai, F. (2015, October). A LSTM-based method for stock returns prediction: A case study of China stock market. In 2015 IEEE international conference on big data (big data) (pp. 2823-2824). IEEE.
55. Choi, E., Schuetz, A., Stewart, W. F., & Sun, J. (2017). Using recurrent neural network models for early detection of heart failure onset. Journal of the American Medical Informatics Association, 24(2), 361-370.
56. Wang, Y., Neves, L., & Metze, F. (2016, March). Audio-based multimedia event detection using deep recurrent neural networks. In 2016 IEEE international conference on acoustics, speech and signal processing (ICASSP) (pp. 2742-2746). IEEE..
57. Chung, J., Gulcehre, C., Cho, K., & Bengio, Y. (2014). Empirical evaluation of gated recurrent neural networks on sequence modeling. arXiv preprint arXiv:1412.3555.
58. Schuster, M., & Paliwal, K. K. (1997). Bidirectional recurrent neural networks. IEEE transactions on Signal Processing, 45(11), 2673-2681.
59. Kramer, M. A. (1991). Nonlinear principal component analysis using autoassociative neural networks. AIChE journal, 37(2), 233-243.
60. Belciug, S., & Gorunescu, F. (2014). Error-correction learning for artificial neural networks using the Bayesian paradigm. Application to automated medical diagnosis. Journal of Biomedical Informatics, 52, 329-337.
61. Sutton, R. S., & Barto, A. G. (2018). Reinforcement learning: An introduction. MIT press.
62. Fahlman, S. E., & Lebiere, C. (1990). The cascade-correlation learning architecture. In Advances in neural information processing systems (pp. 524-532).
63. Russell, S. J., & Norvig, P. (2010). Artificial Intelligence-A Modern Approach, Third International Edition.
64. Goodfellow, I., Bengio, Y., & Courville, A. (2016). 6.5 Back-Propagation and Other Differentiation Algorithms. Deep Learning, 200-220.
65. Zhu, X. J. (2005). Semi-supervised learning literature survey. University of Wisconsin-Madison Department of Computer Sciences.
66. Ratner, A. J., De Sa, C. M., Wu, S., Selsam, D., & Ré, C. (2016). Data programming: Creating large training sets, quickly. In Advances in neural information processing systems (pp. 3567-3575).
67. Rosenthal, S., Farra, N., & Nakov, P. (2017, August). SemEval-2017 task 4: Sentiment analysis in Twitter. In Proceedings of the 11th international workshop on semantic evaluation (SemEval-2017) (pp. 502-518).
68. Hu, Z., Li, X., Tu, C., Liu, Z., & Sun, M. (2018, August). Few-shot charge prediction with discriminative legal attributes. In Proceedings of the 27th International Conference on Computational Linguistics (pp. 487-498).
69. Zhong, H., Guo, Z., Tu, C., Xiao, C., Liu, Z., & Sun, M. (2018). Legal judgment prediction via topological learning. In Proceedings of the 2018 Conference on Empirical Methods in Natural Language Processing (pp. 3540-3549).
70. Luo, B., Feng, Y., Xu, J., Zhang, X., & Zhao, D. (2017). Learning to predict charges for criminal cases with legal basis. arXiv preprint arXiv:1707.09168.
71. He, H., Ganjam, K., Jain, N., Lundin, J., White, R., & Lin, J. (2017, September). An insight extraction system on biomedical literature with deep neural networks. In Proceedings of the 2017 Conference on Empirical Methods in Natural Language Processing (pp. 2691-2701).

72. Fahad, A., Alshatri, N., Tari, Z., Alamri, A., Khalil, I., Zomaya, A. Y., ... & Bouras, A. (2014). A survey of clustering algorithms for big data: Taxonomy and empirical analysis. IEEE transactions on emerging topics in computing, 2(3), 267-279.
73. Choi, E., Bahadori, M. T., Searles, E., Coffey, C., Thompson, M., Bost, J., ... & Sun, J. (2016, August). Multi-layer representation learning for medical concepts. In Proceedings of the 22nd ACM SIGKDD International Conference on Knowledge Discovery and Data Mining (pp. 1495-1504).
74. Pham, T., Tran, T., Phung, D., & Venkatesh, S. (2016, April). Deepcare: A deep dynamic memory model for predictive medicine. In Pacific-Asia Conference on Knowledge Discovery and Data Mining (pp. 30-41). Springer, Cham.
75. Wickramasinghe, N. (2017). Deepr: a convolutional net for medical records. IEEE J Biomed Health Inform.
76. Lv, X., Guan, Y., Yang, J., & Wu, J. (2016). Clinical relation extraction with deep learning. International Journal of Hybrid Information Technology, 9(7), 237-248.
77. Mallya, S., Overhage, M., Srivastava, N., Arai, T., & Erdman, C. (2019). Effectiveness of lstms in predicting congestive heart failure onset. arXiv preprint arXiv:1902.02443.
78. Nie, L., Wang, M., Zhang, L., Yan, S., Zhang, B., & Chua, T. S. (2015). Disease inference from health-related questions via sparse deep learning. IEEE Transactions on knowledge and Data Engineering, 27(8), 2107-2119.
79. Nemati, S., Ghassemi, M. M., & Clifford, G. D. (2016, August). Optimal medication dosing from suboptimal clinical examples: A deep reinforcement learning approach. In 2016 38th Annual International Conference of the IEEE Engineering in Medicine and Biology Society (EMBC) (pp. 2978-2981). IEEE.
80. Choi, E., Bahadori, M. T., Sun, J., Kulas, J., Schuetz, A., & Stewart, W. (2016). Retain: An interpretable predictive model for healthcare using reverse time attention mechanism. In Advances in Neural Information Processing Systems (pp. 3504-3512).
81. Ong, B. T., Sugiura, K., & Zettsu, K. (2016). Dynamically pre-trained deep recurrent neural networks using environmental monitoring data for predicting PM 2.5. Neural Computing and Applications, 27(6), 1553-1566.
82. Che, Z., Purushotham, S., Khemani, R., & Liu, Y. (2015). Distilling knowledge from deep networks with applications to healthcare domain. arXiv preprint arXiv:1512.03542.
83. Jagannatha, A. N., & Yu, H. (2016, June). Bidirectional RNN for medical event detection in electronic health records. In Proceedings of the conference. Association for Computational Linguistics. North American Chapter. Meeting (Vol. 2016, p. 473). NIH Public Access.
84. Bhatt, U., Davis, B., & Moura, J. M. (2019). Diagnostic Model Explanations: A Medical Narrative. In AAAI Spring Symposium: Interpretable AI for Well-being.
85. Kinjo, Y., Sakuma, Y., Kobayashi, T., Sugimoto, C., & Kohno, R. (2019, May). Patient Stress Estimation for Using Deep Learning with RRI Data Sensed by WBAN. In 2019 13th International Symposium on Medical Information and Communication Technology (ISMICT) (pp. 1-4). IEEE.
86. Hu, Y., Chen, F., Cai, Y., & Yuan, Y. A Random Under-sampled Deep Architecture with Medical Event Embedding: Highly Imbalanced Rare Disease Classification with EHR Data. Network, 20(21), 22.
87. Zhao, L., Chen, J., Chen, F., Wang, W., Lu, C. T., & Ramakrishnan, N. (2015, November). Simnest: Social media nested epidemic simulation via online semi-supervised deep learning. In 2015 IEEE International Conference on Data Mining (pp. 639-648). IEEE.
88. Banerjee, I., Li, K., Seneviratne, M., Ferrari, M., Seto, T., Brooks, J. D., ... & Hernandez-Boussard, T. (2019). Weakly supervised natural language processing for assessing patient-centered outcome following prostate cancer treatment. JAMIA open, 2(1), 150-159.
89. Beaulieu-Jones, B. K., Orzechowski, P., & Moore, J. H. (2018, January). Mapping patient trajectories using longitudinal extraction and deep learning in the MIMIC-III Critical Care Database. In PSB (pp. 123-132).
90. World Health Organization. (2000). World Health Organization Collaborating Centre for Drug Statistics Methodology: Guidelines for ATC Classification and DDD Assignment. Oslo, Norway: WHO.

91. Fu, R., Zhang, Z., & Li, L. (2016, November). Using LSTM and GRU neural network methods for traffic flow prediction. In 2016 31st Youth Academic Annual Conference of Chinese Association of Automation (YAC) (pp. 324-328). IEEE.
92. Bradley, A. P. (1997). The use of the area under the ROC curve in the evaluation of machine learning algorithms. Pattern recognition, 30(7), 1145-1159.
93. Google, "Data preprocessing for machine learning." [Online]. Available: https://cloud.google.com/solutions/machinelearning/data-preprocessing-for-ml-with-tf-transform-pt1. [Accessed: 22-Feb-2020]
94. Brownlee, J. (2017). Deep Learning for Natural Language Processing: Develop Deep Learning Models for your Natural Language Problems. Machine Learning Mastery.
95. Manning, C., & Schutze, H. (1999). Foundations of statistical natural language processing. MIT press.
96. Wang, Y., Wang, L., Rastegar-Mojarad, M., Moon, S., Shen, F., Afzal, N., ... & Liu, H. (2018). Clinical information extraction applications: a literature review. Journal of biomedical informatics, 77, 34-49.
97. ATC codes, "World Health Organization Collaborating Centre for Drug Statistics Methodology." [Online]. Available: https://www.whocc.no/atc_ddd_index/ .
98. Sethy, A., & Ramabhadran, B. (2008). Bag-of-word normalized n-gram models. In Ninth Annual Conference of the International Speech Communication Association.
99. Di Nunzio, G. M., & Vezzani, F. (2018). A Linguistic Failure Analysis of Classification of Medical Publications: A Study on Stemming vs Lemmatization. In CLiC-it.
100. Wang, Y., Wang, L., Rastegar-Mojarad, M., Moon, S., Shen, F., Afzal, N., ... & Liu, H. (2018). Clinical information extraction applications: a literature review. Journal of biomedical informatics, 77, 34-49.
101. Savova, G. K., Masanz, J. J., Ogren, P. V., Zheng, J., Sohn, S., Kipper-Schuler, K. C., & Chute, C. G. (2010). Mayo clinical Text Analysis and Knowledge Extraction System (cTAKES): architecture, component evaluation and applications. Journal of the American Medical Informatics Association, 17(5), 507-513.
102. Aronson, A. R., & Lang, F. M. (2010). An overview of MetaMap: historical perspective and recent advances. Journal of the American Medical Informatics Association, 17(3), 229-236.
103. Ferrucci, D., & Lally, A. (2004). UIMA: an architectural approach to unstructured information processing in the corporate research environment. Natural Language Engineering, 1-26.
104. Baldridge, J. (2005). The opennlp project. URL: http://opennlp.apache.org/index.html,(accessed 2 February 2012), 1.
105. Pennington, J., Socher, R. and Manning, C.D., (2014). Glove: Global vectors for word representation. In Proceedings of the 2014 conference on empirical methods in natural language processing (EMNLP) (pp. 1532-1543).
106. Bodenreider, O. (2004). The unified medical language system (UMLS): integrating biomedical terminology. Nucleic acids research, 32(suppl_1), D267-D270.
107. Mikolov, T., Sutskever, I., Chen, K., Corrado, G. S., & Dean, J. (2013). Distributed representations of words and phrases and their compositionality. In Advances in neural information processing systems (pp. 3111-3119).
108. Graves, A., & Schmidhuber, J. (2005). Framewise phoneme classification with bidirectional LSTM and other neural network architectures. Neural networks, 18(5-6), 602-610.
109. Lev, G., Klein, B., & Wolf, L. (2015, June). In defense of word embedding for generic text representation. In International Conference on Applications of Natural Language to Information Systems (pp. 35-50). Springer, Cham.
110. Fawcett, T. (2006). An introduction to ROC analysis. Pattern recognition letters, 27(8), 861-874.
111. Dernoncourt, F., Lee, J. Y., Uzuner, O., & Szolovits, P. (2017). De-identification of patient notes with recurrent neural networks. Journal of the American Medical Informatics Association, 24(3), 596-606.
112. Mikolov, T., Chen, K., Corrado, G., & Dean, J. (2013). Efficient estimation of word representations in vector space. arXiv preprint arXiv:1301.3781.

113. Ratner, A. J., De Sa, C. M., Wu, S., Selsam, D., & Ré, C. (2016). Data programming: Creating large training sets, quickly. In Advances in neural information processing systems (pp. 3567-3575).
114. Wang, Y., Sohn, S., Liu, S., Shen, F., Wang, L., Atkinson, E. J., ... & Liu, H. (2019). A clinical text classification paradigm using weak supervision and deep representation. BMC medical informatics and decision making, 19(1), 1.
115. Fries, J., Wu, S., Ratner, A., & Ré, C. (2017). Swellshark: A generative model for biomedical named entity recognition without labeled data. arXiv preprint arXiv:1704.06360.
116. Hammar, K., Jaradat, S., Dokoohaki, N., & Matskin, M. (2018, December). Deep text mining of instagram data without strong supervision. In 2018 IEEE/WIC/ACM International Conference on Web Intelligence (WI) (pp. 158-165). IEEE.
117. Ratner, A., Bach, S. H., Ehrenberg, H., Fries, J., Wu, S., & Ré, C. (2017, November). Snorkel: Rapid training data creation with weak supervision. In Proceedings of the VLDB Endowment. International Conference on Very Large Data Bases (Vol. 11, No. 3, p. 269). NIH Public Access.
118. Bahdanau, D., Cho, K. and Bengio, Y., (2014). Neural machine translation by jointly learning to align and translate. arXiv preprint arXiv:1409.0473.
119. Leaman, R., & Lu, Z. (2016). TaggerOne: joint named entity recognition and normalization with semi-Markov Models. Bioinformatics, 32(18), 2839-2846.
120. Yang, Z., Yang, D., Dyer, C., He, X., Smola, A., & Hovy, E. (2016, June). Hierarchical attention networks for document classification. In Proceedings of the 2016 conference of the North American chapter of the association for computational linguistics: human language technologies (pp. 1480-1489).
121. Gao, S., Young, M. T., Qiu, J. X., Yoon, H. J., Christian, J. B., Fearn, P. A., ... & Ramanthan, A. (2018). Hierarchical attention networks for information extraction from cancer pathology reports. Journal of the American Medical Informatics Association, 25(3), 321-330.
122. Mullenbach, J., Wiegreffe, S., Duke, J., Sun, J., & Eisenstein, J. (2018). Explainable prediction of medical codes from clinical text. arXiv preprint arXiv:1802.05695.
123. Baumel, T., Nassour-Kassis, J., Cohen, R., Elhadad, M., & Elhadad, N. (2017). Multi-label classification of patient notes a case study on ICD code assignment. arXiv preprint arXiv:1709.09587.
124. Honnibal, M., & Montani, I. (2017). spacy 2: Natural language understanding with bloom embeddings, convolutional neural networks and incremental parsing. To appear, 7(1).
125. Zhang, J., Kowsari, K., Harrison, J. H., Lobo, J. M., & Barnes, L. E. (2018). Patient2vec: A personalized interpretable deep representation of the longitudinal electronic health record. IEEE Access, 6, 65333-65346.
126. Sousa, R. T., Pereira, L. A., Galvao Filho, A. R., & Soares, A. D. S. (2018). MedAttention: A Self-Attentive RNN to Predict Diabetes Complications with Financial Data.
127. Vaswani, A., Shazeer, N., Parmar, N., Uszkoreit, J., Jones, L., Gomez, A. N., ... & Polosukhin, I. (2017). Attention is all you need. In Advances in neural information processing systems (pp. 5998-6008).
128. Devlin, J., Chang, M. W., Lee, K., & Toutanova, K. (2018). Bert: Pre-training of deep bidirectional transformers for language understanding. arXiv preprint arXiv:1810.04805.
129. Wang, A., Singh, A., Michael, J., Hill, F., Levy, O., & Bowman, S. R. (2018). Glue: A multi-task benchmark and analysis platform for natural language understanding. arXiv preprint arXiv:1804.07461..
130. Beltagy, I., Cohan, A., & Lo, K. (2019). Scibert: Pretrained contextualized embeddings for scientific text. arXiv preprint arXiv:1903.10676.
131. Lee, J., Yoon, W., Kim, S., Kim, D., Kim, S., So, C. H., & Kang, J. (2020). BioBERT: a pre-trained biomedical language representation model for biomedical text mining. Bioinformatics, 36(4), 1234-1240..
132. Huang, K., Altosaar, J., & Ranganath, R. (2019). Clinicalbert: Modeling clinical notes and predicting hospital readmission. arXiv preprint arXiv:1904.05342.

Brain Tumor Segmentation and Surveillance with Deep Artificial Neural Networks

Asim Waqas, Dimah Dera, Ghulam Rasool, Nidhal Carla Bouaynaya, and Hassan M. Fathallah-Shaykh

Abstract Brain tumor segmentation refers to the process of pixel-level delineation of brain tumor structures in medical images, such as *Magnetic Resonance Imaging* (MRI). Brain tumor segmentation is required for radiotherapy treatment planning and can diagnosis through surveillance. Automatic segmentation of brain tumors is a challenging problem due to the complex topology of anatomical structures, noise from image acquisition, heterogeneity of signals and spatial/structural variations of tumors. *Machine Learning* (ML) techniques, including *Deep Artificial Neural Networks* (DNNs), have shown significant improvement in classification and segmentation tasks. This chapter provides a comprehensive review of supervised learning models and architectures for image segmentation. A particular emphasis will be placed on *U-Net* and *U-Net* with *Inception* and *dilated Inception* modules for brain tumor segmentation. The performance of the proposed models is evaluated using the multi-modal *BRAin Tumor Segmentation* (BRATS) benchmark dataset. Furthermore, we present a new Bayesian deep learning framework, called *extended Variational Density Propagation* (exVDP), for quantifying uncertainty in the decision of DNNs. In particular, exVDP provides a pixel-level uncertainty map associated with the network's segmentation output. Finally, we present clinical retrospective studies in tumor surveillance using MRI data from patients with glioma and show the advantages accrued from these methods.

Keywords Tumor segmentation · Tumor surveillance · Computer vision · Deep supervised learning · *U-Net* · *Inception* · Uncertainty propagation

A. Waqas (✉) · D. Dera · G. Rasool · N. C. Bouaynaya
Rowan University, Henry M. Rowan College of Engineering, Glassboro, NJ, USA
e-mail: waqasa8@students.rowan.edu; derad6@rowan.edu; rasool@rowan.edu; bouaynaya@rowan.edu

H. M. Fathallah-Shaykh
The University of Alabama at Birmingham, School of Medicine, Birmingham, AL, USA
e-mail: hfshaykh@uabmc.edu

M. Elloumi (ed.), *Deep Learning for Biomedical Data Analysis*,
https://doi.org/10.1007/978-3-030-71676-9_13

1 Introduction

The task of *brain tumor segmentation*, presented in this chapter, is the confluence of multiple techniques usually employed in diverse fields of science such as *Digital Image Processing* (DIP), *Computer Vision* (CV), and *Machine Learning* (ML). ML algorithms, specifically *Deep Artificial Neural Networks* (DNNs), have achieved state-of-the-art accuracy in CV-related tasks, including image segmentation. DNNs are built using large stacks of individual artificial neurons, each of which performs mathematical operations of multiplication, summation, and non-linear operations. One of the key reasons for the success of DNNs is their ability to learn useful features automatically from the data as opposite to manual selection by expert humans [1]. Various architectures of DNNs employed for brain tumor segmentation have been discussed along with a case study of one of those architectures in detail. We discuss a new technique for quantifying uncertainty in the output decision of DNNs. We also present different techniques for tumor surveillance.

The rest of the chapter is organized as follows: In Sect. 2, we touch upon the relevant theoretical background of the techniques involved in solving the tumor segmentation problem. We discuss image segmentation in general and medical image segmentation, particularly, and the concept of surveillance in the medical sphere. Section 3 demonstrates brain tumor segmentation through DNNs. Section 4 presents the particularly suited *Inception* modules in *Deep Learning* (DL) for brain tumor segmentation. Section 5 explains the concept of uncertainty estimation in the decision made by DNNs. Finally in Sect. 6, we discuss tumor surveillance techniques supported by a case study followed by conclusion in Sect. 7.

2 Theoretical Background of the Problem

The task of *brain tumor segmentation* using DNNs inherently involves various tasks. Therefore, it is imperative to imbibe some basic theoretical background about these concepts, which leads up to the task-at-hand.

2.1 Image Segmentation

A picture is worth a thousand words. This is because a picture contains far more information in a few pixels that the human brain can process simultaneously as compared to the numerous words that can express the same amount of information sequentially. Thus, understanding the image and extracting useful information from it bears the central role in the fields of DIP and CV. Classification task, in particular,

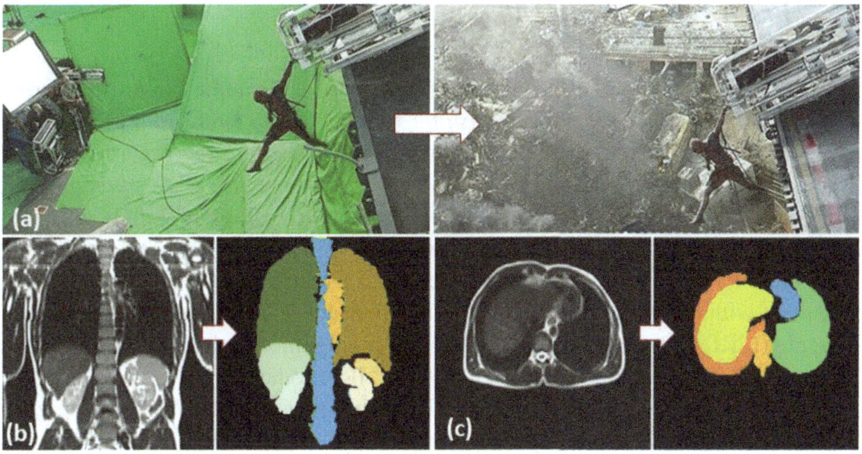

Fig. 1 Use of Image Segmentation in (**a**) *Marvel* movies [9], (**b**), (**c**) Medical imaging: Segmentation of representative organs in thorax and abdomen from CT images [10]

assigns a label or class to an input image. However, image classification does not provide pixel-level information, such as the location of objects in an image, objects' shapes and boundaries, information about which pixel belongs to which object, etc. For this purpose, images are segmented by assigning a specific label to pixels with similar characteristics in an image. *Segmentation* is a technique frequently used in DIP and CV fields for extracting useful information from images [2]. It is the process of partitioning an image into segments (having sets of pixels) representing various objects in the image. The purpose is to modify the representation of an image into a more elaborate format, which is easy to understand anatomically and helpful in extracting meaningful information for analysis. In usual practice, this process is used to locate objects of interest and draw boundaries/shapes conforming to these objects in an image(s). Image segmentation has contributed to many spheres of human life, ranging from the film-making industry to the field of medicine [3]. For example, the green screens used in *Marvel* [4] movies employed segmentation to extract the foreground objects and place them on different backgrounds depicting dangerous real-life scenes, Fig. 1a. An example of medical image segmentation includes the identification of multiple organs in the abdomen and thorax, as shown in Fig. 1b,c. Various techniques are used for classical image segmentation, e.g., *region-based* [5] (*threshold segmentation, regional growth segmentation*), *edge-detection* [6] (*Sobel operator, Laplacian operator*), *clustering-based* [6] (*K-means*), and *weak-supervised learning* [7] methods in CNN. Further details on classical image segmentation techniques are beyond the scope of this chapter, so we will confine ourselves to medical image segmentation, in general, with particular focus on brain tumor [8] segmentation.

2.2 Brain Tumor Segmentation

Brain tumors [8] are masses or growths of abnormal cells in the brain, categorized into *primary* and *secondary* or *metastatic* types. Primary brain tumors originate from brain cells, whereas secondary tumors metastasize into the brain from other organs [11]. The most common type of primary brain tumors are *gliomas* [12], which arise from brain glial cells, and can be of *Low-Grade* (LGG) or *High-Grade* (HGG) sub-types. HGGs are aggressively-growing and infiltrative malignant brain tumors, which usually require surgery or radiotherapy and have poor survival prognosis with the highest mortality rate and prevalence [13]. *Magnetic Resonance Imaging* (MRI) is a crucial diagnostic tool for brain tumor analysis, monitoring and surgery planning. Several complimentary 3D MRI modalities such as T1, *T1 with gadolinium-enhancing Contrast* (T1C), T2-weighted (T2), and *FLuid-Attenuated Inversion Recovery* (FLAIR) are acquired to emphasize different tissue properties and areas of tumor spread. For example, in T1C MRI modality, the contrast agent (e.g., gadolinium) emphasizes hyper-active tumor sub-regions.

Brain tumor segmentation [14] is the technique of labeling tumor pixels in an MRI to distinguish them from normal brain tissues and artifacts. These MRI scans are the representation of the internal structure or function of the brain's anatomic region in the form of an array of picture elements called pixels or voxels. It is a discrete representation resulting from a sampling/reconstruction process that maps numerical values to positions of the space. The number of pixels used to describe the field-of-view of an acquisition modality is an expression of the detail with which the anatomy or function can be depicted. Depiction of the numerical value of pixel depends on the imaging modality, acquisition protocol, reconstruction, and post-processing. MRI scans come in various file formats and standards but the six commonly used are: *Analyze* [15], *Neuroimaging Informatics Technology Initiative* (NIfTI) [16], *Minc* [17], PAR/REC format used by Philips MRI scanners [18], *Nearly Raw Raster Data* (NRRD) [19], and *Digital imaging and communications in medicine* (Dicom) [20]. A comparison of characteristics of these formats is shown in Fig. 2. In clinical practice, the process of separating the tumor pixels from normal brain tissues provides useful information about existence, growth, diagnosis, surveillance and treatment planning. The process of manual delineation requires anatomical knowledge by specially trained persons, whereas such manual practices are expensive, time-consuming, and are prone to errors due to human limitations. The process of automated segmentation of brain tumors from 3D images facilitates in overcoming these shortcomings [21].

2.3 Tumor Surveillance

National Cancer Institute (NCI), part of the U.S. *National Institutes of Health* (NIH), defines tumor surveillance as *closely watching a patient's condition but*

	Dicom	Nifti	Minc	Analyze	NRRD	PAR/REC format
Stands for	Digital Imaging and Communications in Medicine	Neuroimaging Informatics Technology Initiative	Medical Imaging NetCDF Toolkit	Developed by Biomedical Imaging Resource (BIR) at Mayo Clinic	Nearly Raw Raster Data	Philips MRI scanner format
File extension	.dcm	.nii	.mnc	.img and .hdr	.nrrd	.PAR
Data type	Signed, unsigned integer (8, 16-bit, 32-bit only allowed for radiotherapy dose)	Signed, unsigned integer (8-64 bit),float (32-128 bit), complex(64 256 bit)	Signed, unsigned integer (8-32 bit),float (32, 64 bit), complex(32, 64 bit)	Unsigned integer (8-bit), signed integer(16, 32 bit), float(32-64 bit), complex(64 bit)	-	8 or 16 bit unsigned integers
Header Length	Variable length binary format	Fixed-length 352 byte binary format (348 byte in case of data stored as .img and .hdr)	Extensible binary format	Fixed-length: 348 byte binary format	Extensible with attached and detached	8 or 16 bit unsigned integers

Fig. 2 Summary of medical imaging file formats

not treating it unless there are changes in test results. Surveillance is also used to find early signs that a disease has come back. It may also be used for a person who has an increased risk of a disease, such as cancer [22]. The process of surveillance regularly involves (scheduled) medical tests and examinations to track the growth of the tumor. The term has also been used in the realm of public health, wherein collective information of a disease, such as cancer, is recorded in a group of people belonging to a specific category (ethnic, age, gender, regional, etc.) Active surveillance is extremely beneficial, especially for patients with low-risk cancer diagnoses. Apart from the routine biopsy, active surveillance is almost surgically noninvasive. It helps in the delay of more invasive treatments such as surgical removal of a tumor, sparing the patients from burdensome side effects and potential complications for as long as possible. Moreover, by deferring the invasive treatment to the point when the disease worsens, active surveillance enables cancer patients to maintain a quality of life. A case in point is the exceptionally beneficial surveillance of low-risk prostate cancer in men. The reason for this success is that almost 50% of prostate cancer diagnoses are categorized as low-risk with less possibility of spread, and few cases may never require advanced forms of treatment. Such cases do not immediately need to be aggressively treated in the absence of worsened disease while the specialists keep records of the tumor's growth over time. Surveillance allows the specialists to monitor the disease right from the onset, thus leveraging them the liberty to analyze the effects and progress of disease and determine the next course of action [23]. A study by Harvard researchers found that the aggressiveness of prostate cancer at diagnosis appears to remain stable over time for most men. If patients had chosen active surveillance, then this could make them feel more confident in their decision about treatment [24]. Early detection of the tumor through surveillance could assist both the patients and the specialists in taking more considered decisions about treatment.

2.4 Deep Learning Segmentation Task

Computer Vision (CV) is the field of computer science that aims to replicate (to some extent) the complex nature of the human vision system into modern-day computers and machines. It endeavors to enable machines to visually gaining a high-level understanding of objects in the imagery in its quest to mimic the humans. A chronological insight in some of the most active topics of research in computer vision can be found in [25]. Most attractive topics in today's CV tasks include object classification (i.e., categorizing objects in an image), localization (i.e., spatially locating objects in an image), detection/ recognition and segmentation (i.e., identifying the category of each pixel in an image) as shown in Fig. 3.

The commonly used classical image segmentation techniques, described in Sect. 2.1 above have been replaced by their more efficient counterparts in ML because of the former's inherently rigid algorithms and the need for human intervention. Image segmentation is the fundamental component of DL, which is part of a broader family of ML. Compared with other DL algorithms, CNNs have proven to be the more efficient selection for segmentation tasks from imagery. Image segmentation using CNN involves feeding the CNN with the desired image as an input and getting the labels of each pixel, i.e., labeled image as an output. Instead of processing the complete image at once, CNN deals with a fraction of image

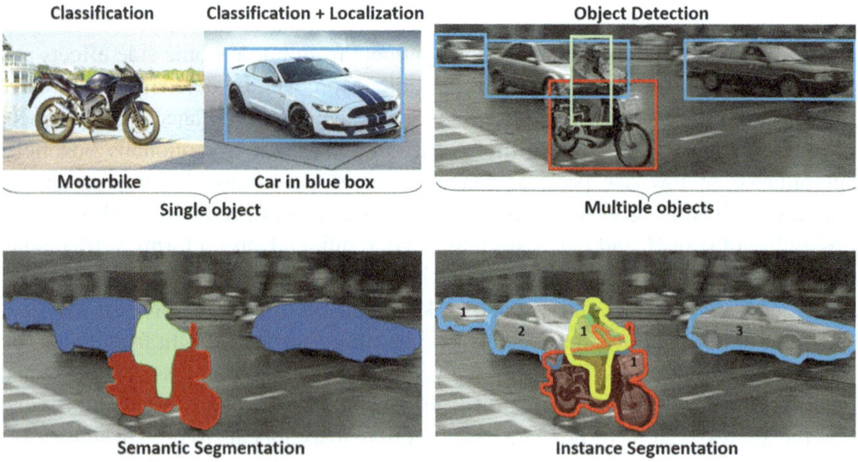

Fig. 3 Important computer vision tasks

conforming to its *filter*, convolving and ultimately mapping over the entire image. To learn more on CNNs, a concise explanation supported by a visualization can be referred to at [26, 27].

2.5 Motivation

A large population suffers from fatalities caused by cancer, and brain tumors are one of the leading causes of death for cancer patients, especially children and adolescents. Brain tumors account for one in every 100 cancers diagnosed annually in the United States [28]. In 2019, the *American Cancer Society* reported that 23,820 new brain cancer cases were discovered in the United States [29]. One of the most frequent primary brain tumors is glioma [30], which affects the glial cells of the brain as well as the surrounding tissues. The HGG or *GlioBlastoMa* (GBM) is the most common and aggressive type with a median survival rate of one to two years [31]. Although neurosurgery may be the only therapy for many brain tumors [32], other treatment methods such as radiotherapy and chemotherapy are also used to destroy the tumor cells that cannot be physically resected or to slow their growth. Before the treatment through chemotherapy, radiotherapy, or brain surgeries, there is a need for medical practitioners to confirm the boundaries and regions of the brain tumor and determine where exactly it is located and the exact affected area. Moreover, all of these invasive treatments face challenging practice conditions because of the structure and nature of the brain. These conditions make it very difficult to distinguish the tumor tissue from normal brain parenchyma for neurosurgeons based on visual inspection alone [33].

Moreover, such manual-visual practices usually involve a group of clinical experts to define the location and the type of the tumor accurately. This lesion localization process is laborious, and its quality depends on the physicians' experience, skills, slice-by-slice decisions, and the results may still not be universally accepted among the clinicians. Treatment protocols for high-grade pediatric brain tumors and general low-grade tumors recommend regular follow-up imaging for up to 10 years. For these longitudinal studies, a comparison of the current MRI with all prior imaging takes a very long time, which is practically infeasible. Automated computer-based segmentation methods present excellent solution to the challenges mentioned above by saving physician's time and providing reliable and accurate results while reducing the diagnosis efforts of surgeons on a single patient [34]. Brain tumor segmentation is motivated by assessing tumor growth, treatment responses, computer-based surgery, treatment of radiation therapy, and developing tumor growth models. Thus, the computer-assisted diagnostic system is meaningful in medical treatments to reduce the workload of doctors and to get accurate results.

2.6 Challenges

Segmentation of gliomas in pre-operative MRI scans—conventionally done by expert board-certified neuro-radiologists and other physicians—provides quantitative morphological characterization and measurement of glioma sub-regions. The quantitative analysis task is challenging due to the high variance in appearance and shape, ambiguous boundaries and imaging artifacts. Although computer-aided techniques have the advantage of fast speed, consistency in accuracy and immunity to fatigue [35], automatic segmentation of brain tumors in multi-modal MRI scans is still one of the most difficult tasks in medical image analysis and applications. Automatic segmentation involves dealing with a complicated and massive amount of data, artifacts due to patient's motion, limited acquisition time, and soft tissue boundaries that are usually not well defined. Moreover, many classes of tumors have a variety of irregular shapes, sizes, and image intensities, especially the surrounding structures of tumors. Numerous attempts have been made in developing ML algorithms for segmenting normal and abnormal brain tissues using MRI images, which will be covered in detail in Sect. 3. However, feature selection to enable automation is challenging and requires a combination of computer engineering and medical expertise. Thus, developing fully-automated brain tumor segmentation remains a challenging task, and a large part of the research community is currently involved in overcoming this challenge in bringing state-of-the-art ideas in this field into reality.

3 Brain Tumor Segmentation Using Deep Artificial Neural Networks

The task of segmenting brain tumor in MRI images has been adopted in DNNs from the image segmentation task in CV. This section focuses on these techniques imported in DL from CV and also gives an overview of the various DL architectures employed on brain MRI datasets.

3.1 Image Segmentation in Computer Vision Realm

Image segmentation is the task of finding groups/clusters of pixels that belong to the same category. It divides an input image into *segments* to simplify image analysis. These segments represent objects or parts of objects and comprise sets of pixels belonging to each part. Practically, the segmentation sorts pixels into larger components, eliminating the need to consider individual pixels as units of observation. In statistics, this problem is known as *cluster analysis* and is a widely studied area with many different algorithms [36–39]. In CV, image segmentation

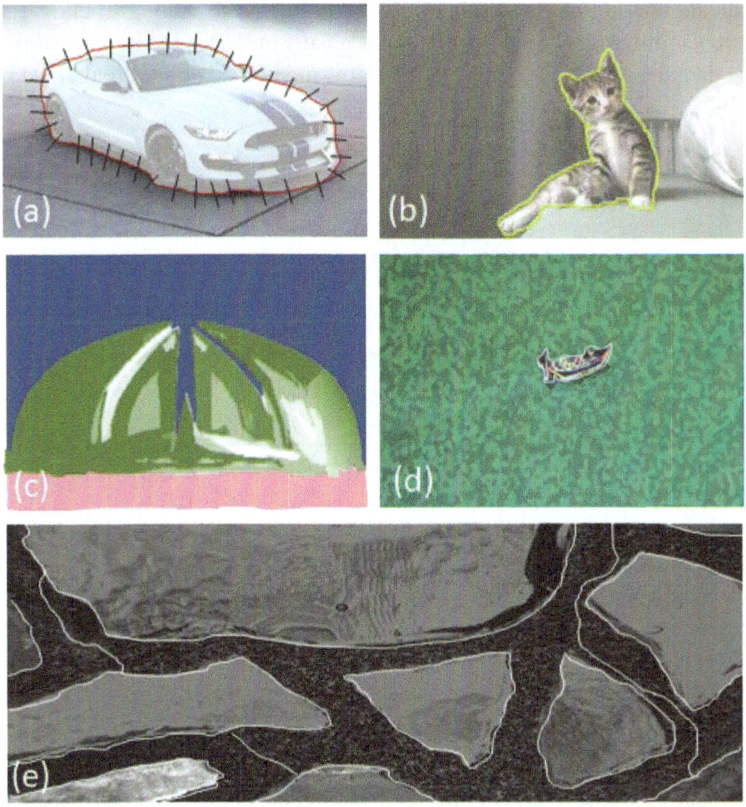

Fig. 4 Few image segmentation techniques in computer vision: (**a**) *active contours* [46]; (**b**) *level sets* [47]; (**c**) *region splitting* and *graph-based merging* [48]; (**d**) *mean shift* (mode finding) [49]; (**e**) *normalized cuts* (splitting based on pixel similarity metrics) [52]

is one of the oldest and most extensively used problems dates back to the 1970s [40–45]. Some of the most extensively known techniques developed for image segmentation are: (a) active contours [46]; (b) level sets [47]; (c) region splitting and graph-based merging [48]; (d) mean shift (mode finding) [49]; (e) normalized cuts (splitting based on pixel similarity metrics), as depicted in Fig. 4. The segmentation process itself has two forms, namely; *semantic*, and *instance* segmentation. The former classifies all the pixels of an image into meaningful or *semantically interpretable* classes of objects and is usually referred to as *dense prediction*. The latter identifies each instance of each object in an image and differs from semantic segmentation in that it does not categorize every pixel. For example, in Fig. 3, semantic segmentation classified all cars, while instance segmentation identifies each one individually. Various metrics are used for performance evaluation of image segmentation including pixel accuracy P_{acc}, mean accuracy M_{acc}, *Intersection-over-Union* (IoU) M_{IU}, frequency weighted IoU F_{IU} and *Dice coefficient* [50]. Let

n_{ij} indicate the number of pixels of class i predicted to belong to class j, where there are n_{cl} different classes, and let $\mathbf{t}_i = \sum_j n_{ij}$ indicates the number of pixels of class i, then the performance evaluation terms mentioned above are defined by:

$$P_{acc} = \frac{\sum_i n_{ii}}{\sum_i t_i}, \tag{1}$$

$$M_{acc} = \frac{1}{n_{cl}} \sum_i \frac{n_{ii}}{t_i}, \tag{2}$$

$$M_{IU} = \frac{1}{n_{cl}} \sum_i \frac{n_{ii}}{t_i + \sum_j n_{ji} - n_{ii}}, \tag{3}$$

$$F_{IU} = \frac{1}{\sum_k t_k} \sum_i \frac{t_i n_{ii}}{t_i + \sum_j n_{ji} - n_{ii}}, \tag{4}$$

Dice Similarity Coefficient (DSC) has also been extensively used for evaluating segmentation algorithms in medical imaging applications [51]. DSC between a predicted binary image P and ground truth binary image G, both of size N x M is given by:

$$DSC(P, G) = 2\frac{\sum_{i=0}^{N-1} \sum_{j=0}^{M-1} P_{ij} G_{ij}}{\sum_{i=0}^{N-1} \sum_{j=0}^{M-1} P_{ij} + \sum_{i=0}^{N-1} \sum_{j=0}^{M-1} G_{ij}}, \tag{5}$$

where i and j represent pixel indices for the height N and width M. The range of DSC is [0, 1], and a higher value of DSC corresponds to a better match between the predicted image P and the ground truth image G.

The application of image segmentation techniques in the medical imaging field opened a new frontier of knowledge with advances in the areas of diabetic retinopathy detection, skin cancer classification, brain tumor segmentation and many more. In this chapter, we will restrict ourselves to brain tumor segmentation only and look at the various techniques employed in *Artificial Neural Networks* (ANNs) for brain tumor segmentation.

3.2 Deep Artificial Neural Networks and Image Segmentation

DNNs have achieved significant milestones in the CV field. DNNs have multiple layers between the input and output layers. The basic element of ANN, i.e., *artificial neuron*, has multiple inputs that are weighted and summed up, followed by a *transfer function* or *activation function*. Then the neuron outputs a scalar value. An example of ANN is illustrated in Fig. 5 [53]. Inspired by biological processes, ANNs use shared-weight architecture where the connectivity pattern between neurons mimics

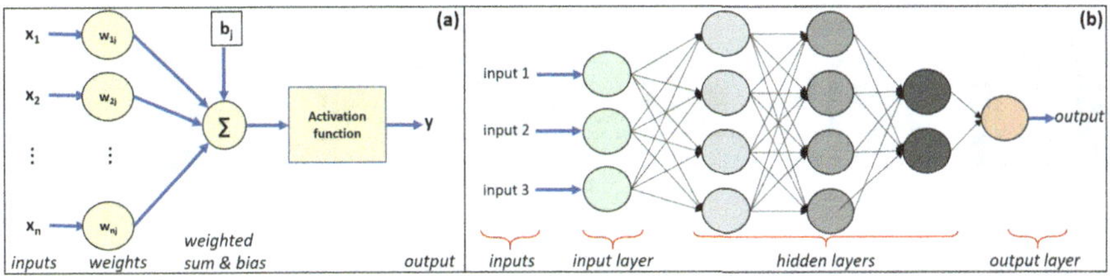

Fig. 5 (**a**) artificial neuron model, (**b**) ANN model

the organization of the brain visual cortex [54, 55]. ANN or DNN imitate the concept of receptive fields where individual cortical neurons respond to stimuli only in a restricted field of view. Because of their shared-weight architecture and translation invariance characteristics, ANNs are shift or space-invariant. Due to the linear operations followed by the non-linear activations, ANNs are capable of extracting higher-level representative features [56] and can compute any function [57].

3.3 DL-Based Image Segmentation Architectures

Most prominent DL architectures used by the CV community include *Convolutional Neural Networks* (CNNs), *Recurrent Neural Networks* (RNNs) and *Long Short Term Memory* (LSTM), *encoder-decoders*, and *Generative Adversarial Networks* (GANs) [58–61]. With our focus on CNNs, let us discuss the three important DL-based image segmentation architectures.

3.3.1 Convolutional Neural Networks

Waibel et al. introduced *Convolutional Neural Networks* (CNN) that had weights shared among temporal receptive fields, and it had back-propagation training for phoneme recognition [62]. LeCun et al. developed a CNN architecture for document recognition as shown in Fig. 6a [58]. The three basic components/ layers of a CNN are: (1) *convolutional layer*, having a kernel (or filter) of weights convolved with the input image to extract features; (2) *nonlinear layer* having an element-wise activation function applied to feature maps; and (3) *pooling layer*, is a down-sampling operation that reduces spatial resolution and replaces appropriate neighborhood of a feature map with some statistical information (mean, max, etc.) [63]. Deep CNNs have performed extremely well on a wide variety of medical imaging tasks, including diabetic retinopathy detection [64], skin cancer classification [65], and brain tumor segmentation [66–70]. Some of the most well-known CNN architectures include *AlexNet* [71], *VGGNet* [72], *ResNet* [73], *GoogLeNet* [74] which use *Inception* modules architecture (explained in detail in Sect. 4), *MobileNet* [75], and *DenseNet* [76].

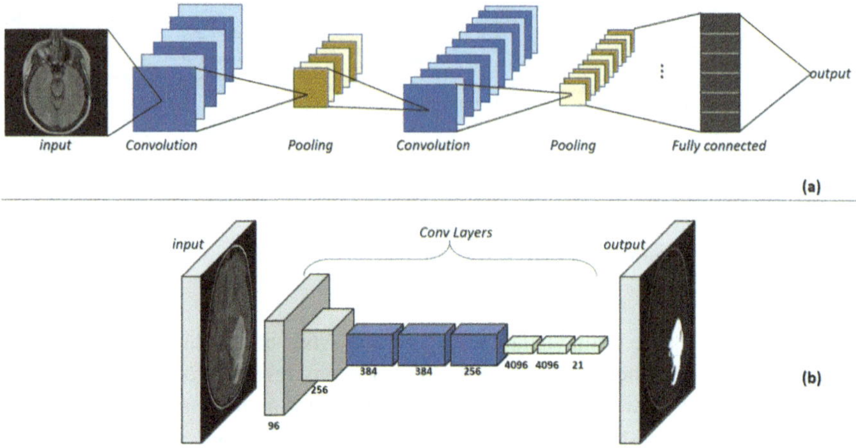

Fig. 6 Basic architecture of (**a**) CNN, (**b**) FCN

3.3.2 Fully Convolutional Networks

Fully Convolutional Networks (FCNs), proposed by Long et al., use convolutional layers to process varying input sizes [50]. It was one of the first DL models for semantic segmentation. As shown in Fig. 6b, the final output layer of FCN has a large receptive field and corresponds to the height and width of the image, while the number of channels corresponds to the number of classes. The convolutional layers classify every pixel to determine the context of the image, including the location of objects. FCNs have been applied to a variety of segmentation problems, such as brain tumor segmentation [68], instance-aware semantic segmentation [77], skin lesion segmentation [78], and iris segmentation [79].

3.3.3 Encoder-Decoder Based Models

Encoder-decoder models are inspired by the FCNs, and the most well-known architecture of these models are *U-Net* and *V-Net* [80, 81]. *U-Net* was proposed for segmenting biological microscopy images, and it used the data augmentation technique to learn from the available annotated images more effectively. The *U-Net* architecture consists of two parts; a *contracting* or *down-sampling* path to capture the context, and a symmetric *expanding* or *up-sampling* path for localization of the captured context. The contracting path has FCN-like architecture that extracts features with 3×3 convolutions while increasing the number of feature maps and reducing their dimensions. Contrarily, the expanding path carries out deconvolutions by reducing the number of feature maps while increasing their dimensions. The feature maps from the contracting path are concatenated to the expanding path to maintain the integrity of pattern information. Finally, a segmentation map is generated from feature maps by 1×1 convolution operation that categorizes each pixel of the input image. *U-Net* was trained on 30 transmitted light microscopy images, and it won the *International Symposium on Biomedical Imaging* (ISBI) cell tracking challenge in 2015 by a large margin. *V-Net* [81] is another well-known FCN-based model proposed for 3D medical image segmentation. It introduced a new objective function based on the Dice coefficient, which enabled the model to deal with strong class imbalance between the number of voxels in the foreground and the background. *V-Net* was trained end-to-end on MRI volumes depicting prostate, and it learned to predict segmentation for the whole volume at once.

3.3.4 Other Deep Learning Models Used in Image Segmentation

In addition to the models described in previous sections, there are many families of DL architectures that are very popular for medical image segmentation. For example, convolutional graphical models (incorporating concepts of *Conditional Random Fields* (CRFs) and *Markov Random Field* (MRFs)), *Multi-scale pyramid network* models (*Feature Pyramid Network* (FPN)) [82], *Pyramid Scene Parsing Network* (PSPN) [83], *Regional CNN* (R-CNN) like *Fast* R-CNN, *Faster* R-CNN, and *Mask-RCNN*, dilated or atrous convolution (*DeepLab Family* [84]), RNN-based models (ReNet [85], *ReSeg* [86]), *Data-Associated* RNNs (DA-RNNs) [87]), and *attention-based* models (OCNet [88], *Expectation-Maximization Attention Network* (EMANet) [89], *Criss-Cross attention Network* (CCNet) [90]). Minaee et al. has presented an elaborate review reference of all of these models [63].

3.4 *Brain Tumor Segmentation Task Challenge*

In this section, we discuss the brain tumor segmentation task in the realm of DL. Internationally held challenges on medical imaging analysis have become the standard for validation of the proposed methods. *Brain Tumor Segmentation* (BraTS) Challenge [91] is one such challenge that is held in conjunction with *Medical Imaging Computing and Computer-Assisted Intervention* (MICCAI) conference [92]. The first challenge workshop was held in 2012, followed by yearly benchmarks held with MICCAI conferences. BraTS challenge evaluates state-of-the-art segmentation methods of brain tumors in MRI scans. It has a publicly available dataset (with accompanying expert delineations), which is used for benchmarking the submitted contenders for segmenting *multi-institutional pre-operative* MRI (mpMRI) scans having intrinsically heterogeneous (in appearance, shape, and histology) brain tumors, namely *gliomas*. In addition, this challenge also encompasses the survival prediction of the patient and evaluates the algorithmic uncertainty in tumor segmentation. The challenge evaluates segmentation of tumor

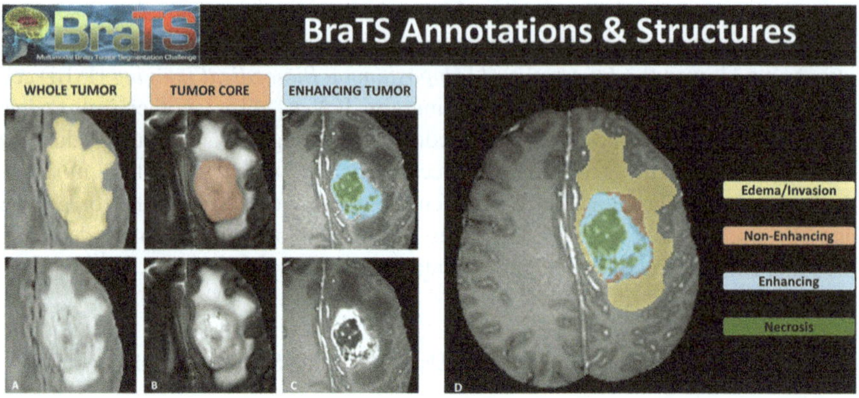

Fig. 7 Image patches with annotated tumor (glioma) sub-regions. (**a**) Whole tumor (yellow) visible in T2-FLAIR, (**b**) Tumor core (orange) visible in T2, (**c**) Enhancing tumor (blue) surrounding the cystic/necrotic core (green) visible in T1c, (**d**) Combined segmentations [94]

sub-regions of *Enhancing Tumor* (ET), *Tumor Core* (TC), and *Whole Tumor* (WT) as shown in Fig. 7. ET are regions of hyper-intensity in T1C when compared to T1, and TC is the bulk of the tumor, that is typically resected. The TC involves ET and the necrotic (fluid-filled) and the non-enhancing (solid) parts of the tumor. WT is the complete extent of the disease, as it is comprised of the TC and the peritumoral *EDema* (ED), depicted by FLAIR. The dataset for the 2018 challenge consisted of a total of 542 patients, with 285 for training, 66 for validation, and 191 for testing scans having 210 *High-Grade Glioma* (HGG) and 75 *Low-Grade Glioma* (LGG) patients with annotations approved from experienced neuro-radiologists through a hierarchical majority vote. The data consists of clinically-acquired 3T multi-contrast MR scans from around 19 institutions, with ground truth labels by expert board-certified neuro-radiologists in NIfTI files (.nii.gz). *Dice coefficient* and *Hausdorff distance* (95%) have been used as evaluation schemes. Apart from these, *Sensitivity* and *Specificity* are also used as metrics. An assessment of state-of-the-art ML methods used for brain tumor segmentation under the BraTS challenge from the period 2012–2018 has been compiled by Bakas et al. [93].

4 Inception Modules in Brain Tumor Segmentation

After assimilating the brain tumor segmentation problem using DNNs, let us now look at one of the architectures that has appreciable accuracy in solving this problem. This architecture is based on *U-Net* with *Inception* modules.

4.1 Brain Tumor Segmentation Using Inception and Dilated Inception modules

Cahall et al. proposed an image segmentation framework for tumor delineation that benefits from two state-of-the-art ML architectures in CV: *Inception* modules and *U-Net* [70, 74, 80]. This new framework includes two learning regimes, i.e., learning to segment intra-tumoral structures (necrotic and non-enhancing tumor core, peritumoral edema, and enhancing tumor) or learning to segment glioma sub-regions (WT, TC, ET). Both learning regimes are described in Sect. 3.4 above. These learning regimes were incorporated into a modified loss function based on the DSC described in Sect. 3.1 Eq. 5 above.

U-Net was originally developed for cell tracking. However, it has been applied recently to other medical segmentation tasks, such as, brain vessel segmentation [95], brain tumor segmentation [96], and retinal segmentation [97]. To tackle different medical imaging segmentation problems, variations and extensions of *U-Net* have also been proposed, such as 3D *U-Net* [98, 99], H-DenseUNet [100], RIC-UNet [101], and Bayesian *U-Net* [102]. Cahall et al. used a cascade learning approach in which three different models were used first to learn the WT, then TC, and finally, ET resulting in a proposed end-to-end implementation for all tumor subtypes [70].

4.2 BraTS Dataset and Pre-processing

BraTS 2018 dataset, described in Sect. 3.4, has been most frequently used, as in these experiments [51, 93, 94, 103]. The dataset contains four sequences for each patient's MRI (T1, T1c, T2, and FLAIR) images. It also contains ground truth in the form of pixel-level manual segmentation markings for three intratumoral structures: necrotic and non-enhancing tumor core (labeled as 1), peritumoral edema (labeled as 2), and enhancing tumor (labeled as 4). The glioma sub-regions have been defined as WT having all three intratumoral structures (labeled as $(1 \cup 2 \cup 4)$), TC containing all except peritumoral edema (labeled as $(1 \cup 4)$), and ET (labeled as 4). Different sequences provide complementary information for identifying the intratumoral structures:

- FLAIR highlights the peritumoral edema.
- T1c distinguishes the ET.
- T2 highlights the necrotic and non-enhancing tumor core.

BraTS images have been pre-processed for skull-stripping, re-sampled to an isotropic 1 mm^3 resolution, and co-registered all four modalities of each patient. Cahall et al. [70] performed additional pre-processing in the following order:

1. Discard excess background pixels from images by obtaining the bounding box of the brain and extracting the selected portion, effectively zooming on the brain.
2. Re-size the cropped image to 128×128 pixels.

3. Drop the images having no tumor regions in the ground truth segmentation.
4. Apply intensity windowing function to each image such that the lowest 1% and the highest 99% pixels were mapped to 0 and 255, respectively.
5. Normalize images by subtracting the mean and dividing by the standard deviation of the dataset.

4.3 Deep Artificial Neural Network Architectures

In medical imaging, semantic segmentation's accuracy depends on the ability to extract the local structural as well as global contextual information from MRI scans while training the model. For this reason, many multi-path architectures in the context of medical imaging have been proposed, and all of them extract the structural and contextual information from input data at multiple scales [98, 104, 105]. This features extraction-aggregation concept at various scales was also done in *Inception* modules [74]. However, the feature extraction mechanism in the *Inception* module is different from the multi-path architectures. The *Inception* module applies filters of various sizes at each layer and concatenates resulting feature maps [74]. Cahall et al. [70] proposed a modified version of *Dilated Residual Inception* (DRI) [106] based on *U-Net* and factorized convolution *Inception* module [74, 80]. DRI's special blocks were inspired from *Inception* module [107] and dilated convolution [108]. DRI has fewer parameters than the original *Inception* module and employs residual connections to alleviate the vanishing gradients problem at a faster convergence rate [73]. *MultiResUNet* combined a *U-Net* with residual *Inception* modules for multi-scale feature extraction, applying the architecture to several multimodal medical imaging datasets [109]. Integration of *Inception* modules with *U-Net* has been evaluated for left atrial segmentation [110], liver and tumor segmentation [111], and brain tumor segmentation [112].

4.3.1 *Inception* Module

The convolutional layer in the proposed *Inception* module [70] in the original *U-Net* was replaced with an *Inception* module having multiple sets of 3×3 convolutions, 1×1 convolutions, 3×3 max pooling, and cascaded 3×3 convolutions as depicted in Fig. 8b. At each layer on the contracting path, the height and width of the feature maps are halved, and the depth is doubled until reaching the bottleneck, i.e., the center of the *U*. On the corresponding expanding path at each layer, the height and width of feature maps are doubled, and the depth is halved until having the segmentation mask as the output. As with *U-Net*, feature maps generated on the contracting path are concatenated to the corresponding expanding path. The authors employed a *Rectified Linear Unit* (ReLU) as the activation function, with batch normalization [113] in each *Inception* module. The architecture setting receives an input image of size $N \times M \times D$ and outputs an $N \times M \times K$ tensor where

$N = M = 128$ pixels, $D = 4$ represents the four MRI modalities (T1, T1c, T2, FLAIR), and $K = 3$ represents the segmentation classes (intra-tumoral structures or glioma sub-regions). The output image of K slices is a binary image representing the predicted segments for the ith class ($0 \leq i \leq K - 1$). Pixel-wise activation functions (sigmoid for *glioma* and *softmax* for intra-tumoral structures ([114])) are used to generate the output binary images.

4.3.2 Dilated Inception *U-Net*

Another useful architecture, called *Dilated Inception U-Net* (DIU-Net), integrates dilated or astrous convolutions [84] and *Inception* modules in the *U-Net* architecture [115] as shown in Fig. 8a. Here, each *dilated Inception* module consists of three 1×1 convolutional operations, followed by one l-dilated convolutional filter (with $l = 1, 2, 3$), as illustrated in Fig. 8c. The 1×1 convolutional filters perform dimensionality reduction, while three l-dilated convolutional filters each of size 3×3 implement atrous convolutions. In dilated convolutions, an image I of size $m \times n$ and a discrete convolutional filter w of size $k \times k$ are convolved by:

$$(I * w)(p) = \sum_s I[p + s]\, w[s]. \tag{6}$$

Simple convolution operation of Eq. 6 can be generalized to l-dilated convolution ($*_l$) as [108]:

$$(I *_l w)(p) = \sum_s I[p + l_s]\, w[s]. \tag{7}$$

For $l = 1$, we get the simple convolutional operation of Eq. 6. For $l > 1$, $l - 1$ zeroes are inserted between each filter element, creating a scaled and sparse filter of size $k_s \times k_s$, where k_s is defined by:

$$k_s = k + (k - 1)(l - 1), \tag{8}$$

$$= l(k - 1) + 1. \tag{9}$$

The scaling s increases the receptive field of the filter by a factor $\frac{k_s}{k}$.

$$\frac{k_s}{k} = \frac{k + (k - 1)(l - 1)}{k}, \tag{10}$$

$$= l + \left(\frac{-l + 1}{k}\right). \tag{11}$$

The receptive field of the filter increases linearly with l, while the number of elements ($k \times k$) remains fixed.

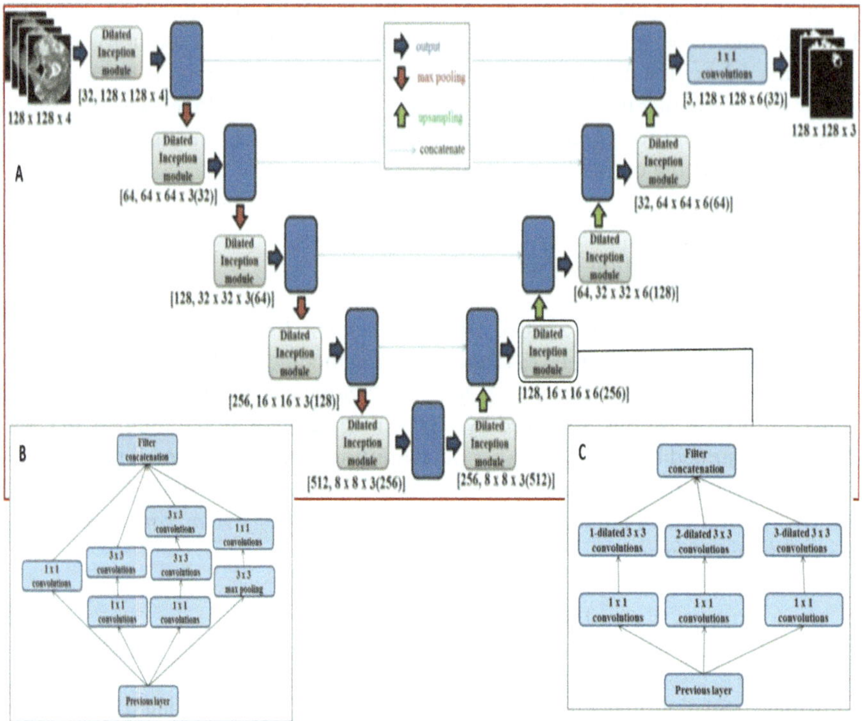

Fig. 8 (**a**) DIU-Net architecture with contracting and expanding path and a bottleneck in the middle. On the contracting path, the multiplication by 3 indicates three *l*-dilated convolutional filters. On expanding path, concatenation of feature maps from contracting path doubles the depth of output feature map, hence the multiplication by 6. (**b**) *Inception* module architecture, (**c**) *Dilated Inception* module with three l-dilated convolutional filters and 1×1 dimensional reduction convolution filters

4.3.3 Modified DSC as Objective/Loss Function

Cahall et al. used a modified version of DSC (Eq. 5) as an objective/ loss function, after incorporating three changes: (1) the sign of DSC was changed to convert it into a minimization problem, (2) a log function was introduced, and (3) a new parameter γ was used to cater for extremely large values of the loss function [70]. From initial experiments, it was empirically observed that $\gamma = 100$ provided the best segmentation performance. Modified DSC as a loss function for a binary class (tumor or not tumor) and multi-class (for K classes) are given in the following two equations:

$$\mathcal{L}_{DSC}(P, G) = -log\left[2\frac{\sum_{i=0}^{N-1}\sum_{j=0}^{M-1} P_{ij}G_{ij} + \gamma}{\sum_{i=0}^{N-1}\sum_{j=0}^{M-1} P_{ij} + \sum_{i=0}^{N-1}\sum_{j=0}^{M-1} G_{ij} + \gamma}\right], \quad (12)$$

$$\mathcal{L}_{DSC}(P, G) = -log\left[\frac{1}{K}\sum_{i=0}^{K-1} DSC(P_i, G_i)\right]. \tag{13}$$

4.4 Experimental Setup and Results

Four different models were trained by Cahall et al. [70], two for the *U-Net* architecture (intra-tumoral structures and glioma sub-regions), and two for the *U-Net* with *Inception* module (intra-tumoral structures and glioma sub-regions). All four models were trained using k-fold *cross-validation* on the dataset that was randomly split into k mutually exclusive subsets of equal or near-equal size. Each algorithm was run k times subsequently, and each time one of the k splits was taken as a validation subset and the rest as the training subset. *Stochastic gradient descent* [116] with an adaptive moment estimator (*Adam* [116]) was used for training all models and their variations [117]. With a batch size of 64 and 100 epochs, the learning rate was initially set to 10^{-4}, which was exponentially decayed every 10 epochs. All learnable parameters (weights and biases) were initialized based on the He initialization method [118]. *Keras* [119] *Application Programming Interface* (API) with *TensorFlow* [120] backend was used for implementation, and all models were trained on a Google Cloud Compute [121] instance with 4 NVIDIA TESLA P100 *Graphical Processing Units* (GPUs).

4.4.1 Results from *Inception* Modules

For intra-tumoral structures, the addition of *Inception* modules to *U-Net* resulted in statistically significant improvements in WT (DSC improved from 0.903 to 0.925, $p < 0.001$), TC (0.938 to 0.952, $p < 0.001$), and ET (0.937 to 0.948, $p < 0.001$). For glioma sub-regions, significant improvements were also noticed in WT (0.898 to 0.918, $p < 0.001$), TC (0.942 to 0.951, $p = 0.001$), and ET (0.942 to 0.948, $p = 0.002$). Changing the objective from intra-tumoral structures to glioma sub-regions learning in the *U-Net* resulted in no difference in performance for WT (0.903 to 0.898, $p = 0.307$), TC (0.938 to 0.942, $p = 0.284$), and ET (0.937 to 0.942, $p = 0.098$). However, *U-Net* with *Inception* modules, which learned the intra-tumoral structures outperformed those which learned glioma sub-regions in WT (0.918 to 0.925, $p = 0.007$), but there was no difference in the performance for TC (0.952 to 0.951, $p = 0.597$) and ET (0.948 to 0.948, $p = 0.402$). This implies that integrating *Inception* modules in the *U-Net* architecture resulted in statistically significant improvement in tumor segmentation performance that was quantified using k-fold cross-validation ($p < 0.05$ for all three glioma sub-regions). The improvement in the validation accuracy can be attributed to the multiple convolutional filters of different sizes employed in each *Inception* module. These filters are able to capture and retain contextual information at multiple scales during the learning process, both in the contracting as well as expanding paths. Cahall et al. also consider that the improvement in the tumor segmentation accuracy is linked to the new loss function based on the modified DSC (Eq. 13). DSC scores for *Inception* modules are comparable or exceed the results of No New-Net [122], which achieved second place in the BraTS 2018 competition, and the ensemble approach proposed in [122–124].

4.4.2 Results from DIU-Net

DIU-Net showed significant improvement in the WT sub-region with an increase in the Dice score from 0.925 to 0.931 with $p < 0.05$. For the TC sub-region, the Dice score improved from 0.952 to 0.957 with $p < 0.05$. However, for the ET, the change was not statistically significant, $p = 0.114$. Interestingly, DIU-Net is computationally more efficient. DIU-Net has 2.5 million fewer parameters than *U-Net* with *Inception* modules. In contrast, DIU-Net achieves significantly better results at a lesser computational cost (15% fewer parameters). The *Dice* scores for each glioma sub-region are comparable or exceed the results of other recently published architectures, including *No New-Net*, SDResU-Net and the ensemble approach proposed in [122–124].

5 Uncertainty Estimation in Brain Tumor Segmentation

As mentioned before, accurate segmentation of brain tumors is crucial for treatment planning and follow-up evaluations. Furthermore, the robustness and trustworthiness of the segmentation results are of particular interest in medical imaging and in the clinic for diagnosis and prognosis due to their link to human health. In this section, we discuss a new DL framework, named *extended Variational Density Propagation* (exVDP), that can quantify uncertainty in the output decision [125]. In exVDP, Dera et al. adopt the *Variational Inference* (VI) [126] framework and propagate the first two moments of the variational distribution through all DNN's layers (convolution, max-pooling and fully-connected) and non-linearities. Dera et al. used the first-order Taylor series linearization [127] to propagate the mean and covariance of the variational distribution through the non-linear activation functions in the DNNs.

exVDP considers a CNN with a total of C convolutional layers and L fully-connected layers, where the convolutional kernels and the weights of the fully-connected layers are random tensors. A non-linear activation function follows every convolutional and fully-connected layer. Moreover, the ANN contains max-pooling layers. The ANN's weights (and biases) are represented by $\boldsymbol{\Omega} = \{\{\{\mathcal{W}^{(k_c)}\}_{k_c=1}^{K_c}\}_{c=1}^{C}, \{\mathbf{W}^{(l)}\}_{l=1}^{L}\}$, where $\{\{\mathcal{W}^{(k_c)}\}_{k_c=1}^{K_c}\}_{c=1}^{C}$ is the set of K_c kernels in the cth convolutional layer, and $\{\mathbf{W}^{(l)}\}_{l=1}^{L}$ is the set of weights in L fully-connected layers. Input tensor is $\mathcal{X} \in \mathbb{R}^{I_1 \times I_2 \times K}$, where I_1, I_2, and K represent image height, width, and number of channels, respectively.

5.1 Variational Learning

Dera et al. introduced a prior distribution over ANN weights, $\boldsymbol{\Omega} \sim p(\boldsymbol{\Omega})$. The convolutional kernels are considered to be independent of each other within a layer as well as across different layers. This independence assumption is desirable as it promotes convolutional kernels to extract uncorrelated features within and across layers. Given the training data $\mathcal{D} = \{\mathcal{X}^{(i)}, \mathbf{y}^{(i)}\}_{i=1}^{N}$ and the prior $p(\boldsymbol{\Omega})$, the posterior $p(\boldsymbol{\Omega}|\mathcal{D})$ is given through the Bayes' rule. However, $p(\boldsymbol{\Omega}|\mathcal{D})$ is typically intractable. VI methods approximate the true posterior $p(\boldsymbol{\Omega}|\mathcal{D})$ with a simpler parametrized variational distribution $q_\phi(\boldsymbol{\Omega})$. The optimal parameters of the variational posterior $\boldsymbol{\phi}^*$ are estimated by minimizing the *Kullback-Leibler* (KL) divergence between the approximate and the true posterior [126, 128].

$$\begin{aligned}\boldsymbol{\phi}^* &= \operatorname{argmin} \mathrm{KL}\left[q_\phi(\boldsymbol{\Omega}) \| p(\boldsymbol{\Omega}|\mathcal{D})\right] \\ &= \operatorname{argmin} \int q_\phi(\boldsymbol{\Omega}) \log \frac{q_\phi(\boldsymbol{\Omega})}{p(\boldsymbol{\Omega}) p(\mathcal{D}|\boldsymbol{\Omega})} d\boldsymbol{\Omega} \\ &= \operatorname{argmin} \mathrm{KL}\left[q_\phi(\boldsymbol{\Omega}) \| p(\boldsymbol{\Omega}\right] - E_{q_\phi(\boldsymbol{\Omega})}\left\{\log p(\mathcal{D}|\boldsymbol{\Omega})\right\}.\end{aligned} \tag{14}$$

The optimization objective is given by the *Evidence Lower BOund* (ELBO) $\mathrm{L}(\phi; \mathbf{y}|\mathcal{X})$:

$$\mathrm{L}(\phi; \mathbf{y}|\mathcal{X}) = E_{q_\phi(\boldsymbol{\Omega})}(\log p(\mathbf{y}|\mathcal{X}, \boldsymbol{\Omega})) - \mathrm{KL}(q_\phi(\boldsymbol{\Omega} \| p(\boldsymbol{\Omega})). \tag{15}$$

ELBO consists of two parts, the expected log-likelihood of the training data given the weights and a regularization term, which can be re-written as:

$$\mathrm{KL}(q_\phi(\boldsymbol{\Omega} \| p(\boldsymbol{\Omega}) = \sum_{c=1}^{C} \sum_{k_c=1}^{K_c} \mathrm{KL}(q_\phi(\mathcal{W}^{(k_c)}) \| p(\mathcal{W}^{(k_c)})) + \sum_{l=1}^{L} \mathrm{KL}(q_\phi(\mathbf{W}^{(l)}) \| p(\mathbf{W}^{(l)})). \tag{16}$$

5.2 Variational Density Propagation

Dera et al. proposed to approximate the true unknown posterior $p(\boldsymbol{\Omega}|\mathcal{D})$ by a variational distribution $q_\phi(\boldsymbol{\Omega})$. *Gaussian* distribution has been defined as a prior over convolutional kernels and weights of the fully-connected layers [125]. The task is now to propagate the moments of the variational distribution $q_\phi(\boldsymbol{\Omega})$ through various

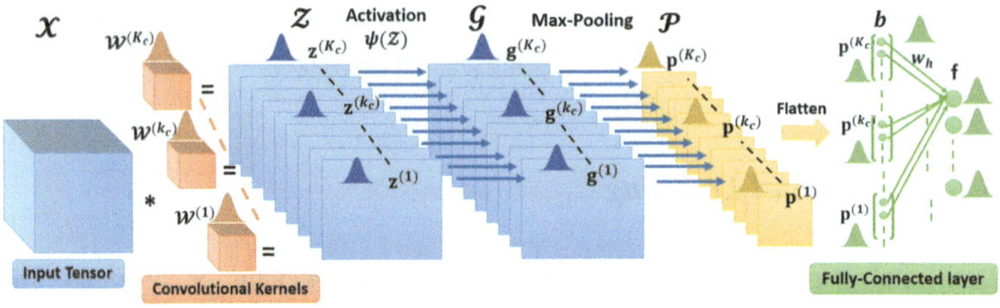

Fig. 9 A schematic layout of the proposed variational density propagation CNN is presented. Dera et al. show the propagation of the mean and covariance of the variational distribution $q_{\phi}(\mathbf{\Omega})$ through multiple layers of a CNN

layers, i.e., convolution, activation, max-pooling, fully-connected, and *softmax*. It is important to note that in these settings, the convolutional kernels, resulting activations, extracted features, logits, and output of the *softmax* function are all random variables. Therefore, instead of performing algebraic operations on real numbers, we are confronted with operations on random variables, including (1) multiplication of a random variable with a constant, (2) multiplication of two random variables, and (3) non-linear transformations [125] operating over random variables. As a result of the multiplication of two Gaussian random variables [127] or non-linear transformation, the resulting random variables may not have *Gaussian distribution* [127]. The goal is to propagate the mean and covariance of the variational distribution and later obtain the mean and covariance of the predictive distribution, $p(\mathbf{y}|\mathcal{X}, \mathcal{D})$. The mean of $p(\mathbf{y}|\mathcal{X}, \mathcal{D})$ represents the ANN's prediction, while the covariance matrix reflects the uncertainty associated with the output decision. An illustration of the proposed variational density propagation CNN with one convolutional layer, one max-pooling and one fully-connected layer is shown in Fig. 9.

5.3 Extended Variational Density Propagation

We start with the mathematical results for the propagation of the mean and covariance of the variational distribution $q_{\phi}(\mathbf{\Omega})$ through convolutional layers, activation functions, max-pooling, fully-connected layers, and the *softmax* function. First-order Taylor series is used for the approximation of the first two moments (mean and covariance) after a non-linear activation function and refer to this method as the exVDP [125].

5.3.1 First Convolutional Layer

The convolution operation between a set of kernels and the input tensor is formulated as a *matrix-vector multiplication*. We first form sub-tensors $\mathcal{X}_{i:i+r_1-1,j:j+r_2-1}$ from the input tensor $\mathcal{X}$, having the same size as the kernels $\mathcal{W}^{(k_c)} \in \mathbb{R}^{r_1 \times r_2 \times K}$. These sub-tensors are subsequently vectorized and arranged as the rows of a matrix $\tilde{\mathbf{X}}$. Thus, convolving $\mathcal{X}$ with the k_cth kernel $\mathcal{W}^{(k_c)}$ is equivalent to the multiplication of $\tilde{\mathbf{X}}$ with $\text{vec}(\mathcal{W}^{(k_c)})$. Let

$$\mathbf{z}^{(k_c)} = \mathcal{X} * \mathcal{W}^{(k_c)} = \tilde{\mathbf{X}} \times \text{vec}(\mathcal{W}^{(k_c)}), \tag{17}$$

where $*$ denotes the convolution operation and $\times$ is a regular matrix-vector multiplication.

Multivariate Gaussian distribution has been defined over the vectorized convolutional kernels, i.e. $\text{vec}(\mathcal{W}^{(k_c)}) \sim \mathcal{N}\left(\mathbf{m}^{(k_c)}, \boldsymbol{\Sigma}^{(k_c)}\right)$. It follows that:

$$\mathbf{z}^{(k_s)} \sim \mathcal{N}\left(\boldsymbol{\mu}_{\mathbf{z}^{(k_c)}} = \tilde{\mathbf{X}}\mathbf{m}^{(k_c)}, \boldsymbol{\Sigma}_{\mathbf{z}^{(k_c)}} = \tilde{\mathbf{X}}\boldsymbol{\Sigma}^{(k_c)}\tilde{\mathbf{X}}^T\right). \tag{18}$$

5.3.2 Non-linear Activation Function

Approximating the mean and covariance after the non-linear activation function ψ using the first-order Taylor series approximation [127]). Let $\mathbf{g}_i^{(k_c)} = \psi[\mathbf{z}_i^{(k_c)}]$ be the element-wise ith output of ψ. We have $\boldsymbol{\mu}_{\mathbf{g}^{(k_c)}}$ and $\boldsymbol{\Sigma}_{\mathbf{g}^{(k_c)}}$:

$$\begin{aligned} \mu_{\mathbf{g}_i^{(k_c)}} &\approx \psi(\mu_{\mathbf{z}_i^{(k_c)}}), \\ \boldsymbol{\Sigma}_{\mathbf{g}^{(k_c)}} &\approx \begin{cases} \sigma^2_{\mathbf{z}_i^{(k_c)}} \left(\dfrac{d\psi(\mu_{\mathbf{z}_i^{(k_c)}})}{d\mathbf{z}_i^{(k_c)}}\right)^2, & \text{if } i = j. \\ \sigma_{\mathbf{z}_i^{(k_c)}\mathbf{z}_j^{(k_c)}} \left(\dfrac{d\psi(\mu_{\mathbf{z}_i^{(k_c)}})}{d\mathbf{z}_i^{(k_c)}}\right)\left(\dfrac{d\psi(\mu_{\mathbf{z}_j^{(k_c)}})}{d\mathbf{z}_j^{(k_c)}}\right), & \text{if } i \neq j. \end{cases} \end{aligned} \tag{19}$$

5.3.3 Max-Pooling Layer

For the max-pooling, $\boldsymbol{\mu}_{\mathbf{p}^{(k_c)}} = \text{pool}(\boldsymbol{\mu}_{\mathbf{g}^{(k_c)}})$ and $\boldsymbol{\Sigma}_{\mathbf{p}^{(k_c)}} = \text{co} - \text{pool}(\boldsymbol{\Sigma}_{\mathbf{g}^{(k_c)}})$, where pool represents the max-pooling operation on the mean and co − pool represents down-sampling the covariance, i.e., we keep only the rows and columns of $\boldsymbol{\Sigma}_{\mathbf{g}^{(k_c)}}$ corresponding to the pooled mean elements.

5.3.4 Flattening Operation

The output tensor $\mathcal{P}$ of the max-pooling layer is vectorized to form the input vector $\mathbf{b}$ of the fully-connected layer such that, $\mathbf{b} = \left[\mathbf{p}^{(1)T}, \cdots, \mathbf{p}^{(K_c)T}\right]^T$. The mean and covariance matrix of $\mathbf{b}$ are given by:

$$\boldsymbol{\mu}_{\mathbf{b}} = \begin{bmatrix} \boldsymbol{\mu}_{\mathbf{p}^{(1)}} \\ \vdots \\ \boldsymbol{\mu}_{\mathbf{p}^{(K_c)}} \end{bmatrix}, \boldsymbol{\Sigma}_{\mathbf{b}} = \begin{bmatrix} \boldsymbol{\Sigma}_{\mathbf{p}^{(1)}} & \cdots & \mathbf{0} \\ \vdots & \ddots & \vdots \\ \mathbf{0} & \cdots & \boldsymbol{\Sigma}_{\mathbf{p}^{(K_c)}} \end{bmatrix}. \tag{20}$$

5.3.5 Fully-Connected Layer

Let $\mathbf{w}_h \sim \mathcal{N}(\mathbf{m}_h, \boldsymbol{\Sigma}_h)$ be hth weight vector of the fully-connected layer, where $h = 1, \cdots, H$, and H is the number of output neurons. We note that f_h is the product of two independent random vectors $\mathbf{b}$ and $\mathbf{w}_h$. Let $\mathbf{f}$ be the output vector of the fully-connected layer, then the elements of $\boldsymbol{\mu}_{\mathbf{f}}$ and $\boldsymbol{\Sigma}_{\mathbf{f}}$ are derived by the following proposition:

Proposition 1

$$\mu_{f_h} = \mathbf{m}_h^T \boldsymbol{\mu}_{\mathbf{b}},$$
$$\boldsymbol{\Sigma}_{\mathbf{f}} = \begin{cases} \operatorname{tr}\left(\boldsymbol{\Sigma}_h \boldsymbol{\Sigma}_{\mathbf{b}}\right) + \mathbf{m}_h^T \boldsymbol{\Sigma}_{\mathbf{b}} \mathbf{m}_h + \boldsymbol{\mu}_{\mathbf{b}}^T \boldsymbol{\Sigma}_h \boldsymbol{\mu}_{\mathbf{b}}, & \\ \mathbf{m}_{h_1}^T \boldsymbol{\Sigma}_{\mathbf{b}} \mathbf{m}_{h_2}, & h_1 \neq h_2, \end{cases} \tag{21}$$

where $h_1, h_2 = 1, \cdots, H$ represent any two weight vectors in the fully-connected layer.

5.3.6 *Softmax* Function

Let the output of the ANN be $\mathbf{y} = \varphi(\mathbf{f})$, where φ is the *softmax* function. Using the first-order Taylor series approximation, the mean and covariance of the output vector, i.e., $\boldsymbol{\mu}_{\mathbf{y}}$ and $\boldsymbol{\Sigma}_{\mathbf{y}}$, are derived as follows [129]:

$$\boldsymbol{\mu}_{\mathbf{y}} \approx \varphi(\boldsymbol{\mu}_{\mathbf{f}}); \boldsymbol{\Sigma}_{\mathbf{y}} \approx \mathbf{J}_{\varphi} \boldsymbol{\Sigma}_{\mathbf{f}} \mathbf{J}_{\varphi}^T, \tag{22}$$

where $\mathbf{J}_{\varphi}$ is the *Jacobian* matrix of φ with respect to $\mathbf{f}$ evaluated at $\boldsymbol{\mu}_{\mathbf{f}}$ [129].

5.3.7 Objective Function

Assuming a diagonal covariance matrix for the variational posterior distribution, N independently and identically distributed (iid) data points and using M Monte Carlo samples to approximate the expectation by a summation, the expected log-likelihood in the ELBO objective function is given as follows:

$$E_{q_\phi(\boldsymbol{\Omega})}(\log p(\mathbf{y}|\mathcal{X}, \boldsymbol{\Omega})) \approx$$

$$-\frac{NH}{2}\log(2\pi) - \frac{1}{M}\sum_{m=1}^{M}\left[\frac{N}{2}\log(|\boldsymbol{\Sigma}_\mathbf{y}|) + \frac{1}{2}\sum_{i=1}^{N}(\mathbf{y}^{(i)} - \boldsymbol{\mu}_\mathbf{y}^{(m)})^T(\boldsymbol{\Sigma}_\mathbf{y}^{(m)})^{-1}(\mathbf{y}^{(i)} - \boldsymbol{\mu}_\mathbf{y}^{(m)})\right] \tag{23}$$

The regularization term in (16) is the KL-divergence between two multivariate Gaussian distributions [127]. If we have a CNN with one convolutional layer followed by the activation function, one max-pooling and one fully-connected layer, thus the regularization term in the ELBO objective function is derived as follows:

$$\mathrm{KL}(q_\phi(\boldsymbol{\Omega}\|p(\boldsymbol{\Omega}) =$$

$$\frac{1}{2}\sum_{k=1}^{K_1}\left(r_1r_2K\sigma_{r_1,k}^2\sigma_{r_2,k}^2\sigma_{K,k}^2 + \|\mathcal{M}^{(k)}\|_F^2 - r_1r_2K - r_1r_2K\left(\log(\sigma_{r_1,k}^2\,\sigma_{r_2,k}^2\,\sigma_{K,k}^2)\right)\right) \tag{24}$$

$$+\frac{1}{2}\sum_{h=1}^{H}\left(n_f\sigma_h^2 + \|\mathbf{m}_h\|_F^2 - n_f - n_f\log\sigma_h^2\right),$$

where $(r_1 \times r_2 \times K)$ is the size of the kernels, K_1 is the number of kernels in the convolutional layer, H is the number of output neurons and n_f is the length of the weight vector $\mathbf{w}_h$ in the fully-connected layer.

5.3.8 Back-Propagation

During back-propagation, we compute the gradient of the objective function $\nabla_\phi \mathrm{L}(\phi; \mathcal{D})$ with respect to the variational parameters:

$$\phi = \left\{\left\{\{\mathcal{M}^{(k_c)}, \sigma_{r_1,k_c}^2, \sigma_{r_2,k_c}^2, \sigma_{K_{c-1},k_c}^2\}_{k_c=1}^{K_c}\right\}_{c=1}^{C}, \{\mathbf{m}_h, \sigma_h^2\}_{h=1}^{H}\right\}, \tag{25}$$

where $(r_1 \times r_2 \times K_{c-1})$ is the size of the k_cth kernel, K_c is the number of kernels in the cth convolutional layer and H is the number of output neurons. Dera et al. used $\nabla_\phi \mathrm{L}(\phi; \mathcal{D})$ to update parameters ϕ using the gradient descent update rule.

5.4 Application to Brain Tumor Segmentation in MRI Images

The performance of the proposed exVDP model on the HGG brain tumor segmentation task using the BraTS 2015 dataset has been evaluated. The dataset consists of 5 classes, i.e. class 0—normal tissue, class 1—necrosis, class 2—edema, class 3—non-enhancing, and class 4—enhancing tumor [94]. The evaluation of segmentation is based on three regions, (1) complete tumor (1, 2, 3 and 4), (2) tumor core (1, 3 and 4), and (3) enhancing tumor (class 4) [94].

Brain tumor segmentation has been formulated as a multi-class classification problem by randomly sampling patches from four MRI modalities, i.e., FLAIR, T1, T2 and T1c [130]. The label of each patch has been manually set to the label of the center pixel. The sampled patches are balanced over all classes, and a total of 100,000 patches of size 33 × 33 are extracted from the BraTS data of 20 patients. These patches are divided into training and validation bins (95% for training and 5% for validation). The test set included randomly sampled 372 images, i.e., 43,264 patches, from each of the four modalities. The proposed exVDP model is compared with a deterministic CNN, presented in [130]. The following CNN architecture has been used: six convolution layers (all kernels were 3×3, and we had 32, 32, 64, 64, 128, 128 kernels in layers one to six, respectively, followed by ReLU activation), two max-pooling layers, and a fully-connected layer. The architecture is shown in Table 1.

DSC has been used to evaluate the segmentation results before and after adding Gaussian noise or targeted adversarial attack (targeted class is class 3, i.e., *non-enhancing tumor*). The evaluation of the proposed model on the BraTS dataset is done without doing any pre-processing or data augmentation techniques.

In Table 2, DSC values for three test cases have been presented, i.e., noise-free, Gaussian, and adversarial noise. We note that for the class of Complete tumor, the DSC value of the proposed model is ($\sim$2.8%) higher than that of the deterministic CNN in the noise-free case, while ($\sim$34%) higher in the case of adversarial attack and ($\sim$13.7%) higher in the case of Gaussian noise. Figure 10 shows segmentation results for exVDP and a deterministic CNN for a representative

Table 1 Architecture of the two models, i.e., exVDP and deterministic CNN

Layer	Type	Filter size	HGG stride	No. kernels	FC units	Input
1	Conv.	3×3	1×1	32	–	33×33×4
2	Conv.	3×3	1×1	32	–	33×33×32
3	Conv.	3×3	1×1	64	–	33×33×32
4	Max-pool.	3×3	2×2	–	–	33×33×64
5	Conv.	3×3	1×1	64	–	16×16×64
6	Conv.	3×3	1×1	128	–	16×16×64
7	Conv.	3×3	1×1	128	–	16×16×128
8	Max-pool.	3×3	2×2	–	–	16×16×128
9	FC	–	–	–	5	6272

Table 2 Segmentation results measured using the DSC for the BraTS test dataset

Method	Tumor regions	No noise	Adversarial noise	Gaussian noise
exVDP	Complete	80.8%	77.4%	80.6%
	Core	74.6%	72.6%	74.5%
	Enhancing	74.0%	69.8%	73.9%
Deterministic CNN	Complete	78.0%	43.4%	66.9%
	Core	65.0%	47.1%	51.9%
	Enhancing	75.0%	43.9%	55.7%

HGG image (with and without adversarial noise). The uncertainty map associated with each segmentation is also presented for the exVDP model. The uncertainty map allows physicians to review the segmentation results quickly and, if needed, make corrections of tumor boundaries in the regions where the uncertainty is high.

6 Tumor Surveillance

As defined in Sect. 2.3, tumor surveillance is the process of monitoring patient's tumor in longitudinal studies to establish severity of the disease and planning treatment accordingly. It helps identifying early signs of tumor occurrence which is critical especially in case of the cancerous tumors.

6.1 Rationale for Tumor Surveillance

Temporal medical imaging data is widely used in oncology as well as radiology for visual comparison of disease over an extended period of time. The 2D medical images (CT or MRI scans) are examined by the physicians to diagnose the disease in 4D (3D tumor volume over time) usually referred to as *change in volume over time*. A detection at an earlier stage of disease is more responsive to treatment, resulting in improved outcomes for the patient. Biological characteristics of various tumor types such as growth, location, and patterns of local as well as metastatic disease are the basis for surveillance scheduling, protocols, and selection of imaging techniques. Usually, low-risk tumor is subjected to active surveillance as part of a patient's treatment plan. An ideal candidate for active surveillance is the one which has one or more of these conditions: (1) disease has not spread, (2) tumor is small and growing slowly, or (3) patient exhibits no symptoms of specific cancer. The data from active surveillance is also used to look for trends and patterns over time in certain regions/groups of people, and to see if preventive measures are making a difference among the sample population. Apart from brain cancer, active surveillance has been widely employed in other forms of cancer diagnosis. It has been shown that use

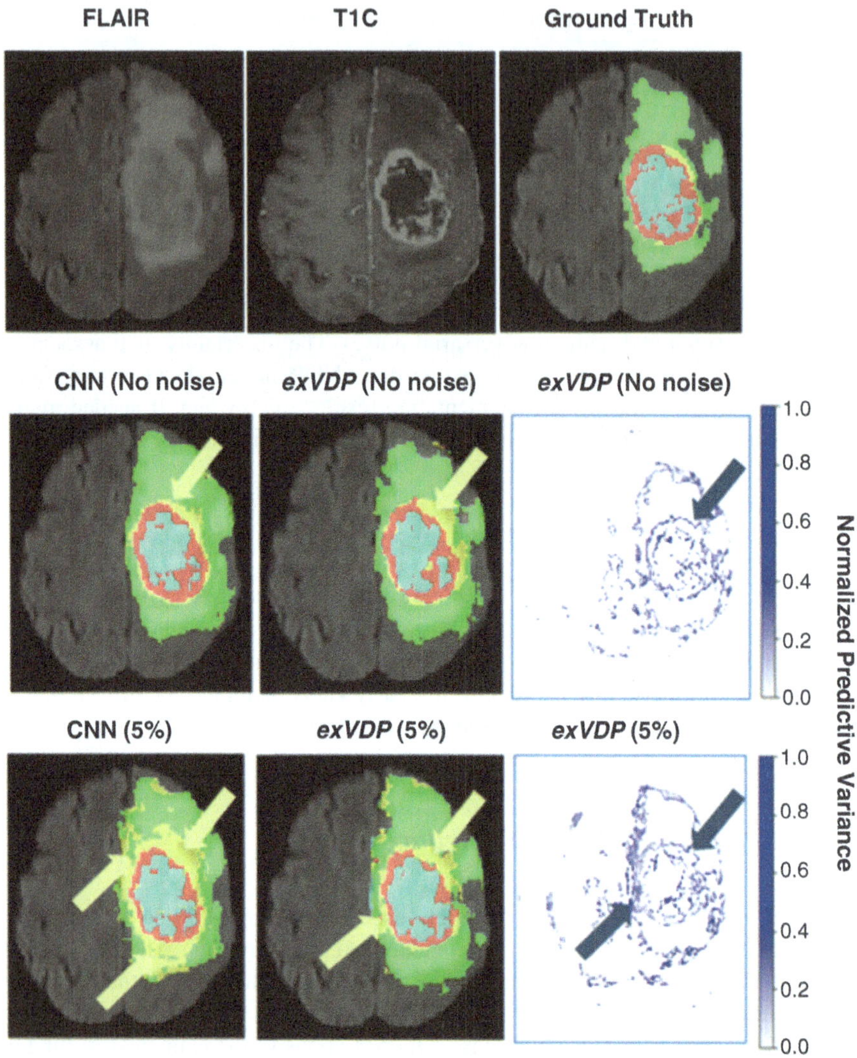

Fig. 10 Example 1: Segmentation results of the proposed exVDP and a deterministic CNN on the BraTS 2015 dataset with and without adding adversarial noise. The uncertainty map associated with each segmentation is also shown for each of the two models. The class label *non-enhancing tumor*, which is the target of the attack, is represented in yellow color in the ground truth image. The green color refers to the edema class, the red color refers to the enhancing tumor, and the blue color refers to the necrosis

of *Prostate-Specific Antigen* (PSA) [131] testing as part of active surveillance of prostate cancer helps in understanding tumor progression and prognosis, enabling the patients diagnosed with lower grade disease feel more comfortable [24].

6.2 Surveillance Techniques

Change-point detection is the classical technique of detecting abrupt changes in sequential data, which focuses predominantly on datasets with a single observable. It has been a long-standing research area in statistics [132, 133], with applications in fields ranging from economics [134], bioinformatics [135, 136], and climatology [137], wherein it is dealt as the problem of detecting abrupt changes in temporal data. The objective is to determine if the observed time series is statistically homogeneous or otherwise to find the point in time when the change happens. There are two variations to the change-point detection technique: posterior and sequential. Posterior tests are done offline after entire data is collected, and a decision of homogeneity or change-point is made based on the analysis of all the collected data. Whereas, sequential tests are done on-the-fly as the data is presented sequentially, and the decisions are made online. *Gleason grade* is another technique used for pathological scoring of the differentiation of prostate cancer, and it has been the most widely used grading system for prostate tumor differentiation and prognostic indicator for prostate cancer progression [138].

6.3 Community-Level Active Surveillance

Apart from the individual tumor surveillance, the term of active surveillance is also often used for collective cancer surveillance data and programs in the United States through *Cancer Registries*. A cancer registry is an information system designed for the collection, storage, and management of data on persons with cancer [139]. Data on cancer in the United States is collected through two types of registries: hospital registries, which are the part of a facility's cancer program, and population-based registries, usually tied to state health departments. Hospital registries provide patient's data on care within the hospital for evaluation. Population-based registries, under state health departments, collect information on all cases diagnosed within a certain geographic area from multiple reporting facilities, including hospitals, doctors' offices, nursing homes, pathology laboratories, radiation and chemotherapy treatment centers, etc. The collected data is used to build statistics like new cancer cases (incidence), death rates (mortality), cancer types related to types of jobs, cancer trends over time to keep an eye on age and racial groups that are most affected by different types of cancer. Registries are staffed under the *Certified Tumor Registrar* (CTR), having pre-defined standards of training, testing, and continuing education, and they compile timely, accurate, and complete cancer information to report to the registry. The major cancer surveillance programs in the United States are the *National Cancer Data Base* (NCDB) [140], *National Cancer Institute*'s (NCI) *Surveillance, Epidemiology and End Results* (SEER) program [139], and *National Program of Cancer Registries* (NPCR) of the *Center for Disease Control and Prevention* (CDC) [141]. *Central Brain Tumor Registry of the United States*

(CBTRUS) is a registry dedicated to collecting and disseminating statistical data on all primary benign and malignant brain tumors [142]. A recent study [143] tries to estimate excess mortality in people with cancer and multi-morbidity in the COVID-19 affected patients through analysis of surveillance data DATA-CAN, the UK National Health Data Research Hub for cancer emergency [144].

6.4 Surveillance of Brain Tumor

Tumors of the *Central Nervous System* (CNS) are the second most common tumors among children after leukemia. Treatment protocols for high-grade pediatric brain tumors recommend regular follow-up imaging for up to 10 years. Based on the surveillance data of high-grade childhood brain tumor patients, a review of maximal time to recurrence and minimal time to radiologically detectable long-term sequel such as secondary malignancies, vascular complications, and white matter disease found that there was no recurrence of the primary brain tumor, either local or distant, 10 years or more after the end of treatment in the reviewed literature and so the results do not justify routine screening to detect tumor recurrence more than 10 years after the end of treatment [145].

Tumor surveillance is being used for building statistical figures in a broad spectrum of ways, including adult glioma incidence and survival by race or ethnicity in the United States [146], county-level glioma incidence and survival variations [147], and accurate population-based statistics on the brain and other central nervous system tumors [148]. A CBTRUS statistical report on the primary brain and *Other Central Nervous System* (OCNS) tumors data diagnosed in the United States in the period 2011–2015 states that brain and OCNS tumors (both malignant and non-malignant) were the most common cancer types in persons age 0–14 years for both males and females. For age 15–39 years, these tumors were the second most common cancer in males and the third most common among females in this age group. For age 40+, these were the eighth most common cancer type, with males having eighth and females having the fifth most common brain cancer. These results were based on the NPCR data of 388,786 brain and OCNS tumors, and 16,633 tumor cases from SEER [149].

Tumor Surveillance among patients also enables the authorities to predict cancer cases and death in advance and respond in time to offset the predicted scores. A recent report by the *American Cancer Society* on cancer statistics in 2020 projected the number of new cancer cases and deaths that will occur in the United States. Incidence data from 2002 to 2017 were collected, and it was estimated that in 2020, 23,890 new cases and 18,020 deaths related to brain and ONS tumors were projected to occur in the United States [150]. Similarly, tumor surveillance is equally important to avoid the side effects of the aggressive forms of treatment. Patients treated for glioma, meningioma, and brain metastases may develop side effects of treatment, including neuropathy (with visual loss), cataracts, hypopituitarism, cognitive decline, increased risk of stroke, and risk of secondary tumor occurring months or even years later. The surgical treatment causes immediate side effects, chemotherapy-caused side effects occur early after treatment (but infertility may not manifest itself until later), and radiotherapy's side effects occur months or even years after treatment. The risks vary depending on the technique used and the area of the brain treated. Surveillance enables the physicians to identify these potential late side-effects earlier which increases the length and quality of life for patients [151].

6.4.1 An Example of Surveillance Study

In this section, we will study an example of tumor surveillance, specifically of patients having low-grade gliomas [152]. Low-grade gliomas, constituting around 15% of all adult brain tumors, significantly affect neurological morbidity by brain invasion. Generally, there is no universally-accepted technique available for the detection of growth of low-grade gliomas in the clinical setting. Clinicians usually consider visual comparisons of two or more longitudinal radiological scans through subjective evaluation for detecting the growth of low-grade gliomas. The paper [152] suggests a *Computer-Assisted Diagnosis* (CAD) method to help physicians detect earlier growth of low-grade gliomas. This method consists of tumor segmentation, computing volumes, and pointing to growth by the online abrupt change-of-point method considering only past measurements. The study suggests that early growth detection of tumor sets the stage for future clinical studies to decide upon the type of treatment-path to be undertaken and whether early therapeutic interventions prolong survival and improve quality of life. Longitudinal (temporal) radiological studies of 63 patients were carried out with a median follow-up period of 33.6 months. These patients were diagnosed with grade 2 gliomas by expert physicians through manual (visual) procedures as well as detection of growth with that of the CAD method, and both detection methods were compared by 7 expert physicians [152] . Each patient had at least 4 MRI scans available for review either after the initial diagnosis or after the completion of chemotherapy with temozolomide (if applicable). The researchers calculated the time to growth detection from the impressions of the radiological reports of these patients from 627 MRI scans. Unexpectedly, the study found large differences in growth detection by visual comparison and by physicians aided by the CAD method. The reasons for missing growth by the visual inspection can be attributed to one or more of these reasons: (1) interpreting a large number of prior studies by physicians takes a very long time, (2) the practice in vogue of comparing the current MRI to a couple of MRI scans immediately preceding it, (3) the lack of determination of baseline MRI, (4) small changes from one scan to the next, and (5) comparing single 2-D images overlooks the growth in the third dimension.

The study [152] showed that the CAD method helped physicians detect growth at earlier times and significantly smaller tumor volumes than the manual standard method. Moreover, physicians aided by the CAD method diagnosed tumor growth in 13 of 22 glioma patients labeled as clinically stable by the standard method.

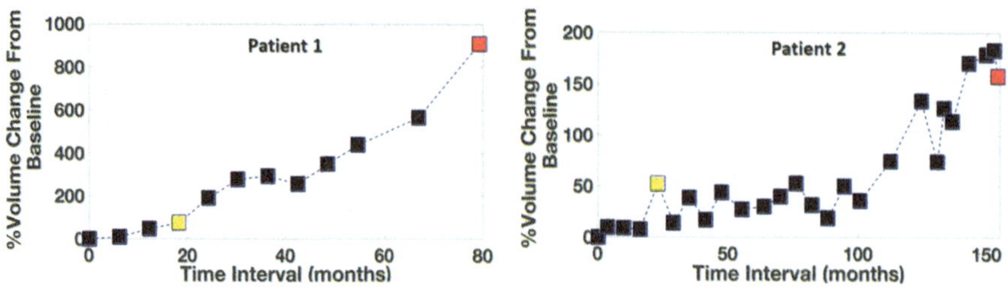

Fig. 11 Volume growth curves of grade 2 gliomas of two patients diagnosed with oligos. x-axis corresponds to the time interval from the baseline MRI, y-axis corresponds to change in volume of tumor from baseline. The volume at the time-to-growth detected by computer-assisted diagnosis is colored in yellow and manual (visual) detection time is colored in red

Figure 11 shows the volume growth curves of grade 2 gliomas of two patients diagnosed with oligos, seen at the University of Alabama at Birmingham clinics between 1 July 2017 and 14 May 2018 [152]. The x-axis corresponds to the time interval from the baseline MRI, and the y-axis corresponds to the change in the volume of tumors from the baseline. The volume at each time step until the growth detected by CAD is colored in yellow, and the manual (visual) detection of change-point time is colored in red. The CAD for patient 1 detected a change-point in 20 months from the baseline, whereas visual detection by a physician was done in 80 months.

Similarly, CAD detection time for patient 2 was also around 20 months, where visual detection was in 150 months, primarily because this tumor did not grow at a faster pace. The detection of tumor volume growth in time enabled the researchers to identify tumors with nonlinear and non-homogeneous growth. Early growth detection holds the potential of lowering the morbidity, and perhaps mortality of patients with low-grade gliomas. The decision to treat a patient would be determined by the rate of growth and proximity to critical areas of the brain, once they have been measured. The study also suggested early interventions for cases where (1) the new growth is in the proximity of key nonsurgical structures like the corpus callosum, (2) the rate of growth is elevated, or (3) the tumor is sensitive to chemotherapy.

7 Conclusion

In this chapter, we have thoroughly reviewed the image segmentation task in the classical CV field and examined various techniques of CV employed in DL frameworks for brain tumor segmentation. We have also assessed multiple DL architectures having varying attributes that make them suit-to-task employment. We have looked into a case study for in-depth analysis of *U-Net* with *Inception* and *dilated Inception* modules in the context of brain tumor segmentation. A new DL framework, called exVDP, that can quantify uncertainty in the output decision of a

DNN, has also been discussed. In the last section, we have discussed the concept of tumor surveillance, its rationale, techniques, and an example study on low-grade gliomas surveillance.

The brain tumor segmentation community has achieved substantial progress in the last decade because of the advances in DL. Although efforts have been made in commercializing the technology for clinicians, there is still a long way to make brain tumor segmentation a reliable and routine tool broadly applied to practical clinical decisions with minimal human interventions. This is due to the lack of existing methods in the face of adversarial examples and research-oriented frameworks that are not suited to production environments. Breakthrough is likely to come with the advent of effective and scalable platforms by the ML community, and direction of research towards adversarial learning.

Acknowledgments This work was supported by the National Science Foundation Awards NSF ECCS-1903466 and NSF DUE-1610911.

References

1. Y. LeCun, Y. Bengio, and G. Hinton, "Deep Learning," *Nature*, vol. 521, no. 7553, p. 436, 2015.
2. A. A. Aly, S. B. Deris, and N. Zaki, "Research review for digital image segmentation techniques," *International Journal of Computer Science & Information Technology*, vol. 3, no. 5, p. 99, 2011.
3. Stanford, "Tutorial 3: Image segmentation," https://ai.stanford.edu/~syyeung/cvweb/tutorial3.html.
4. Marvel, "Marvel movies," https://www.marvel.com/movies.
5. S. Gould, T. Gao, and D. Koller, "Region-based segmentation and object detection," in *Advances in neural information processing systems*, 2009, pp. 655–663.
6. S. Yuheng and Y. Hao, "Image segmentation algorithms overview," *arXiv preprint arXiv:1707.02051*, 2017.
7. X. Xu, G. Li, G. Xie, J. Ren, and X. Xie, "Weakly supervised deep semantic segmentation using cnn and elm with semantic candidate regions," *Complexity*, vol. 2019, 2019.
8. mayoclinic, "Braintumor," https://www.mayoclinic.org/diseases-conditions/brain-tumor/symptoms-causes/syc-20350084.
9. movies.effects, *Instagram*, 2019 (accessed August 27, 2020). [Online]. Available: https://www.instagram.com/p/BuOlFbHhjr7/
10. P. Medicine, "Mipg," https://www.pennmedicine.org/departments-and-centers/department-of-radiology/radiology-research/labs-and-centers/biomedical-imaging-informatics/medical-image-processing-group.
11. A. A. of Neurological Surgeons, "Braintumortypes," https://www.aans.org/en/Patients/Neurosurgical-Conditions-and-Treatments/Brain-Tumors.
12. PennMedicine, "Common types of brain tumors," https://www.pennmedicine.org/updates/blogs/neuroscience-blog/2018/november/what-are-the-most-common-types-of-brain-tumors.
13. S. Bauer, R. Wiest, L.-P. Nolte, and M. Reyes, "A survey of MRI-based medical image analysis for brain tumor studies," *Physics in Medicine & Biology*, vol. 58, no. 13, pp. R97–R129, 2013.

14. A. Bousselham, O. Bouattane, M. Youssfi, and A. Raihani, "Towards reinforced brain tumor segmentation on mri images based on temperature changes on pathologic area," *International journal of biomedical imaging*, vol. 2019, 2019.
15. MayoFoundation, "Biomedical imaging resource," https://analyzedirect.com/.
16. NIH, "Neuroimaging informatics technology initiative," https://nifti.nimh.nih.gov/ and https://brainder.org/2012/09/23/the-nifti-file-format/.
17. R. D. Vincent, P. Neelin, N. Khalili-Mahani, A. L. Janke, V. S. Fonov, S. M. Robbins, L. Baghdadi, J. Lerch, J. G. Sled, R. Adalat, D. MacDonald, A. P. Zijdenbos, D. L. Collins, and A. C. Evans, "Minc 2.0: A flexible format for multi-modal images," *Frontiers in Neuroinformatics*, vol. 10, p. 35, 2016. [Online]. Available: https://www.frontiersin.org/article/10.3389/fninf.2016.00035
18. P. Scanners, "parrec," https://nipy.org/nibabel/reference/nibabel.parrec.html.
19. NRRD, "Nearly raw raster data," http://teem.sourceforge.net/nrrd/.
20. DICOM, "Digital imaging and communications in medicine," https://www.dicomstandard.org/about-home.
21. M. Larobina and L. Murino, "Medical image file formats," *Journal of digital imaging*, vol. 27, no. 2, pp. 200–206, 2014.
22. N. C. Institute, "surveillance," https://www.cancer.gov/publications/dictionaries/cancer-terms/def/surveillance.
23. RCCA, "Active surveillance: Its role in low-risk cancer," https://www.regionalcancercare.org/services/active-surveillance/.
24. K. L. Penney, M. J. Stampfer, J. L. Jahn, J. A. Sinnott, R. Flavin, J. R. Rider, S. Finn, E. Giovannucci, H. D. Sesso, M. Loda *et al.*, "Gleason grade progression is uncommon," *Cancer research*, vol. 73, no. 16, pp. 5163–5168, 2013.
25. R. Szeliski, *Computer vision: algorithms and applications*. Springer Science & Business Media, 2010.
26. V. Dumoulin and F. Visin, "A guide to convolution arithmetic for deep learning," *arXiv preprint arXiv:1603.07285*, 2016.
27. ——, "Convolution arithmetic," https://www.github.com/vdumoulin/conv_arithmetic.
28. C. Sinai, "Brain tumors and brain cancer," https://www.cedars-sinai.org/health-library/diseases-and-conditions/b/brain-tumors-and-brain-cancer.html.
29. R. L. Siegel, K. D. Miller, and A. Jemal, "Cancer statistics, 2019," *CA: a cancer journal for clinicians*, vol. 69, no. 1, pp. 7–34, 2019.
30. E. C. Holland, "Progenitor cells and glioma formation," *Current opinion in neurology*, vol. 14, no. 6, pp. 683–688, 2001.
31. J. C. Buckner, "Factors influencing survival in high-grade gliomas," in *Seminars in oncology*, vol. 30. Elsevier, 2003, pp. 10–14.
32. J. Lemke, J. Scheele, T. Kapapa, S. Von Karstedt, C. R. Wirtz, D. Henne-Bruns, and M. Kornmann, "Brain metastases in gastrointestinal cancers: is there a role for surgery?" *International journal of molecular sciences*, vol. 15, no. 9, pp. 16 816–16 830, 2014.
33. R. C. Miner, "Image-guided neurosurgery," *Journal of medical imaging and radiation sciences*, vol. 48, no. 4, pp. 328–335, 2017.
34. M. E. Karar, D. R. Merk, V. Falk, and O. Burgert, "A simple and accurate method for computer-aided transapical aortic valve replacement," *Computerized Medical Imaging and Graphics*, vol. 50, pp. 31–41, 2016.
35. N. Sharma and L. M. Aggarwal, "Automated medical image segmentation techniques," *Journal of medical physics/Association of Medical Physicists of India*, vol. 35, no. 1, p. 3, 2010.
36. L. Kaufman and P. J. Rousseeuw, *Finding groups in data: an introduction to cluster analysis*. John Wiley & Sons, 2009, vol. 344.
37. A. K. Jain, A. Topchy, M. H. Law, and J. M. Buhmann, "Landscape of clustering algorithms," in *Proceedings of the 17th International Conference on Pattern Recognition, 2004. ICPR 2004.*, vol. 1. IEEE, 2004, pp. 260–263.

38. A. K. Jain, R. P. Duin, and J. Mao, "Statistical pattern recognition: A review, ieee transactions on pattern analysis and machine intelligence," *vol*, vol. 22, 2000.
39. A. K. Jain and R. C. Dubes, *Algorithms for clustering data*. Prentice-Hall, Inc., 1988.
40. C. R. Brice and C. L. Fennema, "Scene analysis using regions," *Artificial intelligence*, vol. 1, no. 3–4, pp. 205–226, 1970.
41. E. M. Riseman and M. A. Arbib, "Computational techniques in the visual segmentation of static scenes," *Computer Graphics and Image Processing*, vol. 6, no. 3, pp. 221–276, 1977.
42. T. Pavlidis, *Structural pattern recognition*. Springer, 1977.
43. R. Ohlander, K. Price, and D. R. Reddy, "Picture segmentation using a recursive region splitting method," *Computer graphics and image processing*, vol. 8, no. 3, pp. 313–333, 1978.
44. A. Rosenfeld and L. S. Davis, "Image segmentation and image models," *Proceedings of the IEEE*, vol. 67, no. 5, pp. 764–772, 1979.
45. R. M. Haralick and L. G. Shapiro, "Image segmentation techniques," *Computer vision, graphics, and image processing*, vol. 29, no. 1, pp. 100–132, 1985.
46. A. Blake and M. Isard, *Active contours: the application of techniques from graphics, vision, control theory and statistics to visual tracking of shapes in motion*. Springer Science & Business Media, 2012.
47. D. Cremers, M. Rousson, and R. Deriche, "A review of statistical approaches to level set segmentation: integrating color, texture, motion and shape," *International journal of computer vision*, vol. 72, no. 2, pp. 195–215, 2007.
48. P. F. Felzenszwalb and D. P. Huttenlocher, "Efficient graph-based image segmentation," *International journal of computer vision*, vol. 59, no. 2, pp. 167–181, 2004.
49. D. Comaniciu and P. Meer, "Mean shift: A robust approach toward feature space analysis," *IEEE Transactions on pattern analysis and machine intelligence*, vol. 24, no. 5, pp. 603–619, 2002.
50. J. Long, E. Shelhamer, and T. Darrell, "Fully convolutional networks for semantic segmentation," in *Proceedings of the IEEE conference on computer vision and pattern recognition*, 2015, pp. 3431–3440.
51. S. Bakas, H. Akbari, A. Sotiras, M. Bilello, M. Rozycki, J. S. Kirby, J. B. Freymann, K. Farahani, and C. Davatzikos, "Advancing the cancer genome atlas glioma mri collections with expert segmentation labels and radiomic features," *Scientific data*, vol. 4, p. 170117, 2017.
52. J. Malik, S. Belongie, T. Leung, and J. Shi, "Contour and texture analysis for image segmentation," *International journal of computer vision*, vol. 43, no. 1, pp. 7–27, 2001.
53. X. Liu, Z. Deng, and Y. Yang, "Recent progress in semantic image segmentation," *Artificial Intelligence Review*, vol. 52, no. 2, pp. 1089–1106, 2019.
54. K. Fukushima and S. Miyake, "Neocognitron: A self-organizing neural network model for a mechanism of visual pattern recognition," in *Competition and cooperation in neural nets*. Springer, 1982, pp. 267–285.
55. Y. LeCun, Y. Bengio *et al.*, "Convolutional networks for images, speech, and time series," *The handbook of brain theory and neural networks*, vol. 3361, no. 10, p. 1995, 1995.
56. J. Gu, Z. Wang, J. Kuen, L. Ma, A. Shahroudy, B. Shuai, T. Liu, X. Wang, G. Wang, J. Cai *et al.*, "Recent advances in convolutional neural networks," *Pattern Recognition*, vol. 77, pp. 354–377, 2018.
57. M. Nielsen, "Ch-4: A visual proof that neural nets can compute any function," http://neuralnetworksanddeeplearning.com/chap4.html.
58. Y. LeCun, L. Bottou, Y. Bengio, and P. Haffner, "Gradient-based learning applied to document recognition," *Proceedings of the IEEE*, vol. 86, no. 11, pp. 2278–2324, 1998.
59. S. Hochreiter and J. Schmidhuber, "Long short-term memory," *Neural computation*, vol. 9, no. 8, pp. 1735–1780, 1997.
60. I. Goodfellow, Y. Bengio, and A. Courville, "Deep learning book," *MIT Press*, vol. 521, no. 7553, p. 800, 2016.
61. I. Goodfellow, J. Pouget-Abadie, M. Mirza, B. Xu, D. Warde-Farley, S. Ozair, A. Courville, and Y. Bengio, "Generative adversarial nets," in *Advances in neural information processing systems*, 2014, pp. 2672–2680.

62. A. Waibel, T. Hanazawa, G. Hinton, K. Shikano, and K. J. Lang, "Phoneme recognition using time-delay neural networks," *IEEE transactions on acoustics, speech, and signal processing*, vol. 37, no. 3, pp. 328–339, 1989.
63. S. Minaee, Y. Boykov, F. Porikli, A. Plaza, N. Kehtarnavaz, and D. Terzopoulos, "Image segmentation using deep learning: A survey," *arXiv preprint arXiv:2001.05566*, 2020.
64. V. Gulshan, L. Peng, M. Coram, M. C. Stumpe, D. Wu, A. Narayanaswamy, S. Venugopalan, K. Widner, T. Madams, J. Cuadros *et al.*, "Development and validation of a deep learning algorithm for detection of diabetic retinopathy in retinal fundus photographs," *Jama*, vol. 316, no. 22, pp. 2402–2410, 2016.
65. A. Esteva, B. Kuprel, R. A. Novoa, J. Ko, S. M. Swetter, H. M. Blau, and S. Thrun, "Dermatologist-level classification of skin cancer with deep neural networks," *nature*, vol. 542, no. 7639, pp. 115–118, 2017.
66. Ö. Çiçek, A. Abdulkadir, S. S. Lienkamp, T. Brox, and O. Ronneberger, "3d u-net: learning dense volumetric segmentation from sparse annotation," in *International conference on medical image computing and computer-assisted intervention*. Springer, 2016, pp. 424–432.
67. F. Isensee, P. Kickingereder, W. Wick, M. Bendszus, and K. H. Maier-Hein, "Brain tumor segmentation and radiomics survival prediction: Contribution to the brats 2017 challenge," in *International MICCAI Brainlesion Workshop*. Springer, 2017, pp. 287–297.
68. G. Wang, W. Li, S. Ourselin, and T. Vercauteren, "Automatic brain tumor segmentation using cascaded anisotropic convolutional neural networks," in *International MICCAI brainlesion workshop*. Springer, 2017, pp. 178–190.
69. L. Sun, S. Zhang, and L. Luo, "Tumor segmentation and survival prediction in glioma with deep learning," in *International MICCAI Brainlesion Workshop*. Springer, 2018, pp. 83–93.
70. D. E. Cahall, G. Rasool, N. C. Bouaynaya, and H. M. Fathallah-Shaykh, "Inception modules enhance brain tumor segmentation," *Frontiers in computational neuroscience*, vol. 13, p. 44, 2019.
71. A. Krizhevsky, I. Sutskever, and G. E. Hinton, "Imagenet classification with deep convolutional neural networks," in *Advances in neural information processing systems*, 2012, pp. 1097–1105.
72. K. Simonyan and A. Zisserman, "Very deep convolutional networks for large-scale image recognition," *arXiv preprint arXiv:1409.1556*, 2014.
73. K. He, X. Zhang, S. Ren, and J. Sun, "Deep residual learning for image recognition," in *Proceedings of the IEEE conference on computer vision and pattern recognition*, 2016, pp. 770–778.
74. C. Szegedy, W. Liu, Y. Jia, P. Sermanet, S. Reed, D. Anguelov, D. Erhan, V. Vanhoucke, and A. Rabinovich, "Going deeper with convolutions," in *Proceedings of the IEEE conference on computer vision and pattern recognition*, 2015, pp. 1–9.
75. A. G. Howard, M. Zhu, B. Chen, D. Kalenichenko, W. Wang, T. Weyand, M. Andreetto, and H. Adam, "Mobilenets: Efficient convolutional neural networks for mobile vision applications," *arXiv preprint arXiv:1704.04861*, 2017.
76. G. Huang, Z. Liu, L. Van Der Maaten, and K. Q. Weinberger, "Densely connected convolutional networks," in *Proceedings of the IEEE conference on computer vision and pattern recognition*, 2017, pp. 4700–4708.
77. Y. Li, H. Qi, J. Dai, X. Ji, and Y. Wei, "Fully convolutional instance-aware semantic segmentation," in *Proceedings of the IEEE Conference on Computer Vision and Pattern Recognition*, 2017, pp. 2359–2367.
78. Y. Yuan, M. Chao, and Y.-C. Lo, "Automatic skin lesion segmentation using deep fully convolutional networks with jaccard distance," *IEEE transactions on medical imaging*, vol. 36, no. 9, pp. 1876–1886, 2017.
79. N. Liu, H. Li, M. Zhang, J. Liu, Z. Sun, and T. Tan, "Accurate iris segmentation in non-cooperative environments using fully convolutional networks," in *2016 International Conference on Biometrics (ICB)*. IEEE, 2016, pp. 1–8.

80. O. Ronneberger, P. Fischer, and T. Brox, "U-net: Convolutional networks for biomedical image segmentation," in *International Conference on Medical image computing and computer-assisted intervention*. Springer, 2015, pp. 234–241.
81. F. Milletari, N. Navab, and S.-A. Ahmadi, "V-net: Fully convolutional neural networks for volumetric medical image segmentation," in *2016 fourth international conference on 3D vision (3DV)*. IEEE, 2016, pp. 565–571.
82. T.-Y. Lin, P. Dollár, R. Girshick, K. He, B. Hariharan, and S. Belongie, "Feature pyramid networks for object detection," in *Proceedings of the IEEE conference on computer vision and pattern recognition*, 2017, pp. 2117–2125.
83. H. Zhao, J. Shi, X. Qi, X. Wang, and J. Jia, "Pyramid scene parsing network," in *Proceedings of the IEEE conference on computer vision and pattern recognition*, 2017, pp. 2881–2890.
84. L.-C. Chen, G. Papandreou, I. Kokkinos, K. Murphy, and A. L. Yuille, "Deeplab: Semantic image segmentation with deep convolutional nets, atrous convolution, and fully connected crfs," *IEEE transactions on pattern analysis and machine intelligence*, vol. 40, no. 4, pp. 834–848, 2017.
85. F. Visin, K. Kastner, K. Cho, M. Matteucci, A. Courville, and Y. Bengio, "Renet: A recurrent neural network based alternative to convolutional networks," *arXiv preprint arXiv:1505.00393*, 2015.
86. F. Visin, M. Ciccone, A. Romero, K. Kastner, K. Cho, Y. Bengio, M. Matteucci, and A. Courville, "Reseg: A recurrent neural network-based model for semantic segmentation," in *Proceedings of the IEEE Conference on Computer Vision and Pattern Recognition Workshops*, 2016, pp. 41–48.
87. Y. Xiang and D. Fox, "Da-rnn: Semantic mapping with data associated recurrent neural networks," *arXiv preprint arXiv:1703.03098*, 2017.
88. Y. Yuan and J. Wang, "Ocnet: Object context network for scene parsing," *arXiv preprint arXiv:1809.00916*, 2018.
89. X. Li, Z. Zhong, J. Wu, Y. Yang, Z. Lin, and H. Liu, "Expectation-maximization attention networks for semantic segmentation," in *Proceedings of the IEEE International Conference on Computer Vision*, 2019, pp. 9167–9176.
90. Z. Huang, X. Wang, L. Huang, C. Huang, Y. Wei, and W. Liu, "Ccnet: Criss-cross attention for semantic segmentation," in *Proceedings of the IEEE International Conference on Computer Vision*, 2019, pp. 603–612.
91. P. S. of Medicine University of Pennsylvania, "Brain tumor segmentation," http://braintumorsegmentation.org/.
92. D. Group, "Medical imaging computing and computerassisted intervention," https://miccai2020.org/en/.
93. S. Bakas *et al.*, "Identifying the Best Machine Learning Algorithms for Brain Tumor Segmentation, Progression Assessment, and Overall Survival Prediction in the BRATS Challenge," *arXiv [Preprint]*, 2018.
94. J. Menze, Bjoern H *et al.*, "The multimodal brain tumor image segmentation benchmark (brats)," *IEEE transactions on medical imaging*, vol. 34, no. 10, pp. 1993–2024, 2014.
95. M. Livne, J. Rieger, O. U. Aydin, A. A. Taha, E. M. Akay, T. Kossen, J. Sobesky, J. D. Kelleher, K. Hildebrand, D. Frey *et al.*, "A u-net deep learning framework for high performance vessel segmentation in patients with cerebrovascular disease," *Frontiers in neuroscience*, vol. 13, p. 97, 2019.
96. H. Dong, G. Yang, F. Liu, Y. Mo, and Y. Guo, "Automatic brain tumor detection and segmentation using u-net based fully convolutional networks," in *annual conference on medical image understanding and analysis*. Springer, 2017, pp. 506–517.
97. F. Girard, C. Kavalec, and F. Cheriet, "Joint segmentation and classification of retinal arteries/veins from fundus images," *Artificial intelligence in medicine*, vol. 94, pp. 96–109, 2019.
98. K. Kamnitsas, C. Ledig, V. F. Newcombe, J. P. Simpson, A. D. Kane, D. K. Menon, D. Rueckert, and B. Glocker, "Efficient multi-scale 3d cnn with fully connected crf for accurate brain lesion segmentation," *Medical image analysis*, vol. 36, pp. 61–78, 2017.

99. P. Sandur, C. Naveena, V. M. Aradhya, and K. Nagasundara, "Segmentation of brain tumor tissues in hgg and lgg mr images using 3d u-net convolutional neural network," *International Journal of Natural Computing Research (IJNCR)*, vol. 7, no. 2, pp. 18–30, 2018.
100. X. Li, H. Chen, X. Qi, Q. Dou, C.-W. Fu, and P.-A. Heng, "H-denseunet: hybrid densely connected unet for liver and tumor segmentation from ct volumes," *IEEE transactions on medical imaging*, vol. 37, no. 12, pp. 2663–2674, 2018.
101. Z. Zeng, W. Xie, Y. Zhang, and Y. Lu, "Ric-unet: An improved neural network based on unet for nuclei segmentation in histology images," *Ieee Access*, vol. 7, pp. 21 420–21 428, 2019.
102. J. I. Orlando, P. Seeböck, H. Bogunović, S. Klimscha, C. Grechenig, S. Waldstein, B. S. Gerendas, and U. Schmidt-Erfurth, "U2-net: A bayesian u-net model with epistemic uncertainty feedback for photoreceptor layer segmentation in pathological oct scans," in *2019 IEEE 16th International Symposium on Biomedical Imaging (ISBI 2019)*. IEEE, 2019, pp. 1441–1445.
103. S. Bakas, H. Akbari, A. Sotiras, M. Bilello, M. Rozycki, J. Kirby, J. Freymann, K. Farahani, and C. Davatzikos, "Segmentation labels and radiomic features for the pre-operative scans of the tcga-lgg collection," *The cancer imaging archive*, vol. 286, 2017.
104. M. Havaei *et al.*, "Brain tumor segmentation with deep neural networks," *Medical image analysis*, vol. 35, pp. 18–31, 2017.
105. S. S. M. Salehi, D. Erdogmus, and A. Gholipour, "Auto-Context Convolutional Neural Network (Auto-Net) for Brain Extraction in Magnetic Resonance Imaging," *IEEE Transactions on Medical Imaging*, vol. 36, no. 11, p. 2319–2330, 2017.
106. S. M. Shankaranarayana, K. Ram, K. Mitra, and M. Sivaprakasam, "Fully Convolutional Networks for Monocular Retinal Depth Estimation and Optic Disc-Cup Segmentation," *IEEE Journal of Biomedical and Health Informatics*, 2019.
107. C. Szegedy, S. Ioffe, V. Vanhoucke, and A. Alemi, "Inception-v4, inception-resnet and the impact of residual connections on learning," *arXiv preprint arXiv:1602.07261*, 2016.
108. F. Yu and V. Koltun, "Multi-scale context aggregation by dilated convolutions," *arXiv preprint arXiv:1511.07122*, 2015.
109. N. Ibtehaz and M. S. Rahman, "MultiResUNet: Rethinking the U-Net Architecture for Multimodal Biomedical Image Segmentation," *arXiv [Preprint]*, 2019.
110. C. Wang, M. Rajchl, A. Chan, and E. Ukwatta, "An ensemble of U-Net architecture variants for left atrial segmentation," in *Medical Imaging 2019: Computer-Aided Diagnosis*, vol. 10950. International Society for Optics and Photonics, 2019, p. 109500M, (San Diego, CA, USA).
111. S. Li and G. K. F. Tso, "Bottleneck supervised u-net for pixel-wise liver and tumor segmentation," *arXiv preprint arXiv:1810.10331*, 2018.
112. H. Li, A. Li, and M. Wang, "A novel end-to-end brain tumor segmentation method using improved fully convolutional networks," *Computers in biology and medicine*, vol. 108, pp. 150–160, 2019.
113. S. Ioffe and C. Szegedy, "Batch normalization: Accelerating deep network training by reducing internal covariate shift," *arXiv preprint arXiv:1502.03167*, 2015.
114. C. Nwankpa, W. Ijomah, A. Gachagan, and S. Marshall, "Activation functions: Comparison of trends in practice and research for deep learning," *arXiv preprint arXiv:1811.03378*, 2018.
115. W. Shi, F. Jiang, and D. Zhao, "Single image super-resolution with dilated convolution based multi-scale information learning inception module," in *2017 IEEE International Conference on Image Processing (ICIP)*. IEEE, 2017, pp. 977–981.
116. S. Ruder, "An overview of gradient descent optimization algorithms," *arXiv preprint arXiv:1609.04747*, 2016.
117. D. P. Kingma and J. Ba, "Adam: A method for stochastic optimization," *arXiv preprint arXiv:1412.6980*, 2014.
118. K. He, X. Zhang, S. Ren, and J. Sun, "Delving Deep into Rectifiers: Surpassing Human-Level Performance on ImageNet Classification," *2015 IEEE International Conference on Computer Vision (ICCV)*, 2015, (Santiago, Chile).
119. F. Chollet *et al.*, "Keras," https://keras.io, 2015.

120. M. Abadi *et al.*, "TensorFlow: A system for large-scale machine learning," in *12th USENIX Symposium on Operating Systems Design and Implementation (OSDI 16)*, 2016, pp. 265–283, (Savannah, GA, USA). [Online]. Available: https://www.usenix.org/system/files/conference/osdi16/osdi16-abadi.pdf
121. Google, "Gpus on compute engine," https://cloud.google.com/compute/docs/gpus.
122. F. Isensee, P. Kickingereder, W. Wick, M. Bendszus, and K. H. Maier-Hein, "No New-Net," in *International MICCAI Brainlesion Workshop*. Springer, 2018, pp. 234–244, (Granada, Spain).
123. P.-Y. Kao, T. Ngo, A. Zhang, J. W. Chen, and B. Manjunath, "Brain Tumor Segmentation and Tractographic Feature Extraction from Structural MR Images for Overall Survival Prediction," in *International MICCAI Brainlesion Workshop*. Springer, 2018, pp. 128–141, (Granada, Spain).
124. J. Zhang, X. Lv, Q. Sun, Q. Zhang, X. Wei, and B. Liu, "Sdresu-net: Separable and dilated residual u-net for mri brain tumor segmentation," *Current Medical Imaging Formerly Current Medical Imaging Reviews*, vol. 15, 08 2019.
125. D. Dera, G. Rasool, and N. Bouaynaya, "Extended variational inference for propagating uncertainty in convolutional neural networks," in *IEEE 29th International Workshop on Machine Learning for Signal Processing (MLSP)*, Oct 2019, pp. 1–6.
126. D. M. Blei, A. Kucukelbir, and J. D. McAuliffe, "Variational inference: A review for statisticians," *Journal of the American Statistical Association*, vol. 112, no. 518, pp. 859–877, 2017.
127. A. Papoulis and S. U. Pillai, *Probability, Random Variables, and Stochastic Processes*, 4th ed. McGraw-Hill Higher Education, 2002.
128. M. Bishop, *Pattern Recognition & Machine Learning*. New York: Springer, Aug. 2006.
129. D. Simon, *Optimal State Estimation: Kalman, H Infinity, and Nonlinear Approaches*. Wiley-Interscience, 2006.
130. S. Pereira, A. Pinto, V. Alves, and C. A. Silva, "Brain tumor segmentation using convolutional neural networks in MRI images," *IEEE Transactions on Medical Imaging*, vol. 35, no. 5, pp. 1240–1251, 2016.
131. N. C. Institute, "Prostate-specific antigen," https://www.cancer.gov/types/prostate/psa-fact-sheet/.
132. M. Basseville, I. V. Nikiforov *et al.*, *Detection of abrupt changes: theory and application*. prentice Hall Englewood Cliffs, 1993, vol. 104.
133. E. Brodsky and B. S. Darkhovsky, *Nonparametric methods in change point problems*. Springer Science & Business Media, 2013, vol. 243.
134. C.-J. Kim, J. C. Morley, and C. R. Nelson, "The structural break in the equity premium," *Journal of Business & Economic Statistics*, vol. 23, no. 2, pp. 181–191, 2005.
135. A. B. Olshen, E. Venkatraman, R. Lucito, and M. Wigler, "Circular binary segmentation for the analysis of array-based dna copy number data," *Biostatistics*, vol. 5, no. 4, pp. 557–572, 2004.
136. A. Futschik, T. Hotz, A. Munk, and H. Sieling, "Multiscale dna partitioning: statistical evidence for segments," *Bioinformatics*, vol. 30, no. 16, pp. 2255–2262, 2014.
137. J. Reeves, J. Chen, X. L. Wang, R. Lund, and Q. Q. Lu, "A review and comparison of changepoint detection techniques for climate data," *Journal of applied meteorology and climatology*, vol. 46, no. 6, pp. 900–915, 2007.
138. O. Andrén, K. Fall, L. Franzén, S.-O. Andersson, J.-E. Johansson, and M. A. Rubin, "How well does the gleason score predict prostate cancer death? a 20-year followup of a population based cohort in sweden," *The Journal of urology*, vol. 175, no. 4, pp. 1337–1340, 2006.
139. N. C. Institute, "Seer program," https://seer.cancer.gov/registries/cancer_registry/index.html, 2020.
140. K. Y. Bilimoria, A. K. Stewart, D. P. Winchester, and C. Y. Ko, "The national cancer data base: a powerful initiative to improve cancer care in the united states," *Annals of surgical oncology*, vol. 15, no. 3, pp. 683–690, 2008.
141. CDC, "Ncpr," https://www.cdc.gov/cancer/npcr/index.htm, 2020.
142. CBTR-US, "Central brain tumor registry of the us," https://cbtrus.org/, 2020.

143. A. G. Lai, L. Pasea, A. Banerjee, S. Denaxas, M. Katsoulis, W. H. Chang, B. Williams, D. Pillay, M. Noursadeghi, D. Linch *et al.*, "Estimating excess mortality in people with cancer and multimorbidity in the covid-19 emergency," *medRxiv*, 2020.
144. HDR-UK, "The health data research hub for cancer," https://www.hdruk.ac.uk/help-with-your-data/our-hubs-across-the-uk/data-can/, 2020.
145. M. Otth and K. Scheinemann, "Surveillance imaging for high-grade childhood brain tumors: What to do 10 years after completion of treatment?" *Pediatric blood & cancer*, vol. 65, no. 11, p. e27311, 2018.
146. Q. T. Ostrom, D. J. Cote, M. Ascha, C. Kruchko, and J. S. Barnholtz-Sloan, "Adult glioma incidence and survival by race or ethnicity in the united states from 2000 to 2014," *JAMA oncology*, vol. 4, no. 9, pp. 1254–1262, 2018.
147. D. J. Cote, Q. T. Ostrom, H. Gittleman, K. R. Duncan, T. S. CreveCoeur, C. Kruchko, T. R. Smith, M. J. Stampfer, and J. S. Barnholtz-Sloan, "Glioma incidence and survival variations by county-level socioeconomic measures," *Cancer*, vol. 125, no. 19, pp. 3390–3400, 2019.
148. C. Kruchko, Q. T. Ostrom, H. Gittleman, and J. S. Barnholtz-Sloan, "The cbtrus story: providing accurate population-based statistics on brain and other central nervous system tumors for everyone," 2018.
149. Q. T. Ostrom, H. Gittleman *et al.*, "CBTRUS Statistical Report: Primary Brain and Other Central Nervous System Tumors Diagnosed in the United States in 2011–2015," *Neuro-Oncology*, vol. 20, 10 2018. [Online]. Available: https://doi.org/10.1093/neuonc/noy131
150. R. L. Siegel, K. D. Miller, and A. Jemal, "Cancer statistics, 2020," *CA: a cancer journal for clinicians*, vol. 70, no. 1, pp. 7–30, 2020.
151. N. I. for Health and C. E. G. Britain), *Brain tumours (primary) and brain metastases in adults*. National Institute for Health and Care Excellence, 2018.
152. H. M. Fathallah-Shaykh, A. DeAtkine, E. Coffee, E. Khayat, A. K. Bag, X. Han, P. P. Warren, M. Bredel, J. Fiveash, J. Markert, N. Bouaynaya, and L. B. Nabors, "Diagnosing growth in low-grade gliomas with and without longitudinal volume measurements: A retrospective observational study," *PLOS Medicine*, vol. 16, no. 5, pp. 1–16, 05 2019. [Online]. Available: https://doi.org/10.1371/journal.pmed.1002810

Index

A

Active learning (AL), 263
Activities of Daily Living (ADL), 84
Acute lymphocytic leukemia (ALL)-Acute myeloid leukemia (AML), 15
Adadelta optimizer, 69
Adam (Adaptive Moment Estimation), 18
Adam optimizer, 69, 71, . 74
AlexNet, 9
Amino acids, 28, 256
Amplicon (AMP) dataset, 45, 50–52
Artificial intelligence (AI)
  DL models
    autoencoder, 282
    convolutional neural network, 281
    multi-layered perceptron, 281
    recurrent neural network, 281–282
  ML algorithms, 282–284
Artificial neural networks (ANNs), 65, 84, 112, 258
  black-box algorithms, 205
  CNN, 201–203
  constructive/growing algorithms, 176
  cross-entropy segmentation, 200
  deep learning, 156
  dice coefficient, 200
  hidden layer, 199
  LIME, 204
  logistic regression, 205
  multi-layer perceptrons, 199, 200–201
  post-hoc analysis, 204
  pruning algorithms, 176
  shallow architectures
    back propagation (BP) algorithm, 173
    classification and regression, 173
    loss function, 173, 175
    non-convergence of learning algorithm, 174
    ReLU, 173, 175
    restricted Boltzmann machine, 175
    sigmoid function, 173
    softplus activation, 175
    training set, 174
  SHAP, 204
  transfer learning, 204
  Universal Approximation Theorem, 199
  weakness
    best neural structure, 176
    interpretability of the obtained results, 177
Attention-based models, 323
Automated computer-based segmentation methods, 317
Automatic Fall Detection (AFD), 83–84
Automatic image retrieval system, 102, 103
Automatic segmentation, 318
Automatic tuberculosis screening techniques, 240

B

Back-Propagation (BP) method, 13
Bacteria classification
  full-length 16S sequences
    frequency chaos game representation, 48–49
    one-hot representation, 44–48
  short reads 16S sequences, spectral representation, 49–52
Basal cancer, 116

M. Elloumi (ed.), *Deep Learning for Biomedical Data Analysis*,
https://doi.org/10.1007/978-3-030-71676-9

Bidirectional encoder representations from transformers (BERT), 302–303
Binary cross-entropy loss function, 68, 71, 74
Binary Relevance (BR), 244
Biological sequence classification, *see* Sequence classification
Biomedical data instantiation and visualization
  conflict zone, 183–184
  massive amount of unlabeled data, 181–182
  visual classification methodology, 182–183
Bladder cancer, 266–267
Blood Pressure (BP), 276
Body Mass Index (BMI), 276
Brain tumor segmentation
  challenges, 318
  deep learning segmentation task, 316–317
  image segmentation, 312–313
    in computer vision realm, 318–320
    convolutional neural networks, 321–322
    deep artificial neural networks and, 320–321
    encoder-decoder based models, 322–323
    fully convolutional networks, 322
  inception modules in
    BraTS dataset and pre-processing, 325–326
    deep artificial neural network architectures, 326–329
    experimental setup and results, 329–330
    tumor delineation, 325
  in MRI Images, 336–337
  task challenge, 323–324
  tumor surveillance, 314–316
    community-level active surveillance, 339–340
    rationale for, 337–338
    surveillance of brain tumor, 340–342
    techniques, 339
  uncertainty estimation in
    extended variational density propagation, 332–335
    variational density propagation, 331–332
    variational learning, 331
Breast cancer, 262–264
Burkitt lymphoma (BL), 16

C

Caenorhabditis elegans (cel), 40, 42, 43
Canadian Health Act, 274
Canadian primary care EMR systems, 277
Canadian Primary Care Sentinel Surveillance Network (CPC-SSN), 274
Central Brain Tumor Registry of the United States (CBTRUS), 339
Change-point detection, 339
Chest Radiograph Anatomical Structure Segmentation (CRASS) dataset, 240
Chest x-rays image analysis, 240
Chromatin, 29, 40, 55, 2565
Clinical data, 257
Closed-Circuit TeleVision (CCTV) cameras, 102
Clustering, 6
Coherence Anti-Stokes Raman Scatttering (CARS), 227
  lung cancer diagnosis, 216–217
COlon ADenocarcinoma (COAD), 264
ColoRectal Adenoma (CRA), 264
Colorectal cancer, 264, 266
Combined feature based CBIR systems, 108
Computer-Aided Diagnosis (CADs) systems, 257, 341
Computer vision (CV), 316
Concatenation-Based Integration (CBI), 258
Conditional Random Fields (CRFs), 323
Content-Based Image Retrieval (CBIR), 105–107
  applications of, 113
  combined feature based CBIR systems, 108
  feature extraction, 106
  feature space generation, 106
  hierarchical feature based CBIR systems, 108–109
  image retrieval, 106
  similarity matching, 106
  single feature based CBIR systems, 107–108
Contrastive Divergence (CD) method, 50
Convolutional kernel, 34
Convolutional neural networks (CNNs)
  convolution layer, 11
  1D-CNNs (*see* 1D Convolutional Neural Networks (1D-CNNs))
  2D-CNNs (*see* 2D Convolutional Neural Networks (2D-CNNs))
  fully connected layer, 11
  imaging applications
    biomedical image analysis, 227
    CARS, 227
    data augmentation, 227
    data management, 225, 228
    diabetic retinopathy, 226
    macular disease detection, 227

melanoma detection, 226
test dataset, 228
max pooling layer, 11
non-linear microscopy
automated ovarian tissue analysis, 218–219
lung cancer diagnosis, 216–217
OCT
non-retinal volumes, 215–216
retinal degeneration in disease states, 210–212
retinal imaging, 212–214
pipeline of CNN architecture, 10
sequence classification, 32, 37, 38, 40, 41, 46–55
TPEF, 218
U-Nets inspired algorithm
architecture, 220
image denoising, 223–225
image segmentation, 220–223
PlotNeuralNet implementation, 220
Convolutional variational autoencoder (CVAE), 172
Convolution layer, 11, 34
Copy number variation (CNV), 267
Cox Proportional Hazard (Cox-PH) model, 267
Criss-Cross attention Network (CCNet), 323

D

Data-associated RNNs (DA-RNNs), 323
Data augmentation, 117–118
Data extraction systems, 275
Data programming, 296
Decision Tree (DT), 6
Deep Belief Networks (DBN), 32, 39, 50–55, 173
Deep Flexible Neural Forest (DFNForest) model, 264
Deep learning based medical image retrieval system, 113–114
Deep Learning Based Skin Cancer Image Retrieval (*see* Deep Learning Based Skin Cancer Image Retrieval)
deep learning models, 114–115
Deep Learning Based Skin Cancer Image Retrieval, 115
data augmentation, 117–118
dataset, 117
inception ResNet V2 model, retrieval using, 120–121
inception V3 model, retrieval using, 118–120
modified VGG-16 model, retrieval using, 124
performance analysis, VGG-16 model, 122–124
skin lesions, survey on, 116–117
VGG-16 model, retrieval using, 121–122
Deep learning (DL) model
CBIR, 112–113
classification task
LeNet architecture, 157
mitosis detection, 158
tissue classification, 159
TPS score, 158
VGG16 architecture, 157, 158
framework, 62
image fusion techniques
average LBF strategy, 146
DenseBlock method, 142
DenseFuse method, 142, 143
mean fusion strategy, 145
training phase, 143–145
microRNAs
Adadelta optimizer, 69
Adam optimizer, 69
alignment and representation layer, 63–65
binary cross-entropy loss function, 68
datasets, 70
dense layer, 68
dropout layer, 65
DSet1, 71–74
DSet2, 74–77
embedding layer, 65
evaluation metrics, 70–71
implementation, 69–70
LSTM layer, 66–67
sigmoid activation function, 68
stochastic gradient descent optimizer, 68
segmentation
cell segmentation, 160–161
Mask-R-CNN architecture, 160
region of interest segmentation, 161
UNet architecture, 159
DeepMirTar, 63, 70, 77
Deep Neural Network 42 (DNN), 62
DenseBlock method, 142
DenseFuse method, 142, 143
Diffuse large B-cell lymphoma (DLBCL), 15
Digital pathology (DP), 154
Dilated Residual Inception (DRI), 326
2-D latent space visualization
identify conflict area, 188–189

2-D latent space visualization (*cont.*)
- learning process of VAE, 188
- reduce conflict area, 189
- result analysis, 190
- training process, 188

DNA sequence, 29–31, 34, 40, 55
Documentation, 275
Dropout layer., 36, 65, 69
Drosophila melanogaster(Dmel), 40, 42, 43
DSet1, 70, 71–74
DSet2, 74–77

E

Early Layer Integration (ELI) model, 260
Electronic Medical Records (EMR)
- adaptation of, 284
- classification of low back pain, 295
- COPD prediction, 295
- CPCSSN dataset, 274, 275
- data preprocessing
  - processing structured and semi-structured EMRs, 288–289
  - processing unstructured text data, 290–292
- for health research, 275
- hybrid models, 298–299
- hypertension, 294
- medical text and signal data, 285–286
- model validation approaches, 293
- PTSD diagnosis, 297–298
- structured, 294
- weak supervision model, 295–296

Emphysema, 240
Ewing sarcoma, 16
Expectation-Maximization Attention Network (EMANet), 323
Extended Variational Density Propagation (exVDP), 330
Extraction, Transfer and Load (ETL), 277

F

Fall Detection Systems (FDSs), 82
Fall detection techniques, 83–84
False-Positive Rate (FPR), 20–22
Faster R-CNN, 323
Fast Healthcare Interoperability Resources (FHIR), 276
Fine-tuning, DBNs, 39
Finite State Machine (FSM) model, 84
Fisher criterion filter approach, 5, 8, 19, 23
Fixed-length sequences representation, 30, 31
Follicular lymphoma (FL), 15
Frequency Chaos Game Representation (FCGR), 31, 32, 37, 48–50, 53
Fully connected layer, 11

G

Gates, 35
Gaussian distribution, 331, 332
Gene expression data
- dataset details, 15–16
- deep learning, 9–10
- filter method, 4, 5
- Fisher criterion filter approach, 8
- format, 4
- *k*-Means clustering algorithm, 8, 9
- minimum redundancy - maximum relevance criterion filter approach, 6–7
- 1D-CNNs (*see* 1D Convolutional Neural Networks (1D-CNNs))
- signal-to-noise-ratio ranking approach, 9
- 2D-CNNs (*see* 2D Convolutional Neural Networks (2D-CNNs))
- wrapper method, 4–5

Generative adversarial network (GAN), 163–164, 177–178
Generative networks (GNs)
- GANs, 177–178
- VAEs, 178–180

Genomic identification of significant targets in cancer (GISTIC 2.0), 267
Gleason grade, 339
Gleason score, 257
GlioBlastoMa (GBM), 317
Global Vectors (GloVe), 292
Google Colab, 69

H

Hematoxylin and Eosin (H&E), 264
Hierarchical feature based CBIR systems, 108–109
Hierarchical Integration Deep Flexible Neural Forest (HI-DFNForest) framework, 263
Histopathological image analysis, DL
- challenges and opportunities
  - annotated data, 161–162
  - GAN, 163–164
  - multiple stainings, 162–163
- classification task
  - LeNet architecture, 157
  - mitosis detection, 158
  - tissue classification, 159

TPS score, 158
VGG16 architecture, 157, 158
DP, 154
GPU cards, 156
gradient backpropagation method, 156
segmentation
cell segmentation, 160–161
Mask-R-CNN architecture, 160
region of interest segmentation, 161
UNet architecture, 159
WSI, 154, 155
Homo sapiens (hsa), 40, 42, 43
Human microbiota, 28

I
Image fusion techniques
color vision, 134
decision-level, 138–139
deep learning
average LBF strategy, 146
DenseBlock method, 142
DenseFuse method, 142, 143
mean fusion strategy, 145
training phase, 143–145
detection and tracking, 133
evaluation results, 148–149
feature-level, 137–138
gray-scale medical imaging, 147–148
imaging modalities, 130
in medical imaging
brain, 132, 147
breast, 132, 148
cardiac, 132
decision labeling, 139
feature extraction, 139
fused image, 140
image registration, 139
knowledge-based methods, 142
lungs, 132
morphological method, 140
neural network-based methods, 141–142
prostate, 132, 149
pyramid-based method, 140
semantic equivalence, 139–140
sparse representation, 141
subspace-based methods, 141
wavelet-based method, 140–141
pixel-level, 136
recognition, 133
remote sensing, 134
spatial domain, 130
surveillance systems, 134–135
training phase analysis, 146–147
transform domain, 130
Image retrieval systems
content-based image retrieval, 105–107
combined feature based CBIR systems, 108
hierarchical feature based CBIR systems, 108–109
single feature based CBIR systems, 107–108
semantic-based image retrieval, 110
classification, 110
deep learning based CBIR, 112–113
machine learning based CBIR, 110–112
training, 110
taxonomy of, 102, 103
text-based image retrieval, 103–105
Image segmentation, 312–313
in computer vision realm, 318–320
convolutional neural networks, 321–322
deep artificial neural networks and, 320–321
encoder-decoder based models, 322–323
fully convolutional networks, 322
Inception modules
BraTS dataset and pre-processing, 325–326
deep artificial neural network architectures
dilated inception module U-Net, 327–328
dilated inception U-net (DIU-Net), 326–327
modified DSC, 328–329
experimental setup and results, 329–330
tumor delineation, 325
Inception ResNet V2 model, retrieval using, 120–121
Inception V3 model, retrieval using, 118–120
Information Extraction (IE), 291
Information gain, 5
Integer encoding, 30
International Skin Imaging Collaboration (ISIC), 115
International Symposium on Biomedical Imaging (ISBI), 323

J
Japanese Society of Radiological Technology (JSRT), 240

K
Keras, 19
$k-$means and signal-to-noise ratio (KM-SNR) filter method, 5, 8

k-mers representation, 37, 40–41, 48, 50–55
k-Nearest Neighbours (KNN), 6
Kullback–Leibler divergence, 33

L
Late layer integration, 261–262
121-layer Deep Convolutional Neural Network (DCNN), 244
Learning functions (LFs), 296
Linker DNA, 29
Linkers, 40, 41, 44, 45, 52
Long Short-Term Memory (LSTM), 32, 35, 37, 38, 41, 46–48, 53, 55, 62, 63, 66–67, 83
Lung cancer, 240
Lung disease detection
  JSRT dataset, 240
  lung nodule detection, 248–250
  lung segmentation and SCR dataset, 240
  performance metrics, 242–243
  thoracic disease detection, 241–242, 244–245
  tuberculosis detection, 245–248
Lung nodule detection
  datasets for, 248–249
  deep feature fusion, 249
  JSRT dataset, 250
  performance metrics for, 249–250

M
Machine learning based CBIR, 110–112
Manitoba Primary Care Research Network (MaPCReN), 277
Markov Random Field (MRFs), 323
Masked Language Model (MLM), 303
Mask-RCNN, 323
Matrix-vector multiplication, 333
Max-pooling layer, 11, 36
Mean Squared Error (MSE), 17, 95
Medical image retrieval system
  deep learning based skin cancer image retrieval, 115
    data augmentation, 117–118
    dataset, 117
    inception ResNet V2 model, retrieval using, 120–121
    inception V3 model, retrieval using, 118–120
    modified VGG-16 model, retrieval using, 124
    performance analysis, VGG-16 model, 122–124
    skin lesions, survey on, 116–117
    VGG-16 model, retrieval using, 121–122
  deep learning techniques, 103
    CBIR/SBIR, applications of, 113
    image retrieval systems (*see* Image retrieval systems)
Medical Imaging Computing and Computer-Assisted Intervention (MICCAI) conference, 323
Medical imaging, fusion techniques
  brain, 132, 147
  breast, 132, 148
  cardiac, 132
  decision labeling, 139
  feature extraction, 139
  fused image, 140
  image registration, 139
  knowledge-based methods, 142
  lungs, 132
  morphological method, 140
  neural network-based methods, 141–142
  prostate, 132, 149
  pyramid-based method, 140
  semantic equivalence, 139–140
  sparse representation, 141
  subspace-based methods, 141
  wavelet-based method, 140–141
Melanoma, 116
Melanoma skin cancer (NMSC), 116
Metabolomic profiling, 28
METABRIC, 262
Metagenomics, 28, 40, 55, 257
MetaMap, 292
Microbial flora, 28
Microbiome, 28, 29, 257
Microbiota, 28, 29, 257
Micro-Controller Unit (MCU), 82
MicroRNAs (miRNAs), 61
  deep learning model
    Adadelta Optimizer, 69
    Adam Optimizer, 69
    alignment and representation layer, 63–65
    binary cross-entropy loss function, 68
    datasets, 70
    dense layer, 68
    dropout layer, 65
    DSet1, 71–74
    DSet2, 74–77
    embedding layer, 65
    evaluation metrics, 70–71
    implementation, 69–70
    LSTM layer, 66–67

sigmoid activation function, 68
stochastic gradient descent optimizer, 68
Middle Integration Deep Neural Network (MI-DNN), 263
Minimum redundancy-maximum relevance (MRMR) filter approach, 6–7
miRNA–mRNA duplex, 64
miRTif, 62
Model-Based Integration (MBI), 258
Model validation approaches, 293
Modified VGG-16 model, retrieval using, 124
Montgomery (MC) dataset, 240
Motivation, 317
Multilayer Perceptron (MLP), 6, 14
Multi-level feature selection (MLFS) approach, 262
Multimodal Deep Neural Network by integrating Multi-dimensional Data (MDNNMD), 262
Multi-Omics data integration
bladder cancer, 266–267
breast cancer, 262–264
colorectal cancer, 264, 266
early layer integration, 259–260
late layer integration, 261–262
middle layer integration, 260
Multi-scale pyramid network models, 323

N
Named entity recognition (NER), 296
National Cancer Data Base (NCDB), 339
National Library of Medicine (NLM), 292
Natural language processing (NLP), 290
Next Generation Sequencing (NGS), 4, 26, 28, 45, 257, 258
Next-Generation Sequencing (NGS) data, 257
Next Sentence Prediction (NSP), 303
Non-linear transformations, 332
Nucleosome identification, 29
Nucleotides, 28, 35, 45

O
Octamer, 29
Omics
defined, 256
multi-omics data integration (*see* Multi-omics data integration)
1D Convolutional Neural Networks (1D-CNNs)
with 3 CNN and 2 MLP layers, 14
vs. 2D-CNNs, 13, 14
gene expression dataset
complete datasets with 70:30 and 80:20 train-test ratio, 20
model creation, 16–18
model testing, 18
model training, 18
preprocessing using fisher ranking approach, 16
for top 200 and top 250 fishers ranked genes with 70:30 and 80:20 train-test ratio, 21, 22
for top 100 fisher ranked genes with 70:30 and 80:20 train-test ratio, 20, 21
for top 150 fisher ranked genes with 70:30 and 80:20 train-test ratio, 21
hyper-parameters for design of, 14
model specifications, 19
with three consecutive hidden CNN layers, 15
One hot encoding, 30, 31, 45–48
Optical coherence tomography (OCT), 206–207
CNN analysis of non-retinal volumes, 215–216
retinal degeneration in disease states, 210–212
retinal imaging, 212–214
Optical imaging methods
challenges and opportunities, 209–210
coherent Raman imaging, 208–209
multiphoton fluorescence microscopy, 207–208
OCT systems, 206–207
SHG, 208

P
Padding, 34
PairWise Error (PWE), 244
Particle Swarm Optimization (PSO), 6
Patients' medical information, 277
Phylum, 45, 47–54
Pneumonia, 240
Post-Traumatic Stress Disorder (PTSD), 275
Pre-training, DBNs, 38
Prostate-Specific Antigen (PSA), 338
Prostate tumor, 15
Proteobacteria phylum, 45
Public datasets, 240
Pyramid Scene Parsing Network (PSPN), 323
Python, 19

Python 3, 70
Python programming, 89
Python TkInter library, 88

Q
Quantitative analysis task, 318
Quantitative Insights Into Microbial Ecology (QIIME), 257

R
Random forest (RF) classifier, 298
Random variables, 332
Rectified Linear Units (ReLU), 17, 34, 38, 42, 49
Recurrent Neural Network (RNN), 32, 34–35, 46, 47
  architectures, 82
  decoder, 244
  fall detection techniques, 83–84, 86–87
    effectiveness of, 96
    embedded, 94–95
    software implementation and training, 89–91
    statistical indicators, 93–94
Restricted Boltzmann machines (RBM), 33
RGB-Depth, 83
Rhabdomyosarcoma, 16
Ribosomal Database Project II (RDP), 45
RNAHybrid, 62
RNA-Seq by Expectation-Maximization (RSEM), 267
RNN-based models, 323
Rontgen, Wilhelm Conrad, 239
Rotational sensors, 83–84

S
Scikit Learn, 19
Second harmonic generation (SHG), 208, 218
Segmentation, 313. *See also* Brain tumor segmentation
Segmented Chest Radiograph (SCR) dataset, 240
Self-attention ANN models, 300–301
Semantic-based image retrieval (SBIR), 110
  applications of, 113
  classification, 110
  deep learning based CBIR, 112–113
  machine learning based CBIR, 110–112
  training, 110
Semi-structured data, 276
Semi-supervised methods, 287
SensorTile®, 82, 83, 94, 96
Sequence classification
  bacteria classification using full-length 16S sequences
    frequency chaos game representation, 48–49
    one-hot representation, 44–48
  bacteria classification using short-reads 16S sequences and spectral representation, 49–52
  deep neural networks
    ANN architecture, 37, 44, 46
    CNN architecture, 38
    convolutional layers, 34
    DBN architecture, 38, 39
    dropout layer, 36
    embedding layer, 36
    LSTM architecture, 38
    max-pooling layer, 36
    recurrent layers, 34–35
    restricted Boltzmann machines, 33
    softmax layer, 36
    unsupervised training, 39
  execution time, 54, 55
  importance of, 28
  nucleosome-linker classification
    k-mers representation, 40–41
    one-hot representation, 41–44
  nucleosome positioning, 29
  sequences representation
    fixed-length sequences, 30, 31
    frequency chaos game representation, 32
    spectral representations, 30–32
Shotgun (SG) dataset, 45, 50–53
Sigmoid activation function, 68
Signal-to-Noise-Ratio (SNR), 9
Single feature based CBIR systems, 107–108
SisFall dataset, 84–85, 87, 90
16S ribosomal RNA (rRNA), 29
Small-Round-Blue-Cell Tumor (SRBCT), 15
Softmax layer, 36
Spectral representations, 30–32, 49–52
Stacked AutoEncoder (SAE), 263
Stacked de-noising Auto-encoder (SdA), 63
State-of-the-art bioinformatics, 28
STM32CubeMX Power Consumption Calculator, 95
Stochastic Gradient Descent (SGD) optimizer, 68
Stride, 34
Supervised methods, 287

Support Vector Machine (SVM), 48, 62, 66
Surface-Enhanced Laser Desorption/Ionization Mass Spectrometry (SELDI-MS), 264

T
Tanh activation function, 47
TargetScan v7.0, 63
TarPmiR, 63
TensorFlow, 19, 95
TensorFlow 1.6, 89
Term Frequency-Inverted Document Frequency (TF-IDF) scores, 291
Text-based Image Retrieval (TBIR), 103–105
The Breast CancerWisconsin dataset
 confusion matrix, 185
 description, 184
 2D-latent space visualization, 187–190
 2D-visual selection method, 190–192
 evaluation metrics, 185–186
 VAE architecture, 186–187
Thoracic disease detection
 Attention Guided CNN (AG-CNN), 243
 ChestX-ray8 dataset, 242, 243
 Indiana dataset, 242, 244, 245
 long-term recurrent convolutional network (LRCN), 245
 LSTMs, 244
 performance metrics, 242–243
Tissue MicroArrays (TMAs), 264
Transfer learning (TL), 267
Transformation-based integration, 258
Transformer
 BERT, 302–303
 network, 302
T-Test, 5
Tuberculosis detection
 Belarus dataset, 246
 Montgomery County dataset, 246, 247
 performance metrics for, 247
 Shenzen dataset, 246–247
Tumor surveillance, 314–316
 of brain tumor, 340–342
 community-level active surveillance, 339–340
 rationale for, 337–338
 techniques, 339
2D Convolutional Neural Networks (2D-CNNs)
 advantages, 12
 complete forward-propagation process, 12–13
 with two convolution and one fully-connected layers, 12
Two-photon excitation fluorescence (TPEF), 218

U
U-Nets inspired algorithm
 architecture, 220
 image denoising, 223–225
 image segmentation, 220–223
 PlotNeuralNet implementation, 220
Unified Medical Language System (UMLS) Meta-thesaurus, 291
Unintentional falls, 82
University of California, Irvine (UCI) ML repository, 5
Unstructured data, 276
Unstructured Information Management Architecture (UIMA) framework, 291
Unsupervised learning methods, 287

V
Variable Length Markov Chain (VLMC), 63
Variant call format (VCF), 267
Variational autoencoder (VAE), 172, 178–180, 186–187
Variational Inference (VI) framework, 330
VGG-16 model
 modified VGG-16 model, retrieval using, 124
 performance analysis, 122–124
 retrieval using, 121–122
Visual classification methodology, 182–183

W
Wearable devices, 83
 fall detection techniques, 83–84
  annotation procedure, 87–89
  dataset and labeling, 85–86
  datasets, 84–85
  embedded, 94–95
  RNN architecture, 86–87
  software implementation and training, 89–91
  statistical indicators, 91–94
  training and inference, 87
Whole-Genome Shotgun (WGS), 29
Whole-slide imaging (WSI), 154, 155
Word embeddings, 292
Word2Vec algorithm, 298

The manufacturer's authorised representative in the EU is Springer Nature Customer Service Centre GmbH, Europaplatz 3, 69115 Heidelberg, Germany. If you have any concerns regarding our products, please contact ProductSafety@springernature.com

Printed and bound by CPI Group (UK) Ltd, Croydon, CR0 4YY
28/09/2026
02224091-0018